Dark Matter

Dark Matter

Evidence, Theory, and Constraints

David J. E. Marsh, David Ellis, and Viraf M. Mehta

PRINCETON UNIVERSITY PRESS
PRINCETON AND OXFORD

Published by Princeton University Press
41 William Street, Princeton, New Jersey 08540
99 Banbury Road, Oxford OX2 6JX

press.princeton.edu

All Rights Reserved

ISBN 978-0-691-24951-3
ISBN (pbk.) 978-0-691-24952-0
ISBN (e-book) 978-0-691-24971-1

Library of Congress Control Number: 2024936836

British Library Cataloging-in-Publication Data is available

Editorial: Ingrid Gnerlich and Whitney Rauenhorst
Production Editorial: Kathleen Cioffi
Cover Design: Wanda España
Production: Jacqueline Poirier
Publicity: William Pagdatoon
Copyeditor: Gráinne O'Shea

Jacket image: Illustration of Dark Matter in a Simulated Universe. © Tom Abel & Ralf Kaehler (KIPAC, SLAC), AMNH

This book has been composed in LaTeX

Printed in the United States of America

10 9 8 7 6 5 4 3 2 1

Contents

This book is accompanied by a series of JUPYTER notebooks, which can be used to reproduce certain figures and numerical exercises. They are available at https://github.com/Dark-Matter-Textbook/Notebook-Exercises. The notebooks will be continually updated, but a rough guide is given here.

- Rotation curves, accompanies Chapter 3. Plot a simple rotation curve, and fit different dark matter profiles to data.
- Lensing, accompanies Chapter 4. A simplified exercise to compute the ellipticity of regions in a sample image, and lens this image.
- WIMP relic density, accompanies Chapter 9. Numerical solution of the Boltzmann equation, produce figures for $Y(T)$.
- Axion relic density, accompanies Chapter 10. Numerical solution of the Klein-Gordon equation, produce figures for $\phi(t)$.
- WIMP direct detection, accompanies Chapter 12. Compute recoil rates and exclusion limits.
- Axion direct detection, accompanies Chapters 14 and 15. Estimate exclusions from haloscopes, helioscopes, and horizontal branch stars.
- PBH microlensing, accompanies Chapter 16. Compute the rate and number of microlensing events, and thus a simple exclusion.

Common Abbreviations

- ΛCDM: Λ cold dark matter, that is, the standard cosmological model
- CMB: cosmic microwave background
- const.: constant
- DM: dark matter
- d.o.f.: degrees of freedom
- dSph: dwarf spheroidal galaxy
- EW: electroweak
- FRW: Friedmann–Robertson–Walker (metric of the expanding Universe)
- FT: Fourier transform
- GR: general relativity
- LHC: the Large Hadron Collider, at CERN
- LIGO: Laser Interferometer Gravitational-Wave Observatory
- MOND: modified Newtonian dynamics
- MW: Milky Way (our galaxy)
- NFW: Navarro–Frenk–White (DM density profile)
- ODE: ordinary differential equation
- PDE: partial differential equation
- PT: phase transition
- QCD: quantum chromodynamics (theory of the strong force)
- QED: quantum electrodynamics
- QFT: quantum field theory
- QM: quantum mechanics
- SIDM: self-interacting DM
- SM: the Standard Model (of particle physics)
- SSB: spontaneous symmetry breaking
- SUSY: supersymmetry
- vev: vacuum expectation value. For a field, $\langle \phi \rangle$
- WIMP: weakly interacting massive particle

Dark Matter

Chapter One

Introduction

1.1 DARK MATTER: THE GREATEST MYSTERY

We live at an extraordinary time in scientific history. Never before have we known so much about the Universe and yet been so certain of our ignorance about it. The twentieth century saw the development and refinement of quantum field theory (QFT) and general relativity (GR). QFT is the language in which the Standard Model is written and provides the tools with which it has been proven with great accuracy. GR allows us to conceive a science of the whole Universe and once again make precise calculations about it that have stood up to every conceivable test.

Cosmology, the large-scale application of GR and its development as a precision observational science, is where the cracks show their widest. On the one hand, GR and QFT can be married perfectly to predict the light element abundances and the existence of the cosmic microwave background (CMB). The precise form of the CMB, as expressed in the peaks and troughs of the anisotropy multipoles, however, tells us that the Standard Model and GR cannot be the whole story. A consistent theory of the CMB and galaxy formation requires that the vast majority of the Universe's mass density is composed of a new form of matter: dark matter (DM).

We live in the post-discovery world of dark matter. The total mass density has been measured to percent-level precision. The time at which it was created in the early Universe can be narrowed down to within a few days or even fractions of a second. CMB lensing measurements map the location of dark matter filaments stretching across billions of light-years. Inside galaxy clusters, lensing reveals vast 'halos' of dark matter moving independently of the baryonic gas. Dark matter provides the cosmological conditions for galaxy formation and star formation. We know where dark matter is, how it moves under gravity and how much of it there is. Part I of this book will convince you that dark matter has been discovered at the macroscopic level.

It is often said that we don't know what dark matter is, but that is not really true. The macroscopic theory works perfectly, and the possibilities for microscopic theories are severely restricted. The list of things that dark matter might be can be broken down into three distinct categories, covered in Part II of this book as 'WIMP-like', 'axion-like' and 'macroscopic' or 'primordial black hole (PBH)-like'. In each category, the space of possibilities is restricted and bounded by precise measurement (we know an awful lot about what dark matter isn't), leaving only finite regions in which the dark matter could be hiding. In Part III of this book we show, for each category, where and how to look for evidence that would favour one category of dark matter candidate over another: how to achieve the 'microscopic' discovery of dark matter.

The science of dark matter has progressed in rigour and complexity in the last forty years to arrive at this almost complete picture. The tools are at hand in almost every case to close those final windows of possibility and discern the microscopic nature of dark matter: a task I am personally confident will be achieved within my lifetime. The twenty-first century will be the century of dark matter. The stage is set, the players are ready, the curtain is about to rise.

In honour of the University of Göttingen, where this book was conceived, we are reminded of the saying of its greatest son:

> Wir müssen wissen; wir werden wissen.
> We must know; we will know.

David Hilbert

1.2 OVERVIEW OF THIS BOOK

This book is designed as a relatively self-contained introduction to the vast and multidisciplinary topic of dark matter. Theoretical background is kept to a minimum, and we cover only some cosmology (chapter 5) and a tour of the Standard Model (chapter 7). You will need a physics background in special relativity, thermodynamics, classical mechanics (including Lagrangian/Hamiltonian dynamics and classical field theory), electromagnetism and quantum mechanics. The mathematics should pose no problem to anyone with a bachelor's degree in physics. Nonetheless, to really get the most out of this book, you will need accompanying introductory texts on cosmology and particle physics. Recommendations are given in the bibliography.

As pertains to dark matter, in every area of this book we attempt some level of balance between topics that has been absent in other reviews and discussion. This means that each topic alone may receive less detail than in specialised reviews, but we hope that the broad scope is nonetheless valuable.

In Part I we cover astrophysical and cosmological evidence and attempt balance between galactic rotation curves (chapter 3), lensing and dynamics of clusters (chapter 4) and cosmological structure formation (chapter 6). Part II focuses on theories for dark matter, giving balance to WIMP-like models (chapter 9), axions (chapter 10) and primordial black holes (chapter 11).

Part III covers constraints on each type of dark matter where we balance not only the treatment by theory, but across laboratory and astrophysical limits. For WIMPs, we cover both direct (chapter 12) and indirect (chapter 13) constraints. Similarly for axions, we cover direct (chapter 14) and indirect (chapter 15) constraints. Only for PBHs are we forced to address only indirect constraints (chapter 16), the reason being that direct collisions of PBHs with Earth that would allow direct detection are thankfully rare. The book closes in chapter 17 with a brief epilogue on the theory zoo beyond standard cold dark matter.

Each chapter in this book aims to build intuition and methods for approximation, rather than teaching technical methods. There are eighteen short quizzes throughout the book to test ongoing understanding (and assumed background knowledge). Each part ends with a set of original problems. Online material includes worked examples of numerical calculations and plotting using JUPYTER notebooks.

We have not written this book as a review article, filled with references. References given in the text are only to those things we relied on directly in writing the material. The bibliography in the appendix gives only bare-bones suggested further reading. It should help a beginner to get started diving into the vast literature on dark matter, but it is only intended as a pointer. The bibliography is not exhaustive, nor is it historically complete.

1.3 HISTORICAL NOTE

The story told in Part I of this book on the evidence for dark matter follows the well-worn and idealised folk history of the subject, as told by teachers to students and used in the introductions to countless theses, papers, books, television programmes and research seminars. This idealised story begins with Fritz Zwicky in the 1930s, moves on to Vera Rubin, stopping briefly at the Bullet Cluster and eventually reaches the crowning glory of the cosmic microwave background anisotropies in the 2000s. Of course, the history of the discovery of dark matter, and its acceptance by the scientific community, is much more complex than this. An excellent telling of this history is given by Bertone and Hooper [1]. The following short section tells this extended history briefly, relying entirely on Bertone and Hooper (where you can find all of the original sources), along with some other important notes on history that do not appear in the rest of this book.

An early analogy to the present situation can be found with Le Verrier, who, by observing the motions of the planets, recognised that a new perturbing gravitational force was needed. This force was provided by a heretofore dark object, the position of which he predicted precisely and which was later observed as the planet Neptune. The case of the orbit of Mercury provides the counterpoint story: the necessary dark matter in this case, the planet Vulcan, was found not to exist, and the mystery was only solved by replacing Newtonian gravity with GR. This story of dark matter versus modified gravity is one we will meet in this book, and we will see how only dark matter can explain the evidence on all scales.

The history of our dark matter begins 30 years prior to Zwicky. Lord Kelvin and Henri Poincaré used the virial theorem (see chapter 2) in our own galaxy to determine that invisible matter accounted for up to half of the local matter density. Given how approximate he knew their calculations to be, Poincaré concluded that DM was not necessary within the uncertainty. Now we know that their 50-50 estimate is actually quite close to the truth about the local DM density (see section 3.4).

The continued application of Kelvin's ideas, the 'theory of gases', to determine the density of matter in our own galaxy was taken up by many more scientists in the following years, including Oort, Jeans, Kapteyn and Linblad. Their estimates of the density of dark objects were of the order of 0.05 solar mass per cubic parsec, a few times 10^{-24} g per cubic centimetre or, in our favoured units, about 1 GeV per cubic centimetre. These estimates are also on the correct scale of current measurements. As early as 1936, Hubble wrote that he considered the problem of missing mass 'real and important'.

In Zwicky's story, there is yet another success compared to modern theory, which came the same year as his famous work on the Coma Cluster. Likely in reference to Oort and others measuring the local density, Zwicky compared this to the expansion rate of the Universe determined by Hubble and determined an overdensity in galaxies compared to the cosmic density of around 10^5. Zwicky's famous work determining the presence of DM in the Coma Cluster (section 2.1.2) was actually pretty inaccurate for DM density in clusters. The immense distance to clusters required use of the Hubble expansion rate when measuring velocities, which at the time was determined inaccurately to be around ten times higher than the value we know today. This error, and the fact that Zwicky could not observe the X-ray emitting gas in the cluster, led to a vast overestimate of the required amount of DM in Coma.

The story in this book then jumps ahead nearly forty years from Zwicky to Rubin, but the time in between was in reality far from scientifically quiet. There were worries that the virial theorem could not be applied: that the systems under study were not in equilibrium. This can be proven by showing that the out-of-equilibrium expansion of clusters would lead to sizes and ages incompatible with those measured (something we will meet in a different context in sections 15.1.2 and 16.3 as providing constraints on 'fuzzy' DM, and PBHs from the out-of-equilibrium heating of star clusters). There were also worries that individual interactions of stars were important, but Chandrasekhar proved that such interactions were negligible.

Rubin's tool, the rotation curve (see chapter 3), was used by many predecessors, of which we note just a few here. As early as 1939, Babcock measured the rotation curve of Andromeda, M31, out to large distances and noted the need for 'absorption in the outer parts', that is, a presence of more mass than he could see light from. A leap forward for M31 came with the advent of radio astronomy and the measurements of the outer rotation curve observed in 21 cm emission in 1957 by van de Hulst et al., and in 1966 by Roberts.

A notable work in this period that foreshadowed the modern interpretation was in 1963 by Arigo Finzi, who saw a commonality between missing mass implied by galactic rotation curves and the virial theorem in clusters. The work is significant because Finzi also proposed many candidates for baryonic dark matter (dwarf planets and so on) and even modified gravity.

Rubin and Ford's famous work came in 1970 as did many other flat rotation curves and mass-to-light ratios that rise with increasing radius from the centre. In particular, Albert Bosma measured twenty-five flat rotation curves. The significance of such measurements as pertains to dark matter became apparent to Morton Roberts and Ken Freeman.

A further shift to the modern era came in 1974 when key figures in Europe and the United States began to take the evidence from clusters, rotation curves and an apparently flat Universe together as a common problem. The connection to the total density of the Universe is related to the key observational number that all DM theories must today predict: the 'relic density' (see sections 5.1.4 and 6.1). Another key advance at this time came due to computing power, which allowed for simulations with hundreds of mass elements ('particles'), which allowed Peebles and Ostriker to show that DM halos were necessary theoretically for the observed fact that stellar disks in galaxies are stable, which they cannot be if they themselves dominate the gravitational potential.

In chapter 6, we meet our modern mainstay: the cosmic history of DM and the precision evidence cosmology provides for its existence, which was pioneered by Peebles and collaborators in the late 1970s and early 1980s. Peebles introduces the concept of the "primeval" (primordial in modern language) power spectrum, which was starting to be calculated also by particle theorists working on inflation (see sections 5.3 and 11.3). It was only at this time that joined up modern thinking about DM emerged. Prior to the 1980s, astronomers thought of the missing mass as astronomical objects, not as particles created in the early Universe. But Peebles saw this, and his work in the 1980s even gives us a correct lower limit to thermal DM: what we call the 'warm DM' bound that the particle mass should be larger than about 1 keV (see section 13.1). Finally, in 1985, further advances in computing by Davis, Efstathiou, Frenk and White allowed for the first cosmological simulations of cold dark matter, which gave rise to a 'cosmic web' consistent with what was observed in the large galaxy surveys that were made around the same time. Thus it is in the 1980s that we can say that the modern era of dark matter begins.

The story of dark matter from the 1980s to the present day is given in a more complete way throughout this book. Here we give just a brief sense of the historical narrative. On the theory side, the 1970s saw each of our main theories written down. Supersymmetry (section 9.2.2) was developed throughout the decade by a large number of scientists. The year 1971 saw the first discussion of PBHs (chapter 11) by Hawking, and 1977 and 1978 saw the proposal of the QCD axion (section 10.2.1) by Peccei, Quinn, Weinberg and Wilczek.

Axions and supersymmetry (SUSY) came into the DM game in the 1980s, with both theories significantly refined. For SUSY, it was in 1981 that Dimopoulos and Georgi constructed the supersymmetric standard model, while for the axion the modern theories appeared between 1979 and 1981. In remarkable synchronicity, yet again the relic densities of both the QCD axion and the SUSY neutralino were computed in 1983 in a series of papers by different authors.

At this stage the connection between theory and observation was so strong that experiments were built to search for each dark matter candidate. The results of these experiments were published in 1987: from the Homestake Mine (section 12.3) and a series of axion haloscopes (section 14.2). How different the scientific landscape would be if these first detection attempts had been lucky enough to succeed!

The years since 1987, the year of my birth, have seen the hunt for dark matter intensify. In cosmology, the 1990s and 2000s saw the dawn and apex (literally: the first acoustic peak) of the measurement of cosmic microwave background anisotropies, determining the presence and necessity of dark matter with exquisite accuracy. The 2000s saw huge leaps forward in the scale and precision of dark matter direct searches for both axions and WIMPs. The dawn of gravitational wave measurements in 2015 saw the great PBH revival. Now, entering the 2020s, we see a newly invigorated programme to close in on dark matter from all sides and on all fronts. This book will give you the tools to join history here and, after the microscopic discovery of dark matter, to continue the path to dark matter precision science.

David J. E. Marsh, December 2022

Warm-up Problems: Units

In this warm-up problem set we discuss units in particle physics and cosmology, deriving various mass scales, time scales and length scales, while briefly introducing some key concepts for later.

Here are the values of fundamental constants, in SI units, and some unit conversions, that you will need.

$$c = 3.00 \times 10^8 \text{ m s}^{-1}; \qquad\qquad \text{speed of light in vacuum,} \quad (1.1)$$

$$\frac{h}{2\pi} = \hbar = 1.05 \times 10^{-34} \text{ J s}; \qquad\qquad \text{reduced Planck's constant,} \quad (1.2)$$

$$k_B = 1.38 \times 10^{-23} \text{ m}^2\text{kg s}^{-2} \text{ K}^{-1}; \qquad\qquad \text{Boltzmann's constant,} \quad (1.3)$$

$$G = 6.67 \times 10^{-11} \text{ m}^3 \text{ kg}^{-1} \text{ s}^{-2}; \qquad\qquad \text{Newton's constant,} \quad (1.4)$$

$$1 \text{ eV} = 1.60 \times 10^{-19} \text{ J}; \qquad\qquad \text{electron volts,} \quad (1.5)$$

$$1 \text{ Mpc} = 3.09 \times 10^{22} \text{ m}; \qquad\qquad \text{megaparsecs.} \quad (1.6)$$

Problem 1. *Natural Units in Particle Physics*

By measuring speeds in units of c, and angular momentum in units of $\hbar$, the fundamental constants related to relativity and quantum mechanics, we can express many dimensionful quantities in units of energy (we will deal with temperature in Problem 3). The unit of energy commonly used is the electron volt.

- The proton mass is $m_p = 1.67 \times 10^{-27}$ kg. What is its mass in eV? In MeV? (1 MeV $= 10^6$ eV.)
- What is the 'nuclear time scale', $1/m_p$, in seconds?
- The Higgs has mass $m_H = 126$ GeV. What is the Higgs mass in kg?
- The maximum energy of the Large Hadron Collider (LHC) is 14 TeV (1 TeV $= 10^9$ eV). Assume that all of this energy can go into pair producing new particles (this is not really the case, since the total energy is distributed among the different constituents of the proton). If the lightest supersymmetric particle (LSP) is *just* accessible via pair production, that is, only two particles are produced in the final state, at the LHC, what is the lightest m_{LSP} can be in kg?

Problem 2. *The Hubble Scale*

The Hubble rate measures the expansion of the Universe. The Hubble rate today is approximately $H_0 = 70 \text{ km s}^{-1} \text{ Mpc}^{-1}$.

- What is H_0 in yr^{-1}? The Hubble time, $1/H_0$, is a good estimate for the age of the Universe; what is it? (We will calculate the age of the Universe more precisely in problem I.9 later.)
- What is H_0 in eV? In kg? Notice how small this is in comparison to typical particle physics energy scales.
- What is the Hubble length, $1/H_0$, in Mpc? This is the length scale associated with the size of the Universe.
- If the Hubble rate as a function of time is given by $H(t) = H_0(t_0/t)$, where t_0 is the age of the Universe, what was the Hubble rate, in eV, when the Universe was just 300,000 years old? What was the Hubble length in Mpc? The CMB (see Problem 3) was first formed at about this time.

Problem 3. *Temperature*

By using units where Boltzmann's constant, k_B, is also set to unity, one can also measure temperature in electron volts. For systems in thermal equilibrium, the temperature sets the scale of the kinetic energy of particles, $\text{KE} \sim k_B T$. Temperatures in SI units are measured on the Kelvin scale, where absolute zero occurs at zero Kelvin, $0\,\text{K} = -273°\text{C}$.

- The CMB is the leftover radiation from the Big Bang. Its temperature was measured by COBE in 1992 to be $T_{CMB} = 2.7\,\text{K}$. What is this in eV? This is the current temperature of the Universe.
- Big Bang nucleosynthesis (BBN) refers to the formation of the light elements (hydrogen, helium and lithium) in the early Universe. This occurs at around an MeV. What is this in kelvin?
- Einstein's famous formula $E = mc^2$ gives the energy of particles at rest, the 'rest energy'. For particles with relativistic momentum $p = \gamma m v$, moving at speed v and 'boost factor' $\gamma = 1/\sqrt{1 - v^2/c^2}$, it is given by

$$E^2 = m^2 c^4 + p^2 c^2. \tag{1.7}$$

The kinetic energy is $\text{KE} = p^2/2m$. Assuming thermal kinetic energy, at what temperature does the kinetic energy equal the rest energy? What is the velocity, in units of c, at this temperature?
- We call particles where the kinetic energy is greater than or equal to the rest energy 'relativistic'. At what temperature, in kelvin, did protons become non-relativistic? What mass of particles are just becoming non-relativistic, out in the cosmos, today?

Problem 4. *The Planck Scale and the String Scale*

Newton's constant is a fundamental constant that characterises the strength of gravity. It appears in general relativity also.

- The 'reduced Planck mass' is defined by $1/M_{\text{Pl}}^2 = 8\pi G$. What is this in GeV? In kg? Notice how much larger this is than energy scales associated with, for example, the LHC. How much larger is M_{Pl} than the maximum energy of the LHC? Measuring masses in units of the reduced Planck mass makes all remaining physical quantities dimensionless.

- Convert the reduced Planck mass to a length, the Planck length ℓ_P. Give your answer in m. This is the length scale associated with quantum gravity.
- What is the Planck time, t_P, in s? Without studying quantum gravity we can know nothing about the evolution of the Universe on time scales shorter than this. What is the current age of the Universe in Planck units?
- String theory has a natural length scale, the string length, l_s. If the mass scale associated with this is $M_s = 1$ TeV, what is the string length in m?
- The extra dimensions of string theory are expected to be on the scale of $l_s \approx 1/M_s$. Experimental limits on the size of extra dimensions are on the scale of microns, where $1 \ \mu m = 10^{-6}$ m. What string scale, M_s, does this correspond to in eV?
- The cosmological constant Λ is given by $\Lambda = \Omega_\Lambda 3H_0^2/8\pi G$, where $\Omega_\Lambda = 0.68$ is approximately the inferred value from cosmological observations. What is this in eV? In Planck units, where $M_{Pl} = 1$? In string units, if $M_s = 1$ TeV? The smallness of this number is a source of 'fine-tuning' in physics, since Λ is expected to be of order one in the fundamental units.

PART I

Evidence for Dark Matter

Chapter Two

Virial Theorem and Spherical Collapse

One of the first, and most famous, pieces of evidence for the existence of dark matter came from Fritz Zwicky in 1933 [2] while observing the Coma galaxy cluster. Applying the virial theorem to the motion of galaxies within the cluster, he noticed a gravitational 'anomaly'. Based on the velocities of the galaxies within the cluster, there must have been much more mass than what was observed to create a strong enough gravitational force to hold the galaxies together within the cluster.

To be able to understand how he achieved this, we first need to derive the previously mentioned virial theorem, one of the most important and useful results in astrophysics.

2.1 THE VIRIAL THEOREM

2.1.1 Derivation

We will derive the virial theorem by considering the phase space of a system, following Ref. [3]. For mechanical systems, such as galaxies bound by gravity, the phase space is a set of generalised coordinates, $\vec{q} = (q_1, \ldots, q_{\mathcal{N}})$, and momenta, $\vec{p} = (p_1, \ldots, p_{\mathcal{N}})$, for each of the N particles (here, galaxies) within the system. For three dimensions of space and no constraints, $\mathcal{N} = 3N$.

The virial theorem is derived in statistical mechanics. The phase space distribution function, $f(\vec{q}, \vec{p})$, is the probability of a state in a 'phase space cell' $d\vec{q}d\vec{p}$, centred on $(\vec{q}, \vec{p})$, given by

$$f(\vec{q}, \vec{p}) = \frac{\exp[-\beta H]}{Z}, \tag{2.1}$$

H is the Hamiltonian, and

$$\beta = \frac{1}{k_B T}, \tag{2.2}$$

where k_B is the Boltzmann factor and T is the temperature of the system. Z is a normalisation factor known as the partition function given by

$$Z = \int \underbrace{d^{\mathcal{N}} q d^{\mathcal{N}} p}_{\text{phase space integral}} \exp[-\beta H], \tag{2.3}$$

in which $\mathcal{N}$ is the number of degrees of freedom. For N particles in three dimensions we have a total of $\mathcal{N} = 3N$ degrees of freedom.

The *distribution function defines averages of observables, $\mathcal{O}$*:

$$\langle \mathcal{O} \rangle = \int \underbrace{\mathrm{d}^N q \, \mathrm{d}^N p \, \mathcal{O} f \, (\vec{q}, \vec{p})}_{\text{Average over phase space with 'weight' } f} \quad . \tag{2.4}$$

The Hamiltonian is given by

$$H = K(\vec{p}) + U(\vec{q}), \tag{2.5}$$

where K and U are the kinetic and potential energies of the system respectively. The total kinetic energy is given by the sum of the kinetic energies of each of the N particles within the system

$$K(\vec{p}) = \frac{1}{2} \sum_{i=1}^{N} m_i |v_i|^2, \tag{2.6}$$

where m_i are the particle masses and v_i are the particle speeds which can be decomposed into the three orthogonal directions:

$$|v_i|^2 = |v_{ix}|^2 + |v_{iy}|^2 + |v_{iz}|^2. \tag{2.7}$$

We can then rewrite the kinetic energy as a sum over each of the velocity components for each particle in the system

$$K = \frac{1}{2} \sum_{\nu=1}^{3N} \alpha_\nu p_\nu^2, \tag{2.8}$$

where $\alpha_\nu = 1/m_i$. We note that

$$\frac{\partial K}{\partial p_\nu} = \alpha_\nu p_\nu = \frac{\partial H}{\partial p_\nu}. \tag{2.9}$$

This allows us to finally write the kinetic energy in terms of the Hamiltonian as

$$K = \frac{1}{2} \sum_{\nu=1}^{3N} p_\nu \frac{\partial H}{\partial p_\nu}. \tag{2.10}$$

Next, we want to calculate the total potential energy within the system. This is done by summing over the potential energy that exists between each pair of particles i and j:

$$U_{\text{tot}} = \sum_{i=1}^{N} \sum_{j=1}^{N} V(r_{ij}), \tag{2.11}$$

where we define a general central potential $V(r_{ij}) = \alpha r_{ij}^n$ with $r_{ij} = |\vec{r}_i - \vec{r}_j|$. Taking the derivative of this general potential with respect to r_{ij}, the total potential energy can then be rewritten as

$$U_{\text{tot}} = \sum_{i=1}^{N} \sum_{j=1}^{N} \alpha r_{ij}^{n},$$

$$= \sum_{i=1}^{N} \sum_{j=1}^{N} \frac{1}{n} \frac{\partial V(r_{ij})}{\partial r_{ij}} r_{ij}. \tag{2.12}$$

We then *choose our generalised coordinates* such that $q_v = r_{ij}$ (the power of Hamiltonian mechanics!). Considering that $n = -1$ for gravity, we obtain our final potential of

$$U_{\text{tot}} = -\sum_{v=1}^{3N} q_v \frac{\partial H}{\partial q_v}. \tag{2.13}$$

Now that we have expressed the kinetic and potential energies of our system in terms of $p \frac{\partial H}{\partial p}$ and $q \frac{\partial H}{\partial q}$ respectively, we can calculate their average values using Eq. (2.4) giving

$$\left\langle u_v \frac{\partial H}{\partial u_v} \right\rangle = \int d^N q \, d^N p \, \frac{e^{-\beta H}}{Z} u_v \frac{\partial H}{\partial u_v}, \tag{2.14}$$

where u_v is either q_v or p_v. Applying the chain rule, we can rewrite this as

$$\left\langle u_v \frac{\partial H}{\partial u_v} \right\rangle = \frac{1}{\beta} \int d^N q \, d^N p \, \frac{\partial}{\partial u_v} \left[\frac{e^{-\beta H}}{Z} \right] u_v. \tag{2.15}$$

Integrating by parts, we find that

$$\left\langle u_v \frac{\partial H}{\partial u_v} \right\rangle = \left[u_v \frac{e^{-\beta H}}{Z} \right]_{-\infty}^{\infty} + \frac{1}{\beta} \int d^N q \, d^N p \, \frac{e^{-\beta H}}{Z}. \tag{2.16}$$

If our system is bound, meaning none of the particles have enough kinetic energy to escape to infinity, then the boundary term (square brackets) vanishes. Furthermore, the total integral of the phase space density (the second term inside the integrand) is normalised to be unity by definition. We are therefore left with

$$\left\langle u_v \frac{\partial H}{\partial u_v} \right\rangle = \frac{1}{\beta} = k_B T. \tag{2.17}$$

This tells us that each of the degrees of freedom contributes $k_B T$ to the average energy (a very general result in statistical mechanics).

Using this result to calculate the average kinetic and potential energies of our system as

$$\langle K \rangle = \frac{1}{2} \sum_{v=1}^{3N} \left\langle p_v \frac{\partial H}{\partial p_v} \right\rangle,$$

$$= \frac{3}{2} N k_B T, \tag{2.18}$$

and

$$\langle U \rangle = -\sum_{\nu=1}^{3N} \left\langle q_\nu \frac{\partial H}{\partial q_\nu} \right\rangle,$$

$$= -3Nk_B T,$$

(2.19)

respectively, we can finally write down the virial theorem which simply states

$$\boxed{2\langle K \rangle = -\langle U \rangle}$$ (the virial theorem).   (2.20)

This is the virial theorem for systems in gravitational equilibrium.

2.1.2 Zwicky and the Coma Cluster

We can now apply the virial theorem to Fritz Zwicky's 1933 observation of the Coma galaxy cluster, with the numbers that were known at the time. The Coma Cluster is a collection of around 800 galaxies spanning approximately 3×10^5 pc. By measuring the light from these galaxies (called 'nebulae' at the time) their mass can be estimated to be around $10^9 M_\odot$ each.

The average potential energy at some radius R from the centre of the cluster is given by

$$\langle U \rangle = -\int_0^R \frac{GM(r)}{r} dm.$$

(2.21)

For simplicity, we will assume that the mass within the cluster is uniformly distributed:

$$M(r) = \frac{4}{3}\pi r^3 \rho \Rightarrow dm = 4\pi r^2 \rho dr,$$

(2.22)

where ρ is the average density of the cluster. This is often called a 'tophat' density profile. Substituting into Eq. (2.21) we find

$$\langle U \rangle = -\int_0^R G \frac{16\pi^3}{3} \rho^2 r^4 dr,$$

$$= -\frac{3}{5} \frac{GM_{\text{tot}}^2}{R},$$

(2.23)

where we have defined the total cluster mass M_{tot} to be

$$M_{\text{tot}} = \frac{4\pi}{3}\rho R^3.$$

(2.24)

To calculate the average kinetic energy we will use the measured Doppler velocity dispersion (explained shortly in chapter 3) and the equation

$$\langle K \rangle = \frac{1}{2} M_{\text{tot}} \sigma_v^2.$$

(2.25)

Note that this is essentially the same as the kinetic energy equation we are most used to but now averaged over an ensemble of objects by considering the collective velocity dispersion σ_v. Now, using the virial theorem with our estimated value for the mass of the 800 galaxies we can see what we would expect the velocity dispersion to be:

$$\sqrt{\frac{3}{5}\frac{GM_{\text{tot}}}{R}} = \sigma_v,$$

$$= 80\text{km s}^{-1}.$$

(2.26)

However, the value that Zwicky observed was actually about 1000 km s^{-1}. Seeing this disagreement between the predicted and observed values, Zwicky realised that the mass of the cluster had been dramatically underestimated.

We can use the observed velocity dispersion to calculate the 'real' cluster mass by rearranging the virial theorem to give

$$M_{\text{tot}} = \frac{5}{3}\frac{R}{G}\sigma_v^2,$$

$$= 1.25 \times 10^{14} M_\odot.$$

(2.27)

This value for the cluster mass is around 200 times greater than that predicted from the visible mass alone. Therefore, based on Zwicky's measurements, only around 1% of the matter within the Coma Cluster is visible. The remaining 99% is unseen dark matter. Astronomers often use the phrase 'mass-to-light ratio' to say how much of the inferred mass of an object comes from stars and other bright things like hot gas ('light') compared to the total. For our treatment of the Coma Cluster here, the mass-to-light ratio would be $99/1 = 99$.

One question we have so far failed to address is what exactly is meant by M and R in the virial theorem. These are usually taken to be the virial mass M_{vir} and the virial radius R_{vir}, that is, the mass and radius for which the virial theorem is exactly true. In astrophysics, it can be difficult to define the total mass and radius of an object, since the object is made up of many diffuse pieces (stars, gas, etc.). This is particularly true when the majority of that object is invisible to us. However, we can make things a little easier by relating the virial mass and radius using the theory of spherical collapse.

2.2 SPHERICAL COLLAPSE

Now we are going to consider the gravitational collapse of a distribution of matter and apply the virial theorem to it. This will give us some guidance on what to expect of gravitationally bound systems like galaxies.

Consider a region of matter that is overdense compared to its otherwise homogeneous surroundings. As before, for simplicity, we will approximate this perturbation to have a tophat density profile given by

$$\rho(r) = \begin{cases} \rho_0 & \text{for } r < R \\ 0 & \text{for } r > R. \end{cases} \tag{2.28}$$

A spherical shell of mass m at a radius R will experience the usual gravitational force of attraction to the centre of the perturbation given by

$$F_{\text{grav}} = -\frac{GMm}{R^2}, \tag{2.29}$$

and hence, from Newton's second law, will have the equation of motion

$$\frac{\mathrm{d}^2 R}{\mathrm{d}t^2} = -\frac{GM}{R^2}. \tag{2.30}$$

This equation for the motion of a shell of matter can equally be derived from the Friedmann equation (see chapter 5) and so applies to lumps of matter in the expanding Universe.

This equation of motion has the parametric solution

$$R = A(1 - \cos\theta),$$

$$t = B(\theta - \sin\theta), \tag{2.31}$$

$$A^3 = GMB^2,$$

where A and B are parameters fixed by the initial conditions. The parameter θ spans from 0 to 2π. As shown in Fig. 2.1, the shell will expand from a point at $t = 0$ to a maximum $R = R_*$ at the point of turnaround when $t_* = \pi B$. Finally, the shell will collapse back to a point at $t_c = 2\pi B$. This can be related to the turnaround radius by

$$R_*^3 = \frac{GM}{\pi^2} t_c^2. \tag{2.32}$$

The full solution is depicted in Fig. 2.1 as the solid line.

For a perfectly homogenous overdensity, if the matter is completely non-interacting, all of the shells will collapse to zero at the exact same time. However, in reality, small irregularities in the clump of matter will prevent it from collapsing to a singularity (unless the density is sufficiently large to form a black hole, for which we would need to use full general relativity rather than this approximation). Instead, the matter will virialise, entering a statistical equilibrium between kinetic and potential energy, and form a self-supported gas: gravity acts inwards, and average kinetic energy acts as an effective pressure outwards. We can use this assumption of virialisation to build a more realistic picture of the overdensity's evolution after turnaround.

The total energy of the system is the sum of its kinetic and potential energy

$$E_{\text{tot}} = K + U. \tag{2.33}$$

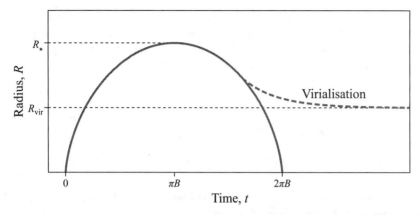

Figure 2.1 Spherical collapse of a tophat density perturbation. The solid line is the exact solution in spherical symmetry. The dashed line shows what happens to a system with small perturbations leading to a state of virial equilibrium at finite radius.

At turnaround, the kinetic energy vanishes $(T = 0)$ and instead all of the system's energy is in the form of gravitational potential energy:

$$E(t = \pi B) = U(t = \pi B) = -\frac{GM}{R_*}. \qquad (2.34)$$

Since energy is conserved, the total energy of the system is equal to that at turnaround $E_{\text{tot}} = E(t = \pi B)$. We expect that the system will later virialise and therefore (by definition) obey the virial theorem. Hence, we can combine Eqs. (2.20) and (2.33) to find

$$E_{\text{vir}} = E_{\text{tot}} = \frac{1}{2}\langle U \rangle. \qquad (2.35)$$

We know the total energy from the potential energy at turnaround, and therefore:

$$-\frac{GM}{R_*} = -\frac{1}{2}\frac{GM}{R_{\text{vir}}}. \qquad (2.36)$$

From this, we see that *the virial radius of the clump should be half of the radius at turnaround*:

$$R_{\text{vir}} = \frac{R_*}{2}. \qquad (2.37)$$

The state of the system going to virialisation is shown by the dashed line in Fig. 2.1.

Now, if we know one of the fitting parameters, A or B, for a clump of matter, we can estimate its final radius. We can therefore also estimate the final density of our virialised system to be

$$\begin{aligned} \rho_{\text{vir}} &= \frac{M}{V} = \frac{3M}{4\pi R_{\text{vir}}^3}, \\ &= \frac{3}{4\pi GB^2}. \end{aligned} \qquad (2.38)$$

To eliminate B we need an additional piece of information which we will obtain by considering how the background density (everything around our clump of matter) evolves. In a Universe containing only matter ($\Omega_m = 1$, see chapter 5), the background density obeys the Friedmann equation

$$\left(\frac{\dot{a}}{a}\right)^2 = \frac{8\pi G}{3}\rho_m,\tag{2.39}$$

and the conservation equation

$$\dot{\rho}_m = -3\left(\frac{\dot{a}}{a}\right)\rho_m,\tag{2.40}$$

where ρ_m is the average (background) matter density in the Universe, a is the scale factor, and overdot is a derivative with respect to time. It can be shown that this pair of equations has the solution

$$\rho_m = \frac{1}{6\pi G t^2}.\tag{2.41}$$

Therefore, at the point in time at which the overdensity becomes virialised, the background density is equal to

$$\rho_m = \frac{1}{24\pi^3 G B^2}.\tag{2.42}$$

Hence, any virialised 'halo' is denser than the cosmic background by a factor Δ_{vir} given by

$$\Delta_{\text{vir}} = \frac{\rho_{\text{vir}}}{\rho} = 18\pi^2.\tag{2.43}$$

The virial factor, Δ_{vir}, in Eq. (2.43), often rounded to be $\Delta_{\text{vir}} \approx 200$, is the most important result of this section. It is the characteristic overdensity of any lump of matter when it is in virial equilibrium. If that lump of matter has some radial density profile $\rho(r)$, then we can define a virial radius, R_{vir}, as the radius at which the average density is $\Delta_{\text{vir}}\rho_m$: all the stuff contained within R_{vir} should be in equilibrium. We can then define the *virial mass* of the object as the mass contained within the virial radius:

$$M_{\text{vir}} = \rho_m \Delta_{\text{vir}}\frac{4\pi}{3}R_{\text{vir}}^3.\tag{2.44}$$

This definition is used to define the masses of galaxies and clusters in astronomy and in theoretical studies of dark matter.

QUIZ

i. In statistical mechanics, we compute averages of observables by integrating the observable multiplied by what other quantity?
   a. The Hamiltonian.
   b. The distribution function.
   c. The temperature.
   d. The virial.

ii. The virial theorem states that for bound systems in statistical equilibrium, the kinetic energy is equal to what constant times the potential energy?
   a. $-1/2$.
   b. $1/2$.
   c. -2.
   d. 2.

iii. The mass element dm for a spherical mass distribution is:
   a. Constant.
   b. Proportional to the radius times the density times dr.
   c. Proportional to the radius squared times the density times dr.
   d. Proportional to the radius cubed times the density times dr.

iv. A system is observed to have a velocity dispersion of 5×10^5 meters per second and a radius of 3 million light years. Applying the virial theorem, what do we expect the mass to be (in kilograms)?
   a. 3×10^{45}.
   b. 2×10^{40}.
   c. 7×10^{46}.
   d. 2×10^{44}.

v. A system is observed to contain 10^{12} Sun-like stars, yet estimates from the virial theorem imply that it has a total mass of 10^{45} kilograms. What is its mass-to-light ratio in units of solar quantities?
   a. 10.
   b. 100.
   c. 500.
   d. 1.

vi. During spherical collapse, a system virialises when what conditions are met? Choose multiple options.
   a. The radius is equal to one-half of the turn around radius.
   b. The radius is equal to the turn around radius.
   c. The kinetic energy is equal to the potential energy.
   d. The kinetic energy is equal to $-1/2$ the potential energy.
   e. The density is about 200 times the cosmic density.

f. The density is about 100 times the cosmic density.

g. The energy is proportional to the virial, Q.

Answers

i. b; ii. a; iii. a; iv. c; v. d; vi. c; vii. a, d, e

Chapter Three

Rotation Curves

Skipping over a lot of the historical development (see discussion in chapter 1), the next part of the story of DM comes in the 1970s when Rubin and Ford [4] measured the rotation curve of the Andromeda Galaxy (M31). Using the Doppler effect to measure the velocity of gas in the galaxy as well as stars, they measured the orbital velocities at a larger distance from the galactic centre than was possible previously. What they realised, much like Zwicky, was that the visible mass was not enough to account for the large velocities, particularly at very large radii.

3.1 MEASURING ORBITAL VELOCITIES

We will assume that 'particles' (stars or gas, or DM) within a galaxy follow circular orbits. Consider a particle of mass m with an orbital radius of R. The total mass M contained within a sphere with a radius R given by the particle's orbital radius exerts the same gravitational force on the particle as a point mass at the origin as shown in Fig. 3.1:

$$F_g = \frac{GM(<R)m}{R^2}. \tag{3.1}$$

We can therefore calculate the body's centripetal acceleration from Newton's second law, which is related to its circular velocity via the equation

$$a = \frac{v_{\text{circ}}^2}{R}. \tag{3.2}$$

The circular velocity is therefore given by

$$v_{\text{circ}}^2 = \frac{GM(<R)}{R}. \tag{3.3}$$

Now that we have related the orbital velocity of an object to the enclosed mass, we can use measurements of the velocities of 'stuff' to effectively 'weigh' the galaxy.

We can calculate the radius of a particle's orbit around its galaxy using the distance between Earth and the galaxy D and the angle θ subtended on the sky by the radius R as shown in Fig. 3.2. To calculate the distance to the galaxy we can use measurements of a particular type of star called a Cepheid variable.

Cepheid variables are stars whose brightness oscillates with a period τ of order days, as sketched in Fig. 3.3.

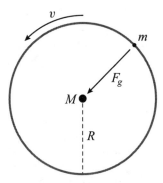

Figure 3.1 Circular orbit of a mass m around a larger mass M with velocity v, bound by the gravitational force F.

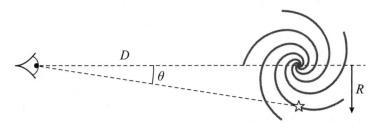

Figure 3.2 Radial distance geometry. The distance to the galaxy is D. The region within the galaxy subtends an angle θ. If the distance D is known, and θ measured, then we can infer the orbital radius R.

The flux of photons $\mathcal{F}$ from the star measured on Earth is related to its luminosity by the inverse square law

$$\mathcal{F} = \frac{L_0}{4\pi r^2}, \tag{3.4}$$

where r is the distance to the star. For stars closer than ~ 100 pc inside the Milky Way (MW), we can measure their distance using parallax. Then, by measuring the flux of light from nearby Cepheids on Earth, we can use Eq. (3.4) to infer the star's absolute luminosity, L_0. It is found that the period τ is directly proportional to the absolute luminosity L_0, and the constant of proportionality can be measured accurately.

Once we have measured the constant of proportionality, we can use this relationship to determine the luminosity of Cepheids from their period. Therefore, measuring the flux and using Eq. (3.4), we can infer their distance, even at large distances where the distance cannot be measured accurately using parallax. If we have a Cepheid in a distant galaxy (for example, Andromeda), then we can infer the distance to Andromeda. The orbital radius of other particles in the galaxy can now be determined using trigonometry from the angular position.

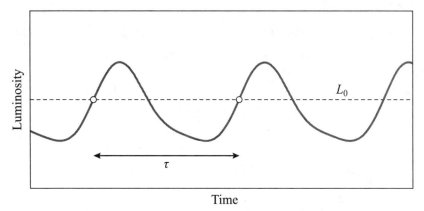

Figure 3.3 Sketch of variable brightness of a Cepheid star. The luminosity varies around an average value L_0 with period τ. For local Cepheids, where the distance is known due to parallax measurements, there is a tight relation between period and absolute luminosity. This can be used to infer the distance to more distant Cepheids by measuring their apparent luminosity and period.

Next, we need to determine the particle's circular velocity, which can be done as long as it gives off light. This is done by measuring the Doppler effect (see Fig. 3.4). As a light source moves towards or away from us, the wavelengths of the photons we receive are compressed (blue-shifted) or expanded (red-shifted) respectively. The wavelength of light measured, λ_r, is related to the wavelength of the source λ_s according to

$$\frac{\lambda_r}{\lambda_s} = \sqrt{\frac{1+\beta}{1-\beta}}, \tag{3.5}$$

where β is given by the ratio of the emitter's speed relative to the observer, v, and the speed of light, c:

$$\beta = \frac{v}{c}. \tag{3.6}$$

How do we know the rest frame wavelength? When electrons move to a lower energy level in an atom, they emit photons of a very specific frequency (recall $v = c/\lambda$). We can measure these wavelengths in the lab allowing us to know λ_s to very high precision. One particular emission line used commonly in astronomy is the 'hyperfine' spin-flip transition of neutral hydrogen, from electron spin parallel to anti-parallel with the proton spin, which has a rest frame wavelength of 21 cm and a frequency of 1.4 GHz. Neutral hydrogen fills galaxies, in the space between stars, and also at very large radii where stars are almost absent. Therefore, from the shift in the frequency of the 21 cm line, we can infer the relative velocity of the neutral hydrogen using the Doppler formula and thus the rotation speed of the host galaxy. For observations of stars, one could also use many other atomic lines of elements known to be inside stars.

As shown in Fig. 3.4, the star's total velocity is a combination of the star's circular velocity v_c, the galaxy's Hubble velocity v_H due to the expansion of the Universe and

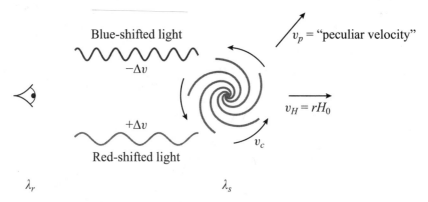

Figure 3.4 Doppler shift.

its 'peculiar velocity' v_p due to local motion. These various velocities are related to the velocity that we observe from the Doppler shift via simple trigonometry.

Rubin and Ford measured the velocity of 43 different regions of the galaxy M31 from the neutral hydrogen gas (in outer regions) and stars (in inner regions) and found a mean of $\bar{v} \approx 300$ km s^{-1}. Since M31 is very close to our own MW galaxy, we can take the Hubble velocity to be zero ($v_H = 0$) enabling us to take the peculiar velocity to be equal to the mean ($v_p = \bar{v}$).

Hence, Rubin and Ford were able to measure the Doppler velocity at increasing distance from the centre of M31 and therefore calculate the orbital velocity of each region as a function of radius.

3.2 MODELLING THE VISIBLE CONTENT

We can estimate what we expect the velocity curves of galaxies to look like by noting that most of the stars, which account for most of the visible mass, are contained within a central bulk. As a first approximation, therefore, we will assume the galaxy has a top-hat density profile (recall Eq. (2.28)) with a radius of R_*, that is, stars with a constant density ρ_0 within R_* and no stars outside R_*. We can integrate this density profile to find the mass contained with a radius $R < R_*$:

$$M(<R) = \int_0^R d^3x\rho(r),$$

$$= 4\pi\rho_0 \int_0^R r^2 dr, \tag{3.7}$$

$$= \frac{4}{3}\pi\rho_0 R^3.$$

For radii greater than R_* the contained mass is constant:

$$M(>R_*) = \frac{4}{3}\pi\rho_0 R_*^3. \tag{3.8}$$

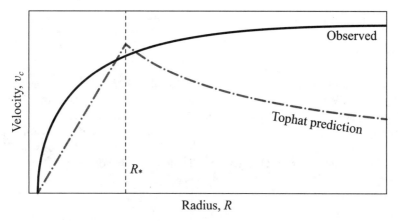

Figure 3.5 Sketch of a velocity curve of a typical galaxy.

We can relate this to the circular velocity at R using Eq. (3.3) to give

$$v_c(R) \sim \begin{cases} R & \text{for } R < R_*, \\ R^{-1/2} & \text{for } R > R_*. \end{cases} \tag{3.9}$$

The dependence of circular velocity v_c on radius R is known as the *rotation curve* of the galaxy.

As shown in Fig. 3.5, this rotation curve does not match the observed velocity curve very well. In particular, we see that the observed velocity of tracers (stars and gas) becomes almost constant at large radii rather than falling with the inverse of the square root of the radius as expected from the density of visible matter (which is roughly constant at small radii and roughly zero at large radii). This implies that there is a large contribution to the mass of galaxies at larger radii, which is unaccounted for in the visible matter. We find that this dark matter surrounds galaxies in a large spherical 'halo'. In fact, since the velocity curve is approximately constant at large radii, we can infer that the internal mass increases proportionally to the radius and therefore $\rho(R) \sim R^{-2}$, and so the enclosed mass goes as $M(R) \propto R$ and is still increasing at large radii. Eventually, DM rotation curves also must turn over (because the amount of mass is finite), but this is at radii larger than usually included in a measurement (at the very largest distances we infer the distribution of DM using statistical properties of the distribution of galaxies, i.e. cosmology; see chapter 5).

To make progress in predicting the distribution of dark matter in galaxies, we need to build a more realistic model of the various galactic components. For example, many galaxies have an optically visible bulge and disk, as well as contributions from gas and the interstellar medium (ISM) and, most importantly for us, dark matter.

First, let's consider the luminosity of galaxies, which is what can be observed (see Ref. [5], 2.162). The luminosity density is $j(\vec{r})$ which has units L pc^{-3}. The surface brightness is $I_b(R)$ and has units luminosity per unit area, that is, L pc^{-2}, projected on a sphere of radius R in a given direction. The surface brightness is defined as the projection of $j(\vec{r})$:

$$I_b(R) = \int_0^R dr\, j(\vec{r}). \tag{3.10}$$

Some textbooks use the units 'per solid angle', that is, $\hat{I}_b(R) = I_b/4\pi$ and $[\hat{I}_b] =$ L pc^{-2} sr^{-1}. We use units like this because telescopes view things on the 'celestial sphere'. Models for $j(\vec{r})$ include the 'Plummer profile'. The surface mass density is Σ (defined as in Eq. (3.10) replacing j with mass density ρ), and a constant 'mass-to-light ratio' means that $M/L = $ const., that is, $\Sigma \propto I_b$. The radial density profile is given by integrating the surface mass profile along the line of sight:

$$\rho(R) = \frac{1}{\pi} \int_R^\infty \frac{\partial \Sigma}{\partial x} \frac{1}{\sqrt{x^2 - R^2}} dx. \tag{3.11}$$

If each component has a mass profile $M_i(R)$, the total orbital velocity is given by

$$v_c(R) = \sqrt{\sum_i v_i^2(R)}, \tag{3.12}$$

where

$$v_i(R)^2 = \frac{GM_i(R)}{R}. \tag{3.13}$$

That is, since mass is additive, the velocity sums in quadrature.

Observations show that galactic disks typically have exponential surface luminosity densities I_b, that is, the brightness of the galaxy falls off exponentially with increasing distance from its centre. Both the bulge and disk can be modelled using the so-called 'Sersic profile' in terms of their half-light radius, $R_{\frac{1}{2}}$, and the luminosity density at that radius, $I_{\frac{1}{2}}$, as

$$I_b(R) = I_{\frac{1}{2}} \exp\left(-b_n\left[\left(\frac{R}{R_{\frac{1}{2}}}\right)^{1/n} - 1\right]\right), \tag{3.14}$$

where n, called the Sersic index, is a free parameter between $1/2$ and 10 which must be fit to the galaxy in question from observations, and b_n is approximately $2n - 1/3$.

The gravitational potential is given from the integral over mass elements:

$$\Phi = -\int \frac{G}{r} dM, \tag{3.15}$$

and the mass element is

$$dM = R'\Sigma(R')dR'd\phi, \tag{3.16}$$

where the coordinates are related to spherical polars as

$$r = \sqrt{R^2 + R'^2 - 2RR'\cos\phi}. \tag{3.17}$$

The total disk mass is $2\pi \Sigma_0 R_d^2$ for surface density $\Sigma(R) = \Sigma_0 e^{-R/R_0}$. The potential gives the circular velocity as

$$v^2 = R \frac{\partial \Phi}{\partial R}. \tag{3.18}$$

Using these definitions one can show that

$$v^2(R) = 2\pi G \int K(R/R') R' \Sigma(R') dR', \tag{3.19}$$

where K is some Kernel (for more details on the necessary integrals, see Ref. [5], Ch. 2). It thus follows that the shape of v^2 follows $R' \Sigma(R')$. The disk rotation curve thus falls sharply beyond the scale radius R_0.

Deviations of stellar motion from circular orbits act as a pressure perpendicular to the plane. When gravity and the velocity dispersion in the z-direction are in equilibrium, the pressure per unit volume is

$$P = c_s^2 \rho, \tag{3.20}$$

where the sound speed c_s^2 is z independent.

If we take the density of the disk to be approximately constant, and the displacement relative to the galactic plane is sufficiently small relative to the radius of the disk, then the acceleration due to gravity along z includes a term which is proportional to that displacement:

$$g_z \sim \alpha z, \tag{3.21}$$

where α is a fitting parameter to be found. The gravitational force per unit volume is $F_g = g_z \rho$ where ρ is the mass density. The pressure force per unit volume is dP/dz, and equilibrium implies that

$$\frac{dP}{dz} = g_z \rho = \alpha \rho z. \tag{3.22}$$

We can estimate v_z, and thus c_s, from thermodynamics. Assuming that the gas is composed of mostly hydrogen,

$$v_z = \sqrt{\frac{k_B T_{\text{gas}}}{2 m_H}}. \tag{3.23}$$

Combining Eqs. 3.20 and 3.23, we can integrate to find the density profile:

$$\rho = \rho_0 \exp\left[-\left(\frac{z}{z_0}\right)^2\right], \tag{3.24}$$

where $z_0 = v_z / \sqrt{\alpha}$. This is known as the Gaussian hydrostatic disk.

Gas and ISM can be modelled with a density profile in terms of both the radius R and the height from the galactic plane z as

$$\rho_g(R, z) = \frac{\Sigma_g}{2 z_g} \exp\left[-\frac{R}{R_g} - \frac{R_m}{R} - \frac{|z|}{z_g}\right], \tag{3.25}$$

where R_g is the gas scale radius, z_g is the gas thickness and R_m accounts for a possible hole in the gas density in the MW at $R \sim 4$ kpc. The ISM is typically much thinner and more extended than the stellar disk with a width of around 80 pc compared to a width of around 300 pc for the stellar disk. All disk components have different thicknesses z_0 given by their velocity dispersion and temperature (see Ref. [6], Table I, for more details). For example, consider the stellar disk with two components, the thin disk and the thick disk:

$$\rho_d(R, z) = \sigma_d e^{-R/R_d} \left(\frac{\alpha_0}{2z_0} e^{-|z|/z_0} + \frac{\alpha_1}{2z_1} e^{-|z|/z_1} \right), \tag{3.26}$$

where $\alpha_0 + \alpha_1 = 1$ by normalisation. In the MW, $\alpha_1 = 0.07\alpha_0$, the thin disk of young stars has $z_0 \approx 300$ pc and the thick disk of older, lower metallicity stars has $z_1 \approx 1$ kpc.

3.3 DARK MATTER FROM $v_C(R)$

Importantly, the surface luminosity density $I(R)$ is measurable for both stars and gas. This allows us to calculate the visible mass component. Since we can infer the mass profile from the rotation curve, $v_c(R)$, we can subtract the visible component to infer the density profile of dark matter:

$$M_{\text{DM}}(R) = \frac{R}{G} \left(v_{\text{tot}}^2 - v_{\text{vis}}^2 \right). \tag{3.27}$$

We first need to find v_{vis}, and we will assume that the surface luminosity density is proportional to the surface mass density. This gives us a free parameter, Υ:

$$\Sigma_0 = \Upsilon I_0. \tag{3.28}$$

This free parameter has to be found separately for the stellar disk, the central bulge and the gas.

Considering all of the visible components together for a moment, we can impose an upper limit of Υ_{max} from

$$v_{\text{tot}} = \sqrt{v_{\text{vis}}^2 + v_{\text{DM}}^2}, \tag{3.29}$$

since this implies $v_{\text{vis}} \leq v_{\text{tot}}$. Then, since $v_{\text{vis}} \propto \Upsilon^{1/2}$, the maximum possible mass-to-light ratio is found by considering the limit $v_{\text{vis}} = v_{\text{tot}}$. The case in which $\Upsilon = \Upsilon_{\text{max}}$ is known as the maximal disk model. A sketch of how the various components combine to form the total that we observe is shown in Fig. 3.6, while an example from real data is shown in Fig. 3.7.

Although we can infer a great deal about the nature of the dark matter density profile, the specific shape is dependent on a number of variables. This gives rise to a level of disk-halo degeneracy. Therefore, while we can set limits on DM, there remains some ambiguity.

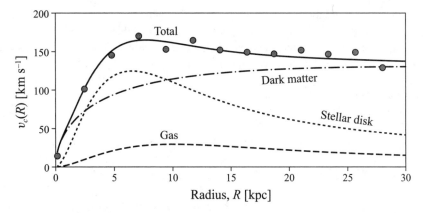

Figure 3.6 Sketch of the total rotation curve for a galaxy containing gas, disk, bulge and DM. See colour insert.

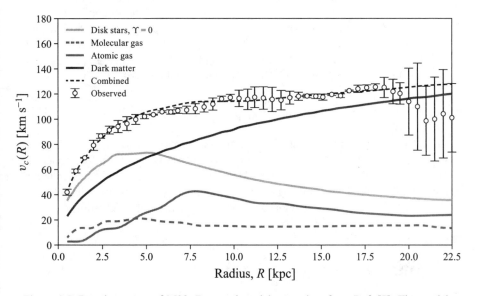

Figure 3.7 Rotation curve of M33. Data and models are taken from Ref. [7]. The model shown takes the mass-to-light ratio as a free parameter. Best-fit NFW parameters are $c = 4.55 \pm 0.01$ and $v_{\text{vir}} = 139.9 \pm 0.5$, where v_{vir} is computed from the virial mass defined in Eq. (2.44) with $\Delta_{\text{vir}} = 200$. See colour insert.

There are a number of models for DM density profiles which are motivated by both observations and simulations. By far the most popular is the Navarro–Frenk–White (NFW) profile given by

$$\rho_{\text{DM}}(r) = \frac{\rho_s}{\left(\frac{r}{r_s}\right)\left(1 + \frac{r}{r_s}\right)^2}, \tag{3.30}$$

where ρ_s and r_s are the 'scale density' and 'scale radius' respectively (this profile is discussed in more detail in section 6.3). We see that the NFW profile has a slope of r^{-1} at small radii and a steeper slope of r^{-3} at larger radii. Importantly, at the scale radius, the NFW profile also gives the previously predicted slope of $\rho \sim r^{-2}$. Other commonly used DM density profiles include the Einasto and Burkert profiles, and the 'generalised' NFW profile.

The mass profile of a halo is

$$M(<R) = \int \rho(r) d^3 x = 4\pi \int r^2 \rho(r) dr. \tag{3.31}$$

For the NFW profile, this integral can be performed analytically (try it!). The mean density within a radius R is $\Delta = M(<R)/(4\pi R^3/3)$. The NFW mass is divergent if we integrate to $R = \infty$, so we regulate the integral using the virial mass and virial overdensity we defined in Eqs. (2.43) and (2.44). We define the virial radius R_{vir} as the radius that encloses a region of mean density equal to $\rho_{vir} = 200\rho_{crit}$: this is the maximum radius within which we should use the NFW profile, and beyond it the halo is not virialised, that is, not gravitationally collapsed and in equilibrium. We define the 'concentration parameter' $c = R_{vir}/r_s$, which parameterises how far the scale radius is separated from the virial radius. For galaxies like the MW, $c \sim 10$, and thus the density at the scale radius is around $10^3 \Delta_{vir}\rho_{crit} \sim 10^5 \rho_{crit}$. Using the analytical result for $M(<R)$ for NFW it is then possible to write ρ_s in terms of c:

$$\rho_s = \rho_{crit}\frac{200}{3}\frac{c^3}{\ln(1+c) - c/(1+c)}. \tag{3.32}$$

The NFW profile can thus be specified entirely in terms of c and $M_{vir} = 4\pi \Delta_{vir}\rho_{crit} R_{vir}^3/3$ or equivalently v_{vir}. This is the most common way to specify the properties of a halo. Even for profiles beyond NFW, it is still possible to specify a c parameter based on the radius when the logarithmic slope of the profile is equal to -2. This provides a universal way to compare halos by virial mass and virial radius and their characteristic size relative to this.

3.4 DARK MATTER IN THE MILKY WAY

3.4.1 General Remarks

In order to search directly for DM (see chapters 12 and 14) we need to know how much of it there is likely to be here on Earth: the local density, ρ_{DM}. We also need to know the DM velocity dispersion, σ_v, and the direction of our motion with respect to the DM halo.

First, let's discuss the location and motion of the Earth and the Sun in the MW. Our motion with respect to the rest frame of the halo has four components: the bulk rotational velocity of the stellar disk (the 'local standard of rest'), $\vec{v}_0$, the peculiar motion of the Sun with respect to this frame, $\vec{v}_{pec}$, the orbit of the Earth around the Sun and our local rotational velocity on the surface of the Earth. Let's neglect the last two, since they are small relative to v_0 (note that these motions in practice can lead to observationally relevant daily and annual modulation of the DM velocity through Earth, although the experimental measurement of such modulation would require a large number of detected DM particles).

There are a number of coordinate systems one can use in the MW. The 'standard galactic coordinate system' takes the Sun as the origin, with galactic longitude, l, and latitude, b, giving any other location on the sky. This is useful for observations to specify the locations of objects on the sky. For fitting properties of the MW, it is more useful to use coordinates with respect to the centre of mass of the DM halo. In spherical polars based on the centre of mass, the solar system is located at a radius of $R_\odot = 8.122 \pm 0.031$ kpc [8]. We can measure angles ϕ with respect to the Sun and define the polar axis relative to the rotation of the stellar disk, with 'galactic north' defined by a standard right-handed coordinate system. As we have already discussed, the stellar disk has a scale height, z (alternatively some polar angle, θ), and the Sun is at $z_\odot \approx 0.025$ kpc [9].

The motion of the Sun is specified in standard Cartesian coordinates, known as UVW coordinates in the MW. Taking the U axis along the radial direction towards the galactic centre and *left* handed, then W gives the vertical motion, and V gives the *right-handed* orbital motion specifying the direction of galactic North. In this coordinate system $\vec{v}_0 = (0, v_0, 0)$, and v_0 is the circular velocity in the gravitational potential of the MW one would find by azimuthally averaging, that is, the value of the rotation curve at $R_\odot$. The direction of V happens to be towards the constellation Cygnus.

There is a supermassive black hole (BH) at the centre of the MW, called 'Sagittarius (Sgr) A*'. Quite recently, the shadow of Sgr A* was spectacularly imaged by the Event Horizon Telescope, confirming that it is indeed a BH [10]. Assuming that Sgr A* is at the centre of mass of the MW, then determining the relative motion of the Sun and Sgr A* (the 'proper motion') gives the motion of the Sun with respect to the MW centre of mass. The circular velocity determined in this way is $v_c = 246 \pm 1$ km s^{-1}. The peculiar motion is $(U, V, W) = (11.1 \pm 1.5, 12.2 \pm 2, 7.3 \pm 1)$ km s^{-1}, and the circular motion and the peculiar motion together give the local standard of rest, $v_0 = 233 \pm 3$ km s^{-1} [11].

For determination of the DM density, we distinguish between two methods: *global* and *local* measurements. An up-to-date review of this subject (as of the time of writing) can be found in Ref. [12].

Global methods measure some average property of the MW and fit a model. An example of a global method is fitting the MW rotation curve. When constructing a rotation curve, we average out the angular data of stellar motion and fit the parameters of spherically or axially symmetric density profiles, $\rho(r, \theta)$. Other global methods include mass modelling of the total distribution function and various moments of it, but we won't discuss these further. We will discuss the rotation curve in a bit more detail below. Global methods have the advantage of giving an easy to interpret 'answer' for $\rho(r)$ 'everywhere'. However, we need to remember that it is only an average. If we want to know the DM density *here on Earth*, we might think we can simply use the value of $\rho(r)$ at $r = R_\odot$, but in reality there could be significant deviations from the average value. Global measurements tend to favour $\rho_{DM} \in [0.3, 0.5]$ GeV cm^{-3} or $[0.008, 0.013] M_\odot$ pc^{-3}.

Deviations from the average could occur due to the history of the MW: for example, there could be remnants of smaller DM halos that went into the formation of the MW still moving around that haven't yet been absorbed into the average spherical profile (these tend to go by the name of 'tidal streams' in the literature). Two other important examples for us come from the properties of the DM itself. If the DM is composed of relatively massive black holes (PBHs, see chapters 11 and 16), then the chance of one of these actually passing through the Earth at any given moment is thankfully very small (see Problem I.6), although there are enough of them on average in a large enough volume to make the dynamics of the MW work correctly. Another example is in the case of axion DM or any light bosonic DM (see chapter 10). In this case the local density of DM varies significantly over the de Broglie wavelength of the particle, which can be astrophysically significant for $m \ll 1$ eV (we will meet this phenomenon in chapter 15).

Due to these possible deviations from the average, local measurements of the DM density are also very important. Local measurements give the average value of ρ_{DM} in a much smaller volume of say a few kpc^3. Methods include vertical kinematics which we met at the end of section 3.2. The smallest scale measurements of ρ_{DM} work within distances of around 100 pc using individual stellar motions. These ultralocal measurements are subject to around 50% uncertainty at present but are important for the reasons mentioned above.

In summary, covering all the measurements reviewed in Ref. [12], we can say that the local density of DM is in the range $\rho_{DM} \in [0.2, 0.7]$ GeV cm^{-3}. This density is roughly equivalent to one proton in one millilitre: compare this to the roughly Avogadro's number (6×10^{23}) of protons in a millilitre of water, and you get a sense of how empty interstellar space is.

3.4.2 The Milky Way Rotation Curve

Now let's focus on the MW rotation curve, which will allow us to determine the average ρ_{DM} and v_c. The circular velocity of the Sun is the sum of the tangential velocity and the peculiar velocity. We describe here the recent measurement of Ref. [13], which makes use of the full 'phase space information' on stars, that is, their complete 3D coordinates and 3D motions given by the *Gaia* satellite, and determines the rotation curve outside 5 kpc using measurements of more than 23,000 red giant stars.

The MW is assumed to be axisymmetric. In order to determine the circular velocity, one needs to assume a model for the radial density profile of the tracer population (i.e. the red giants), which is the dominant source of uncertainty. As we have already seen, the typical choice is an exponential distribution with some scale length. Assuming this density profile and using the measured 3D velocities, the *Jeans equation* (see Ref. [5]) then allows one to determine the gravitational potential and thus the circular velocity.

The rotation curve is shown in Fig. 3.8. The value of the circular velocity at the location of the Sun is $v_c(R_\odot) = 229.0 \pm 0.2$ (stat) ± 2.02 (sys) km s^{-1}. With the circular velocity measured, one then fits the different components with velocities adding in quadrature. Ref. [13] includes the thin disk, the thick disk and the bulge, as described in section 3.2, which are assumed to be well measured and

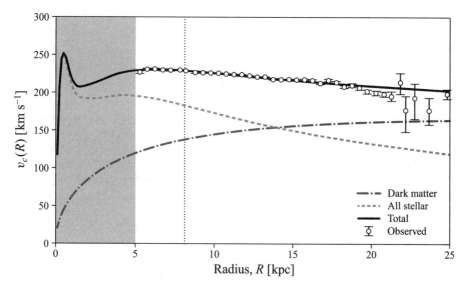

Figure 3.8 Milky Way rotation curve. Data and models are taken from Ref. [13]. Best-fit NFW parameters are $c = 12.8 \pm 0.3$ and $M_{vir} = 7.25 \pm 0.25 \times 10^{11} \, M_\odot$, which imply $\rho_s = 1.06 \pm 0.09 \times 10^7 \, M_\odot \, kpc^{-3}$, $R_{vir} = 189.3 \pm 2.2$ kpc and $R_s = 14.8 \pm 0.4$ kpc. The shaded region denotes the inner galaxy, which is not used in the fit. The vertical dotted line denotes the radial location of the Sun. See colour insert.

given by the fits of Ref. [14] (which give explicit forms for the circular velocity of each component). The DM is assumed to have an NFW profile, and the best fit NFW parameters are $c = 12.8 \pm 0.3$, and $M_{vir} = 7.25 \pm 0.25 \times 10^{11} \, M_\odot$, which imply $\rho_s = 1.06 \pm 0.09 \times 10^7 \, M_\odot \, kpc^{-3}$ $R_{vir} = 189.3 \pm 2.2$ kpc and $R_s = 14.8 \pm 0.4$ kpc. Plugging in the radius of the Sun then leads to a local DM density of $\rho_{DM}(R_\odot) = 0.30 \pm 0.03$ GeV cm^{-3}. We remind ourselves, however, that this is the density on average at all points with this radius, assuming a spherical DM distribution, and may not give the actual value at the true location of the Earth.

3.5 MODIFIED GRAVITY OR DARK MATTER?

It is possible to explain the unexpected velocity curves without requiring that galaxies are full of invisible matter. One alternative theory makes the assumption that it is our understanding of gravity that is incomplete. It is possible to think of ways to modify our models for gravity that would change its behaviour on galactic-length scales while preserving the relatively short-range behaviour that is so well tested on Earth.

Modified Newtonian dynamics (MOND) explains the flat rotation curves by modifying the gravitational force for accelerations below some threshold a_0:

$$a = \begin{cases} \dfrac{GM}{R^2} & \text{for } a \gg a_0, \\[2mm] \dfrac{\sqrt{a_0 GM}}{R} & \text{for } a \ll a_0, \end{cases} \tag{3.33}$$

which then gives a circular velocity,

$$v_c^2 = \begin{cases} \frac{GM}{R} & \text{for } a \gg a_0, \\ \sqrt{a_0 GM} & \text{for } a \ll a_0. \end{cases}$$

(3.34)

There are two problems that immediately arise with this approach. Firstly, the velocity curve can't remain flat indefinitely; it must start decreasing at some point. While this is solved naturally with DM due to the finite extent of the DM halo, MOND falls short. Additionally, a full modified theory of gravity must be embedded into some form of relativistic theory which produces all of the well-tested behaviour of general relativity. MOND, however, does have an advantage. Typically, for dark matter, we have two free parameters for each individual galaxy: the scale radius and scale density. In contrast, MOND requires only a single universal free parameter a_0, which is in theory a universal constant that must be the same for all galaxies (the constancy of a_0 must be demonstrated observationally).

MOND modifies the usual Newtonian acceleration a_N by a single function $\mu(x)$ as follows:

$$\mu\left(\frac{a}{a_0}\right) a = a_N,$$

(3.35)

where a is the modified acceleration. The required behaviour can then be produced with the function

$$\mu(x) = \frac{x}{\sqrt{1+x^2}}.$$

(3.36)

Using such a modification, it is found that MOND can fit galaxy rotation curves very well with parameter $a_0 = 1.21 \times 10^{-10}$ m s^{-2}. MOND also naturally explains the so-called 'Tully-Fisher' relation, which is the observation that the baryonic mass of a galaxy is related to the rotational velocity as $M_{\text{vis}} \propto v^p$ with observed power law $p \approx 3.4$. DM can also explain this relation but requires careful modelling of astrophysics. In clusters, even relativistic completions of MOND require the addition of some dark matter, and these still fall short in explaining cosmology (we will briefly meet such theories in section 6.2.3; see the reading list in the appendix for more discussion).

QUIZ

i. Assuming Newtonian gravity, why does the rotation curve for any mass distribution decay like $R^{-1/2}$ outside its maximum radius?
   a. Because baryonic matter does not gravitate strongly.
   b. It doesn't. Rotation curves are always flat.
   c. Circular $v^2 = GM/R$, and M is constant.
   d. The expansion of the Universe affects the outer region of the rotation curve.

ii. Again in Newtonian gravity, if a rotation curve is flat, what does this imply about the behaviour of the mass density?
   a. It is constant.
   b. It decreases with R^{-2}.
   c. It increases with R.
   d. It must be composed of dark matter.

iii. The surface brightness is (select all that apply):
   a. The luminosity density contained within a sphere.
   b. Equivalent to the luminosity density.
   c. Assumed to be proportional to the surface density of luminous (ordinary) matter.
   d. The luminosity per unit area on the surface of a sphere.

iv. The important components of a galactic rotation curve are (select all that apply):
   a. Bulge.
   b. Supernovae.
   c. Disk.
   d. Gas.
   e. Dark matter.
   f. Planets.

v. What units does the mass-to-light ratio, upsilon, have?
   a. (Solar) mass per (solar) luminosity, $M_\odot L_\odot^{-1}$.
   b. It is dimensionless.
   c. (Solar) mass per unit area, $M_\odot \, \mathrm{pc}^{-2}$.
   d. Electron volts, eV.

vi. In the standard smooth halo model, what is the approximate dark matter density at the location of the Earth in the Milky Way (select all that apply)?
   a. About 1 proton mass per cubic metre.
   b. About half a proton mass per cubic centimetre.

    c. About one elephant on the Earth.

    d. About one squirrel on the Earth.

vii. In MOND, the gravitational force law is modified:

    a. Near black holes.

    b. Due to quantum gravity.

    c. In areas of very low density.

    d. In areas of very high density.

    e. By dark energy.

viii. MOND fits rotation curves without the need for invoking dark matter. It does this:

    a. Very well.

    b. Very poorly.

Answers

i. c; ii. b; iii. c,d; iv. a,c,d,e; v. a; vi. b, d; vii. c; viii. a

Chapter Four

Gravitational Lensing and X-Rays

In chapter 2 we used Zwicky's argument based on the virial theorem to relate σ_v^2 to M in clusters. In this section, we will use X-rays and lensing to do the same, and study a famous object: the 'Bullet Cluster' [15].

4.1 HYDROSTATIC EQUILIBRIUM OF GAS AND CLUSTER TEMPERATURE

We met the concept of equilibrium between gravity and pressure in section 3.2. The equilibrium equation results from equating the force due to pressure, $F_{\text{pressure}} = -\nabla P$ (the minus sign results from pressure acting to oppose gradients), to the inward gravitational force on the gas, $F_{\text{gravity}} = ma$, and demanding $F_{\text{pressure}} + F_{\text{gravity}} = 0$:

$$\frac{dP}{dr} = \rho_{\text{gas}} a(r). \tag{4.1}$$

This applies in a volume element (i.e. P is pressure per unit volume, and ρ is mass density). For an ideal gas we have

$$PV = N k_{\text{B}} T \Rightarrow P = k_{\text{B}} T \frac{\rho_{\text{gas}}}{\mu}, \tag{4.2}$$

where μ is the molecular weight. We wish to solve for a radial distribution of gas so ρ and T are functions of r:

$$\frac{dP}{dr} = \frac{k_{\text{B}} T}{\mu} \frac{d\rho_{\text{gas}}}{dr} + \frac{\rho_{\text{gas}} k_{\text{B}}}{\mu} \frac{dT}{dr}. \tag{4.3}$$

Newton's law for a is

$$a(r) = \frac{-GM(<r)}{r^2}, \tag{4.4}$$

where the minus sign comes from gravity acting inwards, and now the mass is the *total* mass, that is, including any possible pressureless dark matter. Substituting into Eq. (4.1) gives

$$\frac{k_{\text{B}} T}{G\mu} \left[\frac{d \ln \rho_{\text{gas}}}{d \ln r} + \frac{d \ln T}{d \ln r} \right] r = -M(<r). \tag{4.5}$$

Thus, if we can measure $\rho_{\text{gas}}(r)$ and $T(r)$, we can infer the mass profile $M(<r)$. Assuming T is constant for $r > 1$ Mpc in a cluster, while $\rho \sim r^{-p}$ (such that

$d \ln \rho / r \ln r$ is negative and acts outwards as desired) with $1.5 < p < 2$ gives

$$k_B T \approx (1.3\text{--}1.8) \text{ keV} \left(\frac{M}{10^{14} M_\odot} \right) \left(\frac{1 \text{ Mpc}}{r} \right). \tag{4.6}$$

Hot gas emits light with energy fixed by the temperature. The following approximate conversions are useful to keep in mind:

- 1 neV is in MHz, radio frequencies.
- 1 μeV is in GHz, microwave frequencies.
- 1 meV is in THz.
- 1 eV is in the optical frequencies.
- 1 keV is in the X-ray band.
- 100 keV and above are γ-rays.

Let's assume we see a cluster of galaxies, and our X-ray telescope tells us that the gas in the cluster is $T \sim 10$ keV, while we measure the total visible mass of gas and stars $10^{14} M_\odot$. Thus, using Eq. (4.6), the total mass is $6 \times 10^{14} M_\odot$ and *85% of the mass of the cluster must be dark*, that is, non-emitting and pressureless.

Our assumptions were: T constant outside the core (observed); the gas profile $\rho \sim r^{-p}$ which follows from models and observations; $T \sim 10$ keV (observed); and the visible mass content. The visible mass content is stars and gas, but in a galaxy cluster the gas dominates.

Let's investigate the temperature further. Clusters emit due to thermal Bremsstrahlung. 'Bremsstrahlung' is German for 'braking radiation', that is, radiation from slowing down. The gas is ionized, and we say that free electrons 'brem' off the ions. In Feynman diagrams, this is shown in Fig. 4.1.

Momentum is conserved at vertices, so, for example, $q_1 = p_1 - k_1$ for the intermediate electron state. The free electron, e, interacts with the nucleus, N, and loses momentum from p_1 to k_2. Due to momentum conservation it must emit a photon of momentum k_1. The emitted Bremsstrahlung photon is γ_B, while $\gamma_\star$ is a *virtual photon* sourced by the nucleus. Classically, we can think of the electric field E of the charged particle as being composed of virtual photons: this causes deceleration and dipole emission.

Thermal electron energy is of order $k_B T$, and the photon energy is $\hbar \omega$. If the free electron density is n_e and the free ion density is n_Z for ions of charge Z then the emissivity is

$$I = 6.8 \times 10^{-38} T^{-1/2} \underbrace{e^{-\hbar\omega/k_B T}}_{\text{QM cut-off}} \overbrace{n_e n_Z Z^2}^{\text{plasma density}} \underbrace{g_{ff}}_{\text{``Gaunt factor''}} \text{ erg s}^{-1} \text{ cm}^{-3} \text{ Hz}^{-1}, \tag{4.7}$$

where T is measured in kelvin. The *Gaunt form factor* is the equivalent to the *Coulomb logarithm* (see, e.g., BT 1.2), a 'classical correction factor' which arises due to the integral over impact parameter, b, in the scattering, shown in Fig. 4.2.

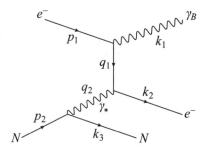

Figure 4.1 Feynman diagram for Bremsstrahlung of an electron, e. The electron loses energy by emitting the photon, γ_B. The exchange of momentum is mediated by the virtual photon, γ_*, which is classically thought of as the electric field around the charged nucleus, N.

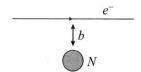

Figure 4.2 Definition of the impact parameter, b, for a scattering event.

The Gaunt factor is given by

$$g_{ff} = \frac{\sqrt{3}}{\pi} \ln \frac{b_{min}}{b_{max}}, \tag{4.8}$$

with

$$b_{max} \approx \frac{v}{4v}, \quad b_{min} \approx \frac{\hbar}{m_e v}. \tag{4.9}$$

Fixing the electron speed v from the kinetic energy $kT = \frac{1}{2}m_e v^2$, we see that g_{ff} depends on $v = \omega/2\pi$ and T and its average value is $\mathcal{O}(1)$. Thus we see that the model Eq. (4.7) allows us to find the gas density given the measured temperature.

4.2 GRAVITATIONAL LENSING

This section is based on Refs. [16, 17]. We work in the *geometric optics* regime, that is, light moves on straight lines independent of frequency. Our general picture of gravitational lensing is shown in Fig. 4.3.

4.2.1 Recap of Special Relativity, and Introduction to General Relativity

In special relativity (SR) we have the invariant distance:

$$s^2 = |\vec{x}|^2 - t^2 \quad \text{(with } c = 1\text{)}. \tag{4.10}$$

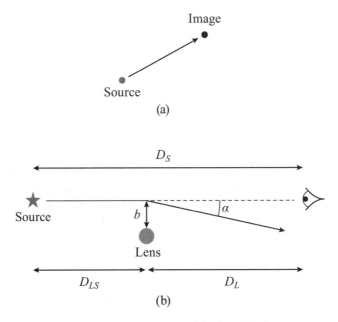

Figure 4.3 *Top*: image plane geometry. *Bottom*: side view. The impact parameter, b, and angle, α, can be thought of as vectors in the lens plane and image plane respectively.

We have contravariant 4-vector $x^\mu = (t, \vec{x})$ with 3-vector $\vec{x} = (x^1, x^2, x^3)$ and time is $t = x^0$. The invariant distance is formed using the Minkowski metric $\eta_{\mu\nu}$ by $s^2 = \eta_{\mu\nu} x^\mu x^\nu = x^\mu \eta_{\mu\nu} x^\nu = x^\mu x_\mu$. In flat space in SR, the covariant 4-vector x_μ has its indices lowered by the Minkowski metric. In most of this book we use the Minkowski metric with 'mostly positive' metric signature.[1] The Minkowski metric can be written out as a matrix:

$$\eta = \begin{pmatrix} -1 & 0 & 0 & 0 \\ 0 & 1 & 0 & 0 \\ 0 & 0 & 1 & 0 \\ 0 & 0 & 0 & 1 \end{pmatrix}. \tag{4.11}$$

The Minkowski spacetime diagram is shown in Fig. 4.4. Light rays move on null geodesics, defined by $s^2 = 0$, while matter with $v < c$ can only propagate in the $s^2 < 0$ region with timelike separations. Regions with $s^2 > 0$ are said to be spacelike separated and are not in causal contact with one another.

In GR we repeat the same activity with a more general *metric tensor*. The invariant separation is

$$s^2 = g_{\mu\nu} x^\mu x^\nu, \tag{4.12}$$

[1] I will let you know when we change signature conventions. The mostly positive case is more convenient in relativity, but the mostly negative case is more convenient in particle physics applications.

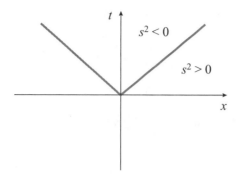

Figure 4.4 Minkowski spacetime diagram with future lightcone.

which we can write differentially as

$$ds^2 = g_{\mu\nu}dx^\mu dx^\nu. \tag{4.13}$$

If we think of dx^μ as providing a basis for the metric, then notice that the 'line element' Eq. (4.13) allows us to write all of the components of the metric tensor out using ds^2 and the basis. (If this way of thinking of dx^μ does not make sense to you, don't worry too much. If you want to know more, take a look in Ref. [18], Chapters 1 and 2. We are only interested here in how to work with tensor indices, not with their formal definition.)

The coordinates have their indices lowered by $g_{\mu\nu}$, as can any 4-vector A^μ such that $A_\nu = g_{\mu\nu}A^\mu$. We define tensors (i.e. multilinear maps) of rank (k, l) with k 'upstairs' contravariant indices and l 'downstairs' covariant indices: $T^{\mu_1\cdots\mu_k}{}_{\nu_1\cdots\nu_l}$. The indices on T can also be lowered with the metric, for example, $T_{\mu_1}{}^{\mu_2\cdots\mu_k}{}_{\nu_1\cdots nu_l} = g_{\mu_1\alpha}T^{\alpha\cdots\mu_k}{}_{\nu_1\cdots\nu_l}$.

Next we define the inverse metric, $g^{\mu\nu}$:

$$g^{\mu\alpha}g_{\alpha\nu} = \delta^\mu_\nu. \tag{4.14}$$

From this definition we find

$$s^2 = g_{\mu\nu}x^\mu x^\nu = g^{\mu\nu}x_\mu x_\nu, \tag{4.15}$$

which implies that the inverse metric 'raises indices': $x^\mu = g^{\mu\nu}x_\nu$, and this again generalises to tensors of arbitrary rank. An important and defining property of the metric is that it is symmetric, $g_{\mu\nu} = g_{\nu\mu}$. The metric has the same 'signature' as $\eta_{\mu\nu}$, which means that locally we can write $g_{\mu\nu}$ at any single point x in spacetime to be equal to $\eta_{\mu\nu}$ by means of a general coordinate transformation.

The partial derivatives $\partial_\alpha = \partial/\partial x^\alpha$ are 'naturally covariant', with $\partial_0 = \partial/\partial t$ and $\partial_i = \nabla_i$ ('an upstairs index downstairs is the same as a downstairs index', or, more formally, partial derivatives are basis vectors for the coordinate basis).

Just like in SR, light moves on *null geodesics* of g such that $s^2 = 0$. For massive particles we can define the proper time along their worldline from $ds^2 = -d\tau^2$, and they move along 4-vectors that satisfy the *geodesic equation*:

$$\frac{d^2x^\mu}{d\tau^2} + \Gamma^\mu_{\alpha\beta}\frac{dx^\alpha}{d\tau}\frac{dx^\beta}{d\tau} = 0. \tag{4.16}$$

This is the GR equivalent of free motion, that is, equivalent to Newton's law $F = m\ddot{x}$ with $F = 0$. The effect of curved spacetime is accounted for by the *Christoffel connection*, $\Gamma^\mu_{\alpha\beta}$, which can be found from the metric:[2]

$$\Gamma^\mu_{\alpha\beta} = \frac{1}{2}g^{\mu\nu}\left(\partial_\alpha g_{\nu\beta} + \partial_\beta g_{\nu\alpha} - \partial_\nu g_{\alpha\beta}\right). \tag{4.17}$$

In gravitational lensing we are interested in the paths of light, that is, null geodesics. In this case we cannot define a proper time (since $ds^2 = 0$) and must use an 'affine parameter' along the path of light. In the geometric optics limit, however, we can 'cheat' the solutions locally simply by using the invariant separation: $ds^2 = g_{\mu\nu}dx^\mu x^\nu = 0$. In Minkowski space you can check that this gives the lightcone solution:

$$ds^2 = -dt^2 + dx^2 = 0 \Rightarrow \left(\frac{dx}{dt}\right)^2 = 1 \Rightarrow x \pm t = \text{const.} \tag{4.18}$$

You are invited to attempt this exercise for the Schwarzschild metric:

$$ds^2 = -\left(1 - \frac{2GM}{r}\right)dt^2 + \left(1 - \frac{2GM}{r}\right)^{-1}dr^2 + r^2(d\theta^2 + \sin^2\theta d\phi^2). \tag{4.19}$$

Consider radial rays and r larger than the Schwarzschild radius, that is, $r > R_s = 2GM$. Solve for $r(t)$ by first finding dr/dt assuming $ds^2 = 0$.

4.2.2 Light Bending by Matter

The statement $ds^2 = 0$ is equivalent to Fermat's principle. We can use it to derive an 'effective refractive index' for a gravitational potential, Φ. The potential is a small perturbation about Minkowski space, $\Phi \ll 1$ (weak field limit), and leads to the metric:

$$ds^2 = -(1 + 2\Phi)dt^2 + (1 - 2\Phi)d\vec{x}^2. \tag{4.20}$$

Using $ds^2 = 0$, this leads to

$$\frac{d\vec{x}}{dt} = \pm\sqrt{\frac{1+2\Phi}{1-2\Phi}} \approx 1 + 2\Phi. \tag{4.21}$$

---

[2] If the seemingly arbitrary index placement on δ, g and Γ concerns you, you should read about the conventions in Ref. [18]. Simply: tensors always have upstairs indices and downstairs indices separate, never above one another. The connection Γ breaks this rule to remind us it is not a tensor, while Kronecker-δ breaks it because for this special tensor the index placement does not matter.

We can think of the left-hand side (LHS) as the effective speed of light in the potential; let's call it c'. Thus the refractive index is[3]

$$n = \frac{c}{c'} = \frac{1}{1+2\Phi} \approx 1 - 2\Phi. \tag{4.22}$$

With this in hand we can do ordinary optics and compute the deflection of light. We consider the image plane shown in Fig. 4.3 and compute the two-dimensional deflection vector $\hat{\vec{\alpha}}$ using the Born approximation (see Ref. [17], Section 1.2). This gives us the *fundamental equation of lensing*:

$$\boxed{\hat{\vec{\alpha}}(b) = 2 \int_{-\infty}^{+\infty} \vec{\nabla}_{\perp} \Phi \, dz}, \tag{4.23}$$

where $\vec{\nabla}_{\perp}$ is the 2D gradient of the potential in the plane perpendicular to the optical axis. The *lens equation* gives the position of an object in the image plane, $\vec{y}$, in terms of its position in the source plane $\vec{x}$ as

$$\boxed{\vec{y} = \vec{x} - \vec{\alpha}}, \tag{4.24}$$

where

$$\vec{\alpha} = \frac{D_L D_{LS}}{\xi_0 D_S} \hat{\vec{\alpha}}. \tag{4.25}$$

The quantity ξ_0 is a length scale in the lens plane such that $\vec{x}, \vec{\alpha}, \vec{y}$ are dimensionless. For a point mass the dimensionless potential is $\Phi = -GM/r$. We use Eq. (4.23) to define the *lensing potential*:

$$\hat{\Psi} = \frac{D_{LS}}{D_L D_S} 2 \int_{-\infty}^{+\infty} \Phi \, dz. \tag{4.26}$$

Normalising we have $\Psi = (D_L^2/\xi_0^2)\hat{\Psi}$, $\nabla_x = \xi \nabla_{\perp}$ and finally

$$\vec{\alpha} = \nabla_x \Psi, \tag{4.27}$$

the deflection angle is the gradient of the lensing potential.

The distortion of an image is found by the Jacobian of the coordinate transformation from source plane to image plane, that is:

$$\frac{\partial y_i}{\partial x_j} \equiv A_{ij} = \delta_{ij} - \frac{\partial \alpha_i}{\partial x_j} = \delta_{ij} - \frac{\partial^2 \Psi}{\partial x_i \partial x_j}. \tag{4.28}$$

The Jacobian matrix A can be decomposed into *convergence*, κ (which measures magnification), and *shear*, γ, as

$$A = \begin{pmatrix} 1 - \kappa - \gamma_1 & -\gamma_2 \\ -\gamma_2 & 1 - \kappa - \gamma_1 \end{pmatrix} = (1-\kappa) \begin{pmatrix} 1 & 0 \\ 0 & 1 \end{pmatrix} - |\gamma| \begin{pmatrix} \cos 2\phi & \sin 2\phi \\ \sin 2\phi & -\cos 2\phi \end{pmatrix}. \tag{4.29}$$

---

[3] Recall that $\Phi < 0$ in a gravitational potential well.

The shear is complex, such that $\gamma = \gamma_1 + i\gamma_2 = |\gamma|e^{i2\phi}$. The convergence and shear can be measured, and thus the gravitational potential Φ can be inferred.

4.2.3 Applying Lensing: The κ Map

We define the surface mass density:

$$\Sigma(\vec{x}) = \int_{-\infty}^{+\infty} \rho \, dz, \tag{4.30}$$

where x is a coordinate in the lens plane. Poisson's equation tells us that $\nabla^2 \Phi = 4\pi G\rho$. Thus if the lens is thin we have

$$\kappa(\vec{x}) = \frac{\Sigma(\vec{x})}{\Sigma_{\mathrm{cr}}}, \quad \Sigma_{\mathrm{cr}} = \frac{1}{4\pi G}\frac{D_S}{D_L D_{LS}}. \tag{4.31}$$

In other words, the convergence is proportional to the surface mass density. A map of κ will give us a map of the projected energy density in the lens plane.

We consider the situation of a foreground galaxy cluster and background galaxies which are lensed by it. The gravitational field of the cluster causes the images of the galaxies to be elongated along the direction perpendicular to the vector $\hat{r}$ pointing from the centre of mass of the cluster to the lensed galaxy.

Ellipticity is defined as

$$|\varepsilon| = \frac{a-b}{a+b}, \tag{4.32}$$

where a and b are the size of the semi-major and semi-minor axes, with $b < a$. The ellipticity of a galaxy is defined from the second moment of the surface brightness, Q. Galaxies are inherently elliptical on the sky, but for a large number of galaxies we expect the orientation axis of the ellipticity to be random. Lensing causes the ellipticity to preferentially align along the shear axis.

For a random distribution of galaxies, $|\varepsilon|$ should have an intrinsic Gaussian distribution. We would like to construct the distribution in the presence of lensing. We consider an image defined by two-dimensional coordinates θ_i with centre of light $\bar{\theta}$. The second moment of the surface brightness is

$$Q_{ij} = \frac{\int d^2\theta \, q_I(\theta_i - \bar{\theta}_i)(\theta_j - \bar{\theta}_j)}{\int d^2\theta q_I}, \tag{4.33}$$

where q_I is an intensity weight function. This example is worked out explicitly in the accompanying JUPYTER notebooks.

Q_{ij} transforms as a tensor under the lensing Jacobian A, that is:

$$Q^{(s)} = AQ^{(i)}A^T, \tag{4.34}$$

where $Q^{(s,i)}$ is Q in the source and image plane respectively. The complex ellipticity defined from Q is

$$\chi = \frac{Q_{11} - Q_{22} + 2iQ_{12}}{Q_{11} + Q_{22}}, \tag{4.35}$$

which transforms to

$$\chi^{(s)} = \frac{\chi - 2g + g^2 \chi^*}{1 + |g|^2 - 2\mathrm{Re}[g\chi^*]}, \tag{4.36}$$

where 'reduced shear' $g(\vec{\theta}) = \gamma(\vec{\theta})/(1 - \kappa(\vec{\theta}))$. Defining $g = g_1 + ig_2$, the reduced shear is related to the convergence by the identity:

$$\nabla \ln(1 - \kappa) = \frac{1}{1 - g_1^2 - g_2^2} \begin{pmatrix} 1 + g_1 & g_2 \\ g_2 & 1 - g_2 \end{pmatrix} \begin{pmatrix} \partial_1 g_1 + \partial_2 g_2 \\ \partial_1 g_2 - \partial_2 g_1 \end{pmatrix}, \tag{4.37}$$

which defines a two-dimensional vector in the image plane using ∇. The algorithm to estimate $\kappa(\vec{\theta})$ is:

1. Measure surface brightness, I.
2. Construct Q.
3. Construct χ from Q.
4. Statistically subtract the intrinsic Gaussian ellipticity distribution.
5. Obtain noisy estimate of g.
6. Integrate identity Eq. (4.37) to obtain noisy estimate of κ.

The κ map thus obtained will clearly identify peaks in the mass distribution from peaks in κ. With a known D_S, D_L, we can also infer the total mass of the lens.

The point of going through the gory details of this process is so you can appreciate how it is in principle possible to infer the convergence, κ, from measuring the ellipticity of galaxies in an image, that is, measuring $Q^{(i)}$, given an assumption about the intrinsic ellipticity in the source plane, $Q^{(s)}$. From the convergence, we directly infer the projected mass density, Σ.

4.3 THE BULLET CLUSTER

Fig. 4.5 shows a composite image of *the Bullet Cluster*. We recall that the visible baryonic matter in a cluster is dominated by hot gas. In Fig. 4.5 gas density is found from an X-ray image, with bright pink colours indicating high gas density. Gas is collisional, and from the presence of the bow shock, we infer that this is in fact an image of two clusters, with the one on the right having punched through the one on the left (hence the name).

The blue colours in the image indicate the reconstruction of the κ map, performed from the lensed images of background galaxies (shown in optical). The κ map has two clear peaks. We note that the *centre of mass in each peak is displaced from the centre of mass of the gas*. Ref. [15] studied the map statistically and showed that the centres are displaced with confidence greater than 8σ. Thus, the centres of gravity of the clusters are displaced from the centre of visible mass: there is a large amount of unseen dark matter in the clusters.

The dark matter is displaced from the visible matter because it is collisionless. In the collision between the clusters the DM in each one passes right through the other. The gas, on the other hand, drags in the collision and thus lags behind the motion

Figure 4.5 *The Bullet Cluster*, composite public domain image from NASA. Inner shaded regions indicate the hot gas, measured using X-rays. Outer shaded regions indicate the mass distribution inferred from the gravitational lensing κ map, measured from the ellipticities of the visible galaxies. (See colour insert.)

of the dark matter. As well as providing striking visual evidence for dark matter, the Bullet Cluster also limits how strongly dark matter can interact with itself and how much pressure it can have. We will return to the consequences of this observation that DM is collisionless and how it affects our fundamental particle physics theories of what DM can and can't be in chapter 17.

QUIZ

i. What is the rough energy range of X-ray frequencies, measured in natural units?
   a. 1 electronvolt.
   b. 10^{-10} Planck masses.
   c. 1 kiloelectronvolt.
   d. 1 microelectronvolt.

ii. What is the physical origin of the thermal Bremsstrahlung process?
   a. Atomic line emission.
   b. Hot electrons scattering off nuclear ions.
   c. Nuclear burning of gravitationally collapsing gas.
   d. Stars in the intercluster medium.

iii. Which statements correctly describe the propagation of light in the general theory of relativity (select multiple answers)?
   a. Light moves on null geodesics of the metric.
   b. Light moves on time like geodesics of the metric.
   c. In a gravitational potential well, light moves with a lower effective speed.
   d. In a gravitational potential well, light moves with a larger effective speed.
   e. Gravitational lensing is only caused by dark matter.
   f. Gravitational lensing can be modelled by an effective refractive index.

iv. In the Born approximation, the lensing deflection angle is given by
   a. The lensing potential.
   b. The Laplacian of the lensing potential.
   c. Proportional to the density of dark matter.
   d. The gradient of the lensing potential.

v. How do we measure dark matter via lensing (select all that apply)?
   a. The intrinsic magnification is compared to known galaxy brightness.
   b. The Jacobian is decomposed into convergence and shear.
   c. Galaxies are elongated perpendicular to the vector towards the centre of mass.
   d. The cluster dynamics are simulated using numerical relativity.
   e. The statistical distribution of galaxy ellipticities is measured.
   f. The ellipticity allows an estimate of the magnification.
   g. Locations of large magnification correspond to peaks in the mass density.
   h. Galaxies are elongated parallel to the vector towards the centre of mass.

vi. The observation of the Bullet Cluster cannot be explained by modified gravity (MOND) because:
a. The cluster is too heavy.
b. MOND is a Newtonian theory, and clusters are regions of large spacetime curvature.
c. MOND can explain the Bullet Cluster.
d. The peaks in the visible mass distribution are displaced from the centre of mass measured by lensing.

Answers

i. c; ii. b; iii. a,c,f; iv. d; v. b,c,e,f,g; vi. d

Chapter Five

Cosmology Toolkit

Some of the most incontrovertible evidence for dark matter these days comes from cosmology. The agreement between the theory of structure formation and the CMB anisotropies can only be explained with a large amount of cold DM. Modified gravity fails completely to explain this concordance. Furthermore, cosmology provides a very precise measurement of exactly how much DM there is in the Universe, the so-called relic density, which is one of the key features a successful theory of DM must reproduce. We don't have time to give a complete account of this story: a full graduate course on cosmology would be needed. This chapter covers the basics of cosmology that you need in order to understand the cosmological evidence for dark matter given in chapter 6. This chapter also gives you the theoretical background to compute the dark matter relic density in Part II.

5.1 THE HOT EXPANDING UNIVERSE

5.1.1 Geometry

We begin with the geometry of the expanding Universe. To a zeroth approximation, the Universe is *homogeneous* (the same at every point in space) and *isotropic* (the same in every direction). Three metrics satisfy this, and they can all be written in the *Friedmann–Robertson–Walker (FRW)* form:

$$ds^2 = -dt^2 + a(t)^2 \left[\frac{dr^2}{1 - kr^2} + \underbrace{r^2(d\theta^2 + \sin^2\theta \, d\phi^2)}_{\text{round metric on the sphere}} \right]. \tag{5.1}$$

The three metrics are distinguished by their *topology*, specified by the sign of the constant k (which cannot change in general relativity):

$$k = \begin{cases} > 0 & \text{positively curved, closed} \\ = 0 & \text{flat} \\ < 0 & \text{negatively curved, open} \end{cases} \tag{5.2}$$

The three cases in two dimensions are depicted in Fig. 5.1. Drawing triangles on each surface, we see that for $k = 0$ the angles add up to 180° as in Euclidean geometry (on $\mathbb{R}^2$ the real plane), while for $k > 0$ the angles add up to more than 180° (S^2,

$k>0$ $k=0$ $k<0$

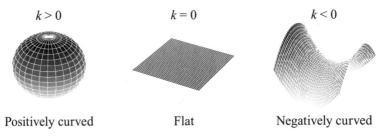

Positively curved   Flat   Negatively curved

Figure 5.1 Two-dimensional curved surfaces as defined by the constant k.

the two-dimensional surface of a sphere, with the standard round metric), and for $k<0$ they add up to less than $180°$ (here depicted on the saddle $z=x^2-y^2$). Drawing parallel lines, on $\mathbb{R}^2$ they never meet, on S^2 they always meet, and on the saddle they diverge. These concepts generalise to three-dimensional geometries, that is, S^3 the surface of the four-dimensional hypersphere, $\mathbb{R}^3$, flat three-dimensional space and three-dimensional hyperbolic space.

The units of the quantities in Eq. (5.1) can be moved around and depend on convention. A common choice is to have the radial coordinate r as dimensionless. The most important quantity in Eq. (5.1) is the function $a(t)$, known as the *scale factor*, which then has unit length. If we normalise $|k|$ to be unity (when it is non-zero), then the absolute value $\sqrt{|a|^2}$ gives the *curvature radius*. If $k=0$, then we can absorb the units of a into those of r. When $k=0$ the value of a is only defined relative to some arbitrarily chosen reference time, normally taken to be the present, t_0, such that $a(t_0)=1$.

A useful set of coordinates is given by (η, χ) where $a^2 d\eta^2 = dt^2$ for *conformal time* η, and

$$d\chi^2 = dr^2/(1-kr^2).\tag{5.3}$$

These are useful, since for radial curves with $d\theta = d\phi = 0$ the spacetime separation is

$$ds^2 = a^2 \underbrace{(-d\eta^2 + d\chi^2)}_{\text{looks like Minkowski}}.\tag{5.4}$$

We say that this space is *conformally flat*, that is, a single transformation by $a(t)$ makes it into flat space. Light travels on null rays $ds^2 = 0 \Rightarrow d\chi^2 = d\eta^2$. Integrating, we find the lightcones:

$$\chi = \pm\eta + \text{const.}\tag{5.5}$$

Then integrating Eq. (5.3) we find how the lightcones depend on the familiar radial coordinate r:

$$\chi = \begin{cases} \sinh^{-1} r; & k=-1 \\ r; & k=0 \\ \sin^{-1} r; & k=+1 \end{cases}.\tag{5.6}$$

5.1.2 Einstein's Equations and Covariant Derivatives

Einstein's equations of GR can be written in one line:

$$\boxed{G_{\mu\nu} = 8\pi G T_{\mu\nu}}.$$
(5.7)

The LHS of Eq. (5.7) is the *Einstein tensor*, which is made up of second derivatives of the metric tensor, $g_{\mu\nu}$. We don't need to know how to compute it here: I will give you the correct expressions. For the definition of $G_{\mu\nu}$ and the derivation of Einstein's equation, see any textbook on GR. The right-hand side (RHS) of Eq. (5.7) is the stress-energy or *energy-momentum tensor*, and G is Newton's constant (which in units $\hbar = c = 1$ has dimensions of $[M]^{-2}$). If we know $T_{\mu\nu}$ for all space and time, then this equation gives a set of sixteen PDEs to solve for $g_{\mu\nu}$. In cosmology, we instead want to solve the initial value problem, where $T_{\mu\nu}$ and $g_{\mu\nu}$ are specified only on some 'initial spacelike hypersurface'. We then need an equation of motion for $T_{\mu\nu}$, which is given by the conservation of energy as

$$\boxed{\nabla_\mu T^\mu{}_\nu = 0}.$$
(5.8)

We now have coupled equations for $T_{\mu\nu}$ and $G_{\mu\nu}$. The quantity ∇_μ in Eq. (5.8) is the *covariant derivative*.

To understand the covariant derivative, first consider the comparison to electrodynamics. Electrons obey the Dirac equation:

$$i\partial_\mu \gamma^\mu \psi - m\psi = 0,$$
(5.9)

where γ^μ are the Dirac matrices. We couple electrons to the electromagnetic vector potential A_μ with the rule

$$\partial_\mu \to D_\mu,$$
(5.10)

$$D_\mu = \partial_\mu + ieA_\mu,$$
(5.11)

where e is the elementary electric charge. The photon field A_μ is the 'gauge connection'.

Why do we make this change from ∂_μ to covariant derivative D_μ? It comes from Maxwell's equations. Maxwell's equations are invariant under 'gauge transformations', $A_\mu \to A_\mu + \partial_\mu \Lambda$ for any $\Lambda(x)$, which ensures that the 4-vector A_μ carries only two physical degrees of freedom: the polarisations of light. The invariance under the gauge transformation is obvious if we consider the sourced Maxwell equation in 4-vector notation:

$$\partial_\mu F^{\mu\nu} = j^\nu,$$
(5.12)

where j^ν is the source and $F^{\mu\nu} = \partial^\mu A^\nu - \partial^\nu A^\mu$ is the Faraday tensor. Substituting $A_\mu \to A_\mu + \partial_\mu \Lambda$ in the Faraday tensor and using the fact that partial derivatives commute, we see that Λ drops out. Using the covariant derivative D_μ we ensure that the entire theory of electrons and photons (i.e. coupled Maxwell and Dirac equations) is invariant under gauge transformations if $\psi \to e^{ie\Lambda}\psi$. If you have not met this idea about gauge transformations before, take a while to work through these ideas: they are fundamental and important throughout theoretical physics.

In GR the Christoffel symbol, Eq. (4.17), plays a similar role to the photon field in the Dirac equation and ensures the invariance of equations of motion like Eq. (5.8) under general coordinate transformations.[1] The rule for covariant derivatives depends on what rank of tensor ∇_μ acts on:

$$\nabla_\sigma T^{\mu_1 \mu_2 \cdots \mu_k}{}_{\nu_1 \nu_2 \cdots \nu_l} = \partial_\sigma T^{\mu_1 \mu_2 \cdots \mu_k}{}_{\nu_1 \nu_2 \cdots \nu_l}$$

$$+ \Gamma^{\mu_1}_{\sigma\lambda} T^{\lambda \mu_2 \cdots \mu_k}{}_{\nu_1 \nu_2 \cdots \nu_l} + \Gamma^{\mu_2}_{\sigma\lambda} T^{\mu_1 \lambda \cdots \mu_k}{}_{\nu_1 \nu_2 \cdots \nu_l}$$

$$+ \cdots$$

$$- \Gamma^{\lambda}_{\sigma\nu_1} T^{\mu_1 \mu_2 \cdots \mu_k}{}_{\lambda \nu_2 \cdots \nu_l} - \Gamma^{\lambda}_{\sigma\nu_2} T^{\mu_1 \mu_2 \cdots \mu_k}{}_{\nu_1 \lambda \cdots \nu_l}$$

$$- \cdots . \tag{5.13}$$

So we have $+\Gamma$ for 'upstairs indices', and $-\Gamma$ for 'downstairs indices', but our rule remains:[2]

$$\partial_\mu \to \nabla_\mu. \tag{5.14}$$

For a scalar, $\nabla_\mu \phi = \partial_\mu \phi$ and life seems simple. However beware: $\nabla_\mu \nabla^\mu \phi \neq \partial_\mu \partial^\mu \phi$ since $\nabla^\mu \phi = \partial^\mu \phi$ is a contravariant vector. Thus the scalar wave equation is

$$\Box \phi \equiv \nabla_\mu \nabla^\mu \phi = \nabla_\mu \partial^\mu \phi = \partial_\mu \partial^\mu \phi + \Gamma^\mu_{\alpha\mu} \partial^\alpha \phi. \tag{5.15}$$

In GR we should also be careful to raise indices using the full metric and beware that the metric components depend on spacetime, such that

$$\partial_\mu \partial^\mu \phi = \partial_\mu g^{\mu\nu} \partial_\nu \phi = g^{\mu\nu} \partial_\mu \partial_\nu \phi + \partial_\nu \phi \partial_\mu g^{\mu\nu}. \tag{5.16}$$

We will meet and use the scalar wave equation on the FRW metric a number of times later in the book.

Returning to the energy-momentum tensor, the symmetries of FRW demand that T must be diagonal and $T^i{}_j \propto \delta^i_j$:

$$T^\mu{}_\nu = \begin{pmatrix} -\rho & 0 & 0 & 0 \\ 0 & P & 0 & 0 \\ 0 & 0 & P & 0 \\ 0 & 0 & 0 & P \end{pmatrix}, \tag{5.17}$$

---

[1] In a similar analogy in electromagnetism, gauge symmetry reduces the four degrees of freedom of A_μ to the two photon polarisations while in GR the propagating degrees of freedom of the metric are reduced to the two polarisations of gravitons or gravitational waves.

[2] Note we have only considered (k,l) tensors, that is, objects carrying indices transforming in vector representations of the Lorentz group $SO(3,1)$. Things are a bit more complicated for fermions or 'spinors'. Spinors arise from the fact that the Lie algebra of SO(3,1), denoted $\mathfrak{so}(3,1)$, is isomorphic to $\mathfrak{su}(2) \times \mathfrak{su}(2)$. We experience the world in the fundamental vector representation of $\mathfrak{so}(3,1)$, while spinors are an $\mathfrak{su}(2)$ representation of it and are hence not very intuitive and behave differently under rotations and boosts than bosons and general (k,l) tensors do. The covariant derivative of a spinor uses the 'vierbeins' and spin connection, which will not concern us here.

where ρ is the energy density and P is the pressure density. Einstein's equations then lead to Friedmann's equations:

$$H^2 \equiv \left(\frac{\dot{a}}{a}\right)^2 = \frac{8\pi G}{3}\rho - \frac{k}{a^2}, \tag{5.18}$$

$$\frac{\ddot{a}}{a} = \frac{-4\pi G}{3}(\rho + 3P). \tag{5.19}$$

Eq. (5.18) defines the Hubble parameter H and is sometimes called the Friedmann equation. It arises from the G_{00} component of Eq. (5.7), and is a first-order constraint for a. Eq. (5.19) is sometimes called the Raychaudhuri equation. It arises from $G_{00} - 3G_{11}$ and is a dynamical equation for the scale factor a.

On the RHS of Friedmann, we notice that it is possible to define 'energy density in curvature', $\rho_k = k/8\pi G a_0^2$, at some reference of the value scale factor a_0 and then rewrite the Friedmann equation as

$$3H^2 M_{\mathrm{Pl}}^2 = \sum_i \rho_i, \tag{5.20}$$

where $M_{\mathrm{Pl}} = 1/\sqrt{8\pi G}$ is the reduced mass. We see that, from Friedmann's equations, if we know the evolution of ρ then we can determine $a(t)$ and the fate of the Universe.

5.1.3 Cosmic Inventory

To get the big picture, we first simplify and assume that all of the components of the Universe are decoupled from one another and evolve independently. Energy-momentum conservation then applies to each component, labelled i, separately. Using the covariant derivative with the FRW metric, $\nabla_\mu T^\mu{}_\nu = 0$, leads to

$$\dot{\rho}_i + 3H(\rho_i + P_i) = 0. \tag{5.21}$$

If we further assume (dropping the subscript labels) $P = w\rho$ where w is a constant *equation of state*, then

$$\dot{\rho} + 3H\rho(1 + w) = 0. \tag{5.22}$$

This equation can be integrated to find $\rho(a)$:

$$\rho = \rho_0 \left(\frac{a}{a_0}\right)^{-3(1+w)}, \tag{5.23}$$

where $\rho_0 = \rho(a_0)$ is a boundary condition, commonly taken as t_0, today.

Current observations are consistent with a flat Universe, that is, $k = \rho_k = 0$, which we will assume for most of this book. The basic cosmic inventory then has just three components with three different equations of state:

$$\text{Radiation:} \quad w = 1/3, \tag{5.24}$$

$$\text{Matter:} \quad w = 0, \tag{5.25}$$

$$\text{Cosmological constant, } \Lambda: \quad w = -1. \tag{5.26}$$

We will derive the equations of state for radiation and matter shortly. The equation of state of Λ follows by definition since it leads to $\rho_\Lambda = $ const.

We can further simplify Friedmann's equation by defining $\Omega_i = \rho_{i,0}/\rho_{\text{crit},0}$ where $\rho_{\text{crit},0}$ is the 'critical density' today defined by $\rho_{\text{crit},0} = 3H_0^2/8\pi G_N$ and H_0 is the Hubble parameter today. The Ω_i measures the fraction of the critical density in each component measured today. Friedmann now reduces to

$$H^2 = H_0^2 \left(\Omega_\Lambda + \Omega_m a^{-3} + \Omega_r a^{-4} \right). \tag{5.27}$$

If only one of the components with equation of state w dominates the energy density, then solutions for $a(t)$ are simple to find

$$a(t) = \left(\frac{t}{t_0} \right)^{\frac{2}{3(1+w)}}, \qquad (w \neq -1), \tag{5.28}$$

$$a(t) \propto e^{H_0 t}, \qquad (w = -1). \tag{5.29}$$

We notice that, if Λ dominates, with $w = -1$ we have accelerated expansion, $\ddot{a} > 0$. From the Raychaudhuri equation, Eq. (5.19), you can see that this is the case whenever $w < -1/3$. In conformal time the solution for any w is

$$a = \left(\frac{\eta}{\eta_0} \right)^{\frac{2}{1+3w}}. \tag{5.30}$$

The range over which η is defined now depends on w. If $w > -1/3$ then $\eta \in [0, \infty]$, while if $w < -1/3$ then $\eta \in [-\infty, 0]$ (if w is constant). In a mixed matter and radiation Universe, the solution in conformal time can be written as

$$a = A\eta + B\eta^2, \tag{5.31}$$

and a suitable choice of reference time can be used to set $A = B = 1$. It is left as an exercise to derive all of these solutions by solving the Friedmann equations. If w changes over time (which it does in the real, multi-component Universe), then we have to either solve the equations numerically or imagine patching together the different solutions. We will return to this in section 5.3.

We can understand the history of the Universe by considering the solution Eq. (5.23) in log-log coordinates as shown in Fig. 5.2. In the expanding Universe, a increases with time, and we can use it as a time coordinate. An alternative time coordinate is the *cosmological redshift*, z, which describes how the wavelength of distant sources is increased by the expansion:

$$\frac{a}{a_0} = \frac{\lambda_e}{\lambda_{\text{obs}}} = \frac{1}{1+z}, \tag{5.32}$$

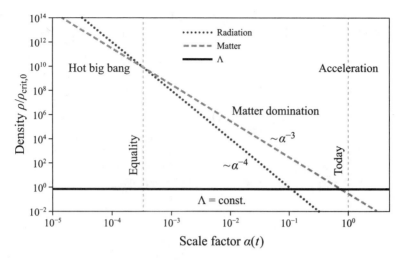

Figure 5.2 Evolution of the cosmic ingredients with scale factor. The vertical dashed lines indicate $a_0 = 1$ (today) and matter-radiation equality at $a_{eq} \approx 3 \times 10^{-4}$.

where λ_e is the wavelength emitted at the source and λ_{obs} is what we observe today. The redshift is $z = 0$ today and increases as we look out into the Universe, effectively looking back in time due to the finite speed of light.

As we look back into the history of the Universe or consider its evolution from the past, the relative behaviour of the components and what dominates the contents of the Universe are unavoidably fixed by the equations of state. In the distant future, the cosmological constant must take over. Sometime in the not-so-distant past, the matter content dominated the Universe. At some other point, matter and radiation must have had equal densities: so-called matter-radiation equality, which occurred when $z = z_{eq} \approx 3400$ (as measured by the CMB). Finally, our distant past must have been dominated by radiation in the *hot Big Bang*. The role of cosmology is to measure the Ω's today and determine what happened during these three epochs in the history of the Universe and maybe what came before or what might come after.

5.1.4 The Dark Matter Relic Density

The part of the cosmic inventory we are concerned with is the DM, and in particular how much of it there is. This is called the *relic density*, and here we define a useful measure of this. Take the Friedmann equation,

$$3H^2 M_{Pl}^2 = \sum_i \rho_i. \tag{5.33}$$

Using today's Hubble constant,

$$H_0 = 100\, h \mathrm{kms}^{-1} \mathrm{Mpc}^{-1} = 2.13 \cdot 10^{-33}\, he\mathrm{V} \equiv M_H h, \tag{5.34}$$

where h ('little h') is a constant to be measured and we have defined the characteristic Hubble mass scale $M_H = 2.13 \times 10^{-33}$ eV. We define the critical density

$$\rho_{\text{crit}} = 3H_0^2 M_{\text{Pl}}^2. \tag{5.35}$$

If all of the components making up the total energy density of the Universe are $\sum_i \rho_i = \rho_{\text{crit}}$, not including curvature ρ_k, then the Universe is **flat**, which is consistent with observations (see section 6.1).

Each component in the cosmic inventory contributes Ω_i, as defined above, to the energy budget where the normalising critical density, ρ_{crit}, depends on the measured value of the Hubble parameter today, H_0. The Ω_i are intuitive, since they represent fractions of the critical density and must add to unity. Due, however, to the uncertainty in the measured Hubble parameter, as encoded in h, it's often best to use the 'physical density':

$$\rho_i = \Omega_i h^2 \times \left(\frac{\rho_{\text{crit}}}{h^2} \right) \tag{5.36}$$

with

$$\frac{\rho_{\text{crit}}}{h^2} = 3M_H^2 M_{\text{Pl}}^2 = 8.1 \times 10^{-11} \text{ eV}^4. \tag{5.37}$$

The quantity $\Omega_i h^2$ simply measures the density in some particularly useful units, in terms of one scale we have simply defined, M_H, and the fundamental constant, M_{Pl}.

The CMB power spectrum (see section 6.2.3) is *separately* sensitive to both,

$$h, \Omega_{\text{DM}} h^2, \tag{5.38}$$

and therefore, we can make a good measurement of $\Omega_{\text{DM}} h^2$ largely independent of, for example, distance measure systematics affecting the determination of h. For instance, the Planck CMB power spectrum is best fit with

$$\boxed{\Omega_{\text{DM}} h^2 = 0.1200 \pm 0.0012.} \tag{5.39}$$

The cosmic DM density is known to 1% precision. In section 6.2 we will learn how this measurement is made. We tend to use the notation $\Omega_{\text{DM}} h^2$, as we will meet many DM candidates, and they should all have $\Omega_{\text{DM}} h^2 \approx 0.12$ if they are to compose the majority of the DM. However, you will commonly meet $\Omega_c h^2$ with 'c' for CDM, and we sometimes use this notation also, as successful DM candidates should be close to CDM on cosmological scales where $\Omega_{\text{DM}} h^2$ is measured.

5.1.5 Cosmic Thermodynamics

Our simple picture above assumed that our cosmic ingredients were totally independent of each other. The matter was already created and coupled in no way to the radiation. In the hot early Universe, we need to think about where the matter, and importantly the DM, came from. We will need to consider thermodynamics, scattering, thermal equilibrium and departures from equilibrium in cosmology.

The tool to use is statistical mechanics, and quantities are defined by integrals over *phase space distribution functions* (which we met in chapter 2 in the context of the virial theorem), $f(\vec{x}, \vec{p}, t)$, which give the number of particles occupying an infinitesimal phase space cell:

$$\text{Number density:} \quad n = g \int \frac{d^3 p}{(2\pi)^3} f(\vec{p}), \qquad (5.40)$$

$$\text{Energy density:} \quad \rho = g \int \frac{d^3 p}{(2\pi)^3} E f(\vec{p}), \qquad (5.41)$$

$$\text{Pressure:} \quad P = g \int \frac{d^3 p}{(2\pi)^3} \frac{|\vec{p}|^2}{3E} f(\vec{p}), \qquad (5.42)$$

where g is the number of degrees of freedom of the particle described by f (e.g. one for a real scalar boson, two for the photon polarisations and so on). What the above equations tell us is that quantities of interest are moments of the distribution function, averaged over momentum. In *kinetic equilibrium* the distribution function is

$$f(\vec{p}) = \frac{1}{\exp\left[(E - \mu)/k_B T\right] \pm 1}, \qquad (5.43)$$

where μ is the chemical potential (describing changes in particle number), the $+$ is for fermions (such that $f \leq 1$ and no phase space cell can be multiply occupied in accordance with the Pauli exclusion principle), and the $-$ is for bosons. In most of what follows, we set $k_B = 1$ and measure T in units eV.

Consider a four-particle scattering interaction:

$$1 + 2 \longleftrightarrow 3 + 4. \qquad (5.44)$$

In such a case, the number density of each particle species i evolves according to the *collisional Boltzmann equation*. For particle 1, this looks like the following:

$$a^{-3} \frac{d(n_1 a^3)}{dt} = \prod_{i=1}^{4} \overbrace{\int \frac{d^3 p_i}{(2\pi)^3 2E_i}}^{\text{LIPS}}$$

$$\underbrace{(2\pi)^4 \delta^{(3)}(p_1 + p_2 - p_3 - p_4)\delta(E_1 + E_2 - E_3 - E_4)}_{\text{energy-momentum conservation}}$$

$$|\mathcal{M}|^2 \left\{ \underbrace{f_3 f_4 [1 \pm f_1][1 \pm f_2]}_{\text{backward scattering}} - \underbrace{f_1 f_2 [1 \pm f_3][1 \pm f_4]}_{\text{forward scattering}} \right\}, \qquad (5.45)$$

where 'LIPS' is the Lorentz invariant phase space familiar from particle physics, and $\mathcal{M}$ is the matrix element for the scattering interaction, which one would compute using Feynman diagrams (we will not do such calculations explicitly). The plus signs apply for bosons and represent *Bose enhancement*, while minus signs apply for fermions and represent *Pauli blocking*. Backward scattering creates particle types 1 and 2 (hence a plus sign, leading to a positive derivative for n_1), and forward scattering creates particle types 3 and 4 (reducing the number of particles of type 1, leading to a negative contribution to the derivative). The right-hand side of the Boltzmann equation is known as the collision term, $C[f]$. The collisional Boltzmann equation is one of the most important equations in particle cosmology, and probably in all of physics. Remember it, and make sure you understand what each term means!

The phase space integral for ρ, Eq. (5.41), is evaluated with the following tricks:

$$d^3 p = p^2 dp \, \sin\theta \, d\theta \, d\phi \quad \text{(spherical polars in momentum space)},$$

(5.46)

$$E^2 = p^2 + m^2 \Rightarrow E dE = p dp \quad \text{('on-shell' condition)}, \tag{5.47}$$

$$\Rightarrow \int \frac{d^3 p}{(2\pi)^3} E = \int \frac{dp}{2\pi^2} p^2 E = \int \frac{dE}{2\pi^2} p E^2 f, \tag{5.48}$$

$$\Rightarrow \rho = g \int \frac{dE}{2\pi^2} \frac{(E^2 - m^2)^{1/2} E^2}{\exp[(E-\mu)/T] \pm 1}. \tag{5.49}$$

This integral can be performed in the limit $T \gg m$, $T \gg \mu$, that is, ultrarelativistic radiation:

$$\rho = \begin{cases} \left(\frac{\pi^2}{30}\right) g T^4 & \text{(bosons)}, \\ \frac{7}{8} \left(\frac{\pi^2}{30}\right) g T^4 & \text{(fermions)}. \end{cases} \tag{5.50}$$

In the radiation-dominated Universe, we can thus write Friedmann's equation as

$$\boxed{3 H^2 M_{\text{Pl}}^2 = \frac{\pi^2}{30} g_{\star,R} T_\gamma^4}, \tag{5.51}$$

where the photon temperature T_γ is taken as the temperature, and we have defined the *effective number of relativistic species*, $g_{\star,R}$:

$$g_{\star,R} = \sum_{i=\text{bosons}} g_i \left(\frac{T_i}{T_\gamma}\right)^4 + \sum_{i=\text{fermions}} \frac{7}{8} g_i \left(\frac{T_i}{T_\gamma}\right)^4. \tag{5.52}$$

Similarly, the integrals can be evaluated for pressure and number density, which in the ultrarelativistic limit yields

$$n \propto T^3, \tag{5.53}$$

$$P = \frac{\rho}{3}, \tag{5.54}$$

which gives the equation of state for radiation $w = 1/3$ as promised.

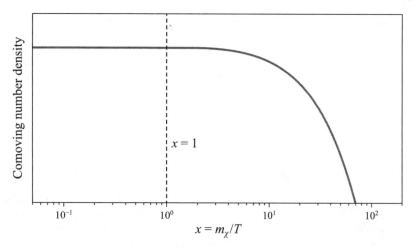

Figure 5.3 Equilibrium number density per comoving volume as a function of $x = m/T$ where m is particle mass. This solution follows directly from the integral of the distribution function, Eq. (5.40). When relativistic, $T \gg m$, the number density per comoving volume is constant. Once the particles become non-relativistic the number density is Boltzmann suppressed as $e^{-m/T}$.

In the non-relativistic limit, $m \gg T$, the results are the same for fermions and bosons and are

$$n = g \left(\frac{mT}{2\pi} \right)^{3/2} \exp \left[-\frac{(m - \mu)}{T} \right], \tag{5.55}$$

$$\rho = mn, \tag{5.56}$$

$$P = nT \ll \rho. \tag{5.57}$$

The exponential suppression in n can be interpreted as follows. When the photon energy is not large enough to produce the particles, the number must be Boltzmann suppressed. If particles and anti-particles are both counted by f, as is the case for electrons and positrons, then they are kept in equilibrium with the photons by annihilation. When $T < m$, the forwards process, annihilation, is more probable than the backwards process, pair creation. Thus we sometimes refer to the time when $T < m_e$ as the time of electron-positron annihilation.

The equilibrium number density we have just derived, $n_{\rm eq}(T)$, is plotted in Fig. 5.3, normalised to a comoving volume, which is equivalent to normalisation by entropy density s (see below).

The evolution of $g_{\star,R}$ in the Standard Model is shown in Fig. 5.4. For most of the history of the Standard Model, the temperature of all of the components is equal to the photon temperature. However, at late times, the neutrinos and the photons have different temperatures:

$$\frac{T_\gamma}{T_\nu} = \left(\frac{11}{4} \right)^{1/3}, \tag{5.58}$$

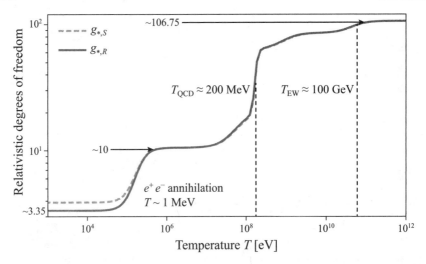

Figure 5.4 Relativistic degrees of freedom in the Standard Model. Fit taken from Ref. [19].

which leads to the final value of $g_{*,R} \approx 3.35$. To derive this result and the relationship between temperature and time, we need to consider *entropy*, S.

The second law of thermodynamics in differential form is

$$TdS = d(\rho V) + PdV. \tag{5.59}$$

Using commuting partial derivatives in the form

$$\frac{\partial^2 S}{\partial T \partial V} = \frac{\partial^2 S}{\partial V \partial T}, \tag{5.60}$$

we find

$$dP = \frac{\rho + P}{T} dT, \tag{5.61}$$

$$dS = d\left[\frac{(\rho + P)V}{T} + \text{const.}\right], \tag{5.62}$$

and thus define

$$\boxed{S = a^3 \frac{\rho + P}{T}}, \tag{5.63}$$

the entropy per comoving volume. The first law of thermodynamics is

$$d[(\rho + P)V] = VdP. \tag{5.64}$$

Substituting we find

$$\boxed{dS = d\left[\frac{(\rho + P)V}{T}\right] = 0,} \tag{5.65}$$

that is, *the entropy per comoving volume is conserved.* The entropy density is

$$s \equiv \frac{S}{V} = \frac{\rho + P}{T}. \tag{5.66}$$

Since $P \ll \rho$ for non-relativistic matter, we find that radiation is the dominant contribution to the entropy density of the Universe. Using the formulae above for ρ and P of relativistic fermions and bosons, we can write

$$s = \frac{2\pi^2}{45} g_{\star,S} T_\gamma^3 \propto n_\gamma, \tag{5.67}$$

$$g_{\star,S} = \sum_{i=\text{bosons}} g_i \left(\frac{T_i}{T_\gamma}\right)^3 + \sum_{i=\text{fermions}} \frac{7}{8} g_i \left(\frac{T_i}{T_\gamma}\right)^3, \tag{5.68}$$

which defines $g_{\star,S}$, the relativistic degrees of freedom in the entropy, which is also shown in Fig. 5.4. Here we made sure to be explicit that we define everything relative to the photon temperature T_γ, but we will henceforth drop the subscript and use T understood as *the* temperature where there is no ambiguity.

The conservation of entropy, S, implies that $s \propto a^{-3}$ or alternatively:

$$\boxed{a \propto g_{\star,S}^{-1/3} T^{-1}}. \tag{5.69}$$

Eq. (5.69) follows from the conservation of entropy and allows us to relate temperature to the scale factor and to time. As long as $g_{\star,S}$ is approximately constant, we find that $a \propto T^{-1}$: the temperature is red-shifted by the expansion of the Universe. However, this simple relationship does not hold when $g_{\star,S}$ evolves with temperature, and this occurs when particles become non-relativistic and drop out of the contribution to entropy, leading to steps in the value of $g_{\star,S}$.

To conserve the entropy, the massive particle transfers entropy to the relativistic degrees of freedom. This transfer heats the relativistic degrees of freedom and causes T to decrease slightly more slowly. Importantly, only particles in thermal equilibrium with the photons share in the entropy transfer and heating. For any decoupled particles, $T_i \propto a^{-1}$ with no additional heating. This leads to the different temperatures of neutrinos and photons. Neutrinos decouple at $T \sim 1$ MeV (see Problem II.5), before electrons become non-relativistic at $T = m_e = 500$ keV (this epoch is referred to as e^+e^- annihilation since this is the mechanism by which the electron number density becomes exponentially suppressed).[3] Thus the photons are heated relative to the neutrinos by the entropy production as electrons become non-relativistic, leading to $T_\gamma > T_\nu$ (the derivation of the factor in Eq. (5.58) is left as an exercise). Finally, because $g_{\star,R}$ has a term proportional to $(T_\nu/T_\gamma)^4$, while $g_{\star,S}$ has a term proportional to $(T_\nu/T_\gamma)^3$, this heating of photons relative to neutrinos

[3] This would be exactly true if there were an equal number of electrons and positrons to annihilate. We know, however, that the Universe contains a leftover amount of matter (electrons) compared to antimatter. This asymmetry is also present for baryons and is known as the baryon asymmetry, which is equivalent to the baryon-to-photon ratio we will meet in section 6.1, and is a very small number. The origin of the asymmetry is called baryogenesis, which we will not discuss further in this book.

leads to the differing values of $g_{*,R}$ and $g_{*,S}$ after e^+e^- annihilation seen in Fig. 5.4 (thus note that sometimes we drop the 'S' or 'R' in the subscript when there is no ambiguity).

5.1.6 Photon Decoupling and Recombination

Recombination is the process by which the hot plasma in the early Universe becomes neutral and leads to the formation of the CMB. To study it, we begin by rewriting the collisional Boltzmann equation, Eq. (5.45). We will also use exactly the same tricks when we compute the abundance of thermal DM, so it is good to pay attention now.

The number density is

$$n_i = g_i e^{\mu_i/T} \int \frac{d^3 p}{(2\pi)^3} e^{-E_i/T}. \tag{5.70}$$

We denote the equilibrium number density $n_i^{(0)} = n_i(\mu_i = 0)$, that is, no chemical potential, $\mu = 0$, implies no change in particle number, hence equilibrium. Thus we can write the chemical potential term as

$$e^{\mu_i/T} = n_i/n_i^{(0)}. \tag{5.71}$$

Next we define the *thermally averaged cross section*:

$$
\langle \sigma v \rangle = \frac{1}{n_1^{(0)} n_2^{(0)}} \left[\prod_{i=1}^4 \int \frac{d^3 p_i}{(2\pi)^3 2E_i} \right] e^{-(E_1+E_2)/T}
$$
$$
(2\pi)^4 \delta^{(3)}(p_1 + p_2 - p_3 - p_4) \delta(E_1 + E_2 - E_3 - E_4) |\mathcal{M}|^2, \tag{5.72}
$$

where the averaging is with respect to the equilibrium phase space density of species 1 and 2 in the scattering process. To understand why we use 'σv' (and what 'v' this refers to) here, see section 9.1.3 and Refs. [20, 21].

Now we can write the Boltzmann equation in a compact form:

$$a^{-3} \frac{dn_1 a^3}{dt} = \underbrace{n_1^{(0)} n_2^{(0)} \langle \sigma v \rangle}_{\text{reaction rate}} \left[\frac{n_3 n_4}{n_3^{(0)} n_4^{(0)}} - \frac{n_1 n_2}{n_1^{(0)} n_2^{(0)}} \right]. \tag{5.73}$$

We can estimate the LHS parametrically:

$$\frac{dn_1}{dt} \sim \frac{n_1}{t} \sim n_1 H, \tag{5.74}$$

where we have used the fact that $t^{-1} \sim H$ in an expanding Universe. If the reaction rate is larger than the Hubble rate, $n_2^{(0)} \langle \sigma v \rangle \gg H$, then, dividing both sides of Eq. (5.73) by $n_1^{(0)} n_2^{(0)} \langle \sigma v \rangle$, we find that this implies that the quantity in square brackets must vanish. This is the condition for *chemical equilibrium*, also known as the

Saha equation:

$$\boxed{\frac{n_1 n_2}{n_1^{(0)} n_2^{(0)}} = \frac{n_3 n_4}{n_3^{(0)} n_4^{(0)}}.}$$

(5.75)

Using the relation between the number densities and the chemical potential, Eq. (5.71), this implies that in chemical equilibrium:

$$\mu_1 + \mu_2 = \mu_3 + \mu_4.$$

(5.76)

For photons, we have $\mu_\gamma = 0$ (the number of photons is not conserved for a black-body spectrum; the Gibbs free energy and μ are only defined for systems with conserved particle number). At late times, neutrinos have decoupled. If Coulomb scattering is strong enough to keep

$$e^- + p \longleftrightarrow H + \gamma$$

(5.77)

in equilibrium then Eq. (5.75) implies

$$\frac{n_e n_p}{n_H} = \frac{n_e^{(0)} n_p^{(0)}}{n_H^{(0)}}.$$

(5.78)

This occurs as long as photons with $E \gg 1$ eV (approximate H binding energy) are abundant and can easily ionise the hydrogen atoms.

The free electron fraction, X_e, is

$$X_e = \frac{n_e}{n_e + n_H} = \frac{n_p}{n_p + n_H}.$$

(5.79)

Using the fact that the plasma is electrically neutral, such that $n_e = n_p$, we have $n_e + n_H = n_p + n_H = n_b$ where n_b is the total number of baryons (we neglect the helium fraction for now), and thus $n_e = n_b X_e$. Next, we use the fact that the baryon number is conserved in a comoving volume, and so $n_b a^3 = $ const. Performing the phase space integrals, the Boltzmann equation for the free electron fraction is

$$\frac{dX_e}{dt} = \underbrace{(1 - X_e)\beta}_{\text{ionization rate}} - \underbrace{X_e^2 n_b \alpha^{(2)}}_{\text{recombination rate}},$$

(5.80)

where

$$\alpha^{(2)} = \langle \sigma v \rangle \approx 9.78 \frac{\alpha^2}{m_e^2} \left(\frac{\epsilon_0}{T} \right)^{1/2} \ln \frac{\epsilon_0}{T},$$

(5.81)

$$\beta = \alpha^{(2)} \left(\frac{m_e T}{2\pi} \right)^{3/2} e^{-\epsilon_0/T},$$

(5.82)

where $\epsilon_0 \approx 13.6$ eV is the ionization energy of hydrogen.

Eq. (5.80) can be solved analytically in the 'Saha approximation' or numerically. *Recombination* is defined as the temperature when X_e drops to below 10%. The temperature of recombination is found to be around 3000 K (the exact number depends on a detailed calculation and on the cosmological parameters). Using the Friedmann equation and entropy conservation, this can be converted into a redshift, given by $z_{rec} \approx 1100$ (every cosmologist should remember this number).

When recombination happened at 3000 K, all of a sudden the Universe became neutral, and photons could move freely. When we look out into the Universe, at some point we will see it as it was when it was 3000 K, and the photons from that time have travelled freely to us. These photons are the CMB. We say that $z = 1100$ *is the redshift from which the CMB is emitted* (since we can think of redshift as a time coordinate). The redshift factor tells us that the CMB should be observed today at around $T = 3000/1100 \approx 3$ K. In 1948 the CMB was predicted by Alpher and Hermann, and it was subsequently observed in 1964 by Penzias and Wilson. This success provides observational confirmation of the theory of the early Universe we have been studying in this section: confirmation of the hot Big Bang phase in the evolution of the Universe.

QUIZ

i. The FRW metric assumes what about the Universe (select all that apply)?
   a. It is flat.
   b. It is homogeneous.
   c. It is isotropic.
   d. It is radiation dominated.

ii. In a negatively curved Universe, the angles of a triangle:
   a. Add up to less than π.
   b. Add up to exactly π.
   c. Add up to more than π.
   d. Cannot be measured by any static observer.

iii. Coordinate-covariant derivatives, ∇_μ, involve what quantity in addition to the partial derivative?
   a. The electromagnetic vector potential.
   b. The Ricci curvature.
   c. The Christoffel connection.
   d. The dark matter field.

iv. What do the Friedmann equations determine?
   a. The curvature of the Universe.
   b. The time evolution of the cosmic scale factor.
   c. The growth of cosmic perturbations.
   d. The topology of the Universe.

v. What is the effective equation of state of each component?
   a. Radiation.
   b. Non-relativistic matter.
   c. The cosmological constant.
   d. Curvature.

vi. Place these events in cosmic history in the correct time order (earliest first):
   a. Today.
   b. Matter-radiation equality.
   c. The electroweak phase transition.
   d. Cosmological constant domination.

vii. Match the energy scale to the cosmic event. Events:
   a. Today.
   b. Electroweak phase transition.
   c. QCD phase transition.

    d. Recombination and photon decoupling.

    e. Electron-positron annihilation.

Energy scales:

    a. 100 GeV.

    b. 3 Kelvin.

    c. 200 MeV.

    d. 1 eV.

    e. 1 MeV.

viii. In the ultra-relativistic limit the energy density is proportional to:

    a. T.

    b. T^2.

    c. T^3.

    d. T^4.

ix. Photons are hotter than neutrinos because:

    a. Electron-positron annihilation occurs after neutrino decoupling.

    b. Electron-positron annihilation occurs before neutrino decoupling.

    c. They are the same temperature.

    d. Neutrinos have mass.

x. The entropy density of the Universe (select all that apply):

    a. Is dominantly composed of radiation.

    b. Is dominantly composed of dark matter.

    c. Is given by the energy density divided by temperature.

    d. Is conserved.

    e. Is proportional to T^3.

    f. Is not a well-defined quantity.

xi. Photon decoupling occurs when what physical condition is met?

    a. Electrons and positrons annihilate.

    b. The Coulomb scattering rate becomes larger than the Hubble expansion rate.

    c. Hydrogen burns into helium during nucleosynthesis.

    d. The Coulomb scattering rate becomes smaller than the Hubble expansion rate.

Answers

i. b,c; ii. a; iii. c; iv. c; v. b; vi. 1/3,0,-1,1/2; vi. EWPT, matter dom., cc. dom. today; vii. 3K, 100 GeV, 200 MeV, 1 eV, 1 MeV; viii. d; ix. a,e; x. a,e; xi. d

5.2 COSMOLOGICAL PERTURBATION THEORY

The treatment here closely follows Ref. [22]. Our goal is to derive the equations of motion for fluctuations in DM, photons, baryons and massless neutrinos and their coupling to the gravitational potentials.

5.2.1 Metric Perturbations and Einstein Equations

GR is coordinate invariant, but we need to adopt a coordinate system to compute in. This is referred to as a choice of gauge, and we choose the *Newtonian gauge*. The Newtonian gauge only considers *scalar metric perturbations* and is quite intuitive due to its similarity to Newtonian gravity. The metric components are

$$g_{00}(t, \vec{x}) = -1 - 2\Psi(t, \vec{x}), \tag{5.83}$$

$$g_{0i} = 0, \tag{5.84}$$

$$g_{ij}(t, \vec{x}) = a^2 \delta_{ij}[1 + 2\Phi(t, \vec{x})], \tag{5.85}$$

where I have written out explicitly that these functions depend on time and space (unlike the FRW scale factor which depends only on time). The sign convention for the metric potential Ψ in Eq. (5.83) ensures that an overdensity has $\Psi < 0$, the same as the standard Newtonian potential. The metric potential Φ in Eq. (5.85) represents perturbations to the spatial curvature. Note that we have the convention that the potential is dimensionless.

The Newtonian gauge is not the only gauge. Another common gauge is the *synchronous gauge* (see e.g. Ref. [23] for gauge transformations). The Newtonian gauge is 'restricted' in the sense that it only allows for scalar perturbations and is also *gauge fixed*, which implies Ψ and Φ are physical and there is no remaining coordinate freedom.

In the non-relativistic limit, which we will define more carefully later, we have $\Psi = \Phi = $ 'the Newtonian potential' (recall lensing in section 4.2).

The Einstein equation is

$$G_{\mu\nu} = 8\pi G T_{\mu\nu}. \tag{5.86}$$

The Einstein tensor is

$$G_{\mu\nu} = R_{\mu\nu} - \frac{1}{2} g_{\mu\nu} R + \Lambda g_{\mu\nu}, \tag{5.87}$$

where Λ is the cosmological constant. I take the convention to absorb Λ into the energy-momentum tensor by subtracting off both sides of the Einstein equation, so

$$G_{\mu\nu} \to G_{\mu\nu} - \Lambda g_{\mu\nu}; \quad 8\pi G_N T_{\mu\nu} \to 8\pi G_N T_{\mu\nu} - \Lambda g_{\mu\nu}. \tag{5.88}$$

$R_{\mu\nu}$ is the Ricci tensor, $R = g^{\mu\nu} R_{\mu\nu}$ is the Ricci scalar, and $g^{\mu\nu}$ is the inverse metric, which, like the metric itself, is symmetric, that is: $g^{\mu\nu} = g^{\nu\mu}$. The Einstein tensor and energy-momentum tensor are thus also symmetric, $G_{\mu\nu} = G_{\nu\mu}$, $T_{\mu\nu} = T_{\nu\mu}$. The Ricci tensor is given in terms of derivatives of the Christoffel connection:

$$R_{\mu\nu} = \partial_\alpha \Gamma^\alpha_{\mu\nu} - \partial_\nu \Gamma^\alpha_{\mu\alpha} + \Gamma^\alpha_{\beta\alpha}\Gamma^\beta_{\mu\nu} - \Gamma^\alpha_{\beta\nu}\Gamma^\beta_{\mu\alpha}. \tag{5.89}$$

Finally, we remind ourselves of the Christoffel symbols in Eq. (4.17). Since Γ depends on the first derivatives of g and R depends on the first derivatives of Γ, the Einstein equations lead to second-order equations of motion for the components of g, that is, Ψ and Φ, with $T_{\mu\nu}$ components as source terms.

How many equations of motion are there? $G_{\mu\nu}$ is a four-by-four symmetric matrix, so in general there are ten equations. However, since we only consider the scalar degrees of freedom, we only need to use two independent equations. We work in perturbation theory to linear order in Ψ and Φ and Taylor expand the inverse metric:

$$g^{\mu\alpha} g_{\nu\alpha} = \mathbb{1}_{4\times4}, \tag{5.90}$$

$$g_{00} = -(1 + 2\Psi) \Rightarrow g^{00} = -\frac{1}{(1+2\Psi)} = -1 + 2\Psi + \mathcal{O}(\Psi)^2. \tag{5.91}$$

It is convenient to work in Fourier space, where we have the rules

derivative $\rightarrow \times ik$, multiplication $\rightarrow$ convolution (i.e. integrate over 'dummy' k). (5.92)

Since our equations will be linear in first-order perturbation theory, there will never be a multiplication of two components that both depend on x, that is, we never have $\Psi \times \Psi$. Thus we never have to do the convolutions. This has the advantage that *all k modes evolve independently*. In particle physics language, there is no exchange of momentum at linear order, and we are really computing 'only the propagators'.

Our Fourier convention is

$$\phi(\vec{x}) = \int \frac{d^3k}{(2\pi)^3} e^{i\vec{k}\cdot\vec{x}} \tilde{\phi}(\vec{k}), \tag{5.93}$$

where ϕ is any field and the Fourier conjugate variable k is the field 'momentum' (in analogy to quantum mechanics). I also drop the tilde on $\tilde{f}$, and we recognise the Fourier transform (FT) only by the use of the argument $\vec{k}$. In cosmology, all of our variables should be considered *classical stochastic fields*, and we are interested in their correlation functions. 'Statistical isotropy' guarantees that our functions depend only on the magnitude of $\vec{k}$, that is, $k = \sqrt{k_i k^i}$, and not on the direction in Fourier space. It is important to note that the definition Eq. (5.93) means that the FT changes the mass dimension of the field, for example:

$$[\phi] = 1 \Rightarrow [\tilde{\phi}] = [k]^{-3} = [M]^{-3} = [L]^3. \tag{5.94}$$

Finally, note that *momentum space is Euclidean* (except in some exotic theories of quantum gravity), that is, $k_i = k^i$ and for velocities $v_i = v^i$: regardless of the curvature, we can raise and lower indices and take scalar products on momentum space using the Kronecker delta.

We are now in a position to write down the Einstein tensor in Fourier space. We compute it in the form $G^{\mu}{}_{\nu}$, and the details are given in Ref. [22]. We find

$$G^0{}_0 = g^{00}\left[R_{00} - \frac{1}{2}g_{00}R\right], \tag{5.95}$$

$$= (-1 + 2\Psi)R_{00} - \frac{1}{2}R. \tag{5.96}$$

$$R_{00} = -3\frac{\ddot{a}}{a} - \frac{k^2}{a^2}\Psi + 3H\left(\dot{\Psi} - 2\dot{\Phi}\right) - 3\ddot{\Phi}, \tag{5.97}$$

$$R = \bar{R} + \delta R, \tag{5.98}$$

$$\bar{R} = 6H^2 + 6\frac{\ddot{a}}{a}, \tag{5.99}$$

$$\delta R = -12\Psi\left(H^2 + \frac{\ddot{a}}{a}\right) + 2\frac{k^2}{a^2}\Psi + 6\ddot{\Phi} - 6H\left(\dot{\Psi} - 4\dot{\Phi}\right) + 4\frac{k^2}{a^2}\Phi. \tag{5.100}$$

Note that in these equations 'k's go with $1/a$': this arises from the cosmological expansion and the sense in which the expansion redshifts the momentum of the field. This gives rise to the notion of 'comoving momentum', which is unchanged by the expansion. The overdots on fields represent derivatives with respect to coordinate time t (this notation differs from Ref. [22], which reserves overdot specifically for derivatives with respect to 'conformal time' η defined by $a\,d\eta = dt$).

A barred quantity is the background FRW solution, which depends only on t, and we solve order by order. Using this notation, we can write $R_{00} = \bar{R}_{00} + \delta R_{00}$ and see $\bar{R}_{00} + \bar{R}/2 = 3H^2$, which we recognise as giving the Friedmann equation. Since the background equations are assumed to be satisfied, we can subtract all of the barred components and find the equation of motion for perturbations only. The LHS of the relevant Einstein equation is

$$\boxed{\delta G^0{}_0 = -6H\left(\dot{\Phi} - H\Psi\right) - 2\frac{k^2}{a^2}\Psi}. \tag{5.101}$$

The spatial components of the Einstein tensor take the form

$$G^i{}_j = A\delta^i_j + k_i k_j \frac{(\Psi + \Phi)}{a^2}, \tag{5.102}$$

where A is some function we now project out. The *longitudinal traceless* component is found by projecting with $\hat{k}_i\hat{k}_j - \delta^i_j/3$, which equals zero when acting on δ^i_j, and gives our second Einstein equation:

$$\boxed{\left[\hat{k}_i\hat{k}_j - \frac{1}{3}\delta^i_j\right]G^i{}_j = \frac{2}{3}\frac{k^2}{a^2}(\Psi + \Phi)}. \tag{5.103}$$

5.2.2 Matter Equations of Motion

The *collisional Boltzmann equation*, Eq. (5.45), is derived from the following equation for the phase space density, f:

$$\frac{df}{dt} = C[f], \qquad (5.104)$$

where the LHS is the collisionless evolution and the RHS is the collision term, involving products of f's. The Boltzmann equation in the form of Eq. (5.104) is an equation on phase space $(\vec{x}, \vec{p})$, and the total derivative is

$$\frac{df}{dt} = \frac{\partial f}{\partial t} + \frac{\partial f}{\partial \vec{x}} \cdot \underbrace{\frac{d\vec{x}}{dt}}_{p/m} + \frac{\partial f}{\partial \vec{p}} \cdot \underbrace{\frac{d\vec{p}}{dt}}_{\text{EOM}}. \qquad (5.105)$$

This equation needs evaluating now using the perturbed metric $g_{\mu\nu}$. We work with the 4-vectors x^μ and

$$P^\mu = dx^\mu / d\lambda, \qquad (5.106)$$

where λ is any 'affine parameter' that increases along the particle worldline. This defines an eight-dimensional phase space, reduced to the usual six by the 'mass-shell condition' for real particles, which is imposed as a constraint. We now need to work out what the Boltzmann equation looks like for our cosmic ingredients. Our goal is to reduce the non-linear partial differential equation on phase space into a series of linear ordinary differential equations where the variables depend only on time.

5.2.2.1 Boltzmann Equation for Photons

The mass-shell condition is $P^\mu P_\mu = g_{\mu\nu} P^\mu P^\nu = 0$. The 4-momentum is $P = (P^0, \vec{P})$, and so using the perturbed metric we have

$$(-1 + 2\Psi)(P_0)^2 + \underbrace{g_{ij} P^i P^j}_{\equiv p^2} = 0, \qquad (5.107)$$

$$\Rightarrow P^0 = \frac{p}{\sqrt{1 + 2\Psi}} = p(1 - \Psi) + \mathcal{O}(\Psi)^2, \qquad (5.108)$$

where Eq. (5.107) defined the total 3-momentum p implicitly using g_{ij}. Eq. (5.108) is the equivalent of the equation $E = p$, or perhaps more familiarly, $E = h\nu$, generalized to perturbed FRW. Our convention for the metric in Eqs. (5.83–5.85) has overdensities defined to give $\Psi < 0$, thus $(1 - \Psi) > 1$. If the 3-momentum p is constant, then P^0 (the energy) must decrease as a photon leaves a potential well/overdensity, as sketched in Fig. 5.5. Thus, we have just derived *gravitational redshift*.

We work in perturbation theory in the distribution function f as well as in the metric perturbations, that is,

$$f = f_0 + \delta f ; \quad \delta f \sim \mathcal{O}(\Psi, \Phi) \sim \mathcal{O}(\epsilon), \qquad (5.109)$$

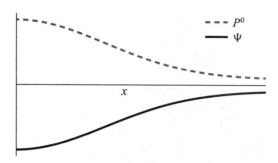

Figure 5.5 Gravitational redshift: an overdensity in the x-direction creates a negative potential, Ψ. According to Eq. (5.108) a photon with energy P^0 and constant 3-momentum p leaving this potential well must lose energy, decreasing P^0.

where ϵ is the perturbative expansion parameter. For photons f_0 is the Bose–Einstein distribution, $f_0 = f_{\rm BE}(E)$, and $E = p$ for on-shell photons. We normalise the 3-momentum such that $\vec{p} = C\hat{p}$ and $\hat{p}^i\hat{p}^j\delta_{ij} = 1$, with C a constant to be solved for below. Eq. (5.105):

$$\frac{df}{dt} = \frac{\partial f}{\partial t} + \frac{\partial f}{\partial x^i}\frac{dx^i}{dt} + \frac{\partial f}{\partial \vec{p}}\cdot\frac{d\vec{p}}{dt} + \overbrace{\frac{\partial f}{\partial \hat{p}^i}}^{\sim\mathcal{O}(\Psi)}\underbrace{\frac{d\hat{p}^i}{dt}}_{\partial\delta f/\partial\hat{p}^i}\ , \qquad (5.110)$$

from which we see that the final term is second order in perturbations and so can be dropped. Next we substitute Eq. (5.106):

$$\frac{dx^i}{dt} = \frac{dx^i}{d\lambda}\frac{d\lambda}{dt} = \frac{P^i}{P^0}, \qquad (5.111)$$

since $P^0 = dt/d\lambda$ (see Eq. (5.106)). Our normalised 3-momentum is

$$p^2 = g_{ij}P^iP^j = C^2a^2(1+2\Phi)\ \underbrace{\delta_{ij}\hat{p}^i\hat{p}^j}_{=1,\ \text{unit vector}}\ , \qquad (5.112)$$

which allows us to find the normalisation constant $C = p/(a\sqrt{1+2\Phi}) = p(1-\Phi)/a + \mathcal{O}(\Phi^2)$. Thus:

$$\frac{dx^i}{dt} = \frac{\hat{p}^i p(1-\Phi)}{ap(1-\Psi)} = \frac{\hat{p}^i}{a}(1+\Psi-\Phi) + \mathcal{O}(\epsilon^2). \qquad (5.113)$$

This term multiplies $\partial f/\partial x^i = \partial\delta f/\partial x^i$ which is already first order, and so the potential terms in Eq. (5.113) appear only at second order in Eq. (5.110). To linear order in perturbations, ϵ, Eq. 5.110 reduces to

$$\frac{df}{dt} = \frac{\partial f}{\partial t} + \frac{\hat{p}^i}{a}\frac{\partial f}{\partial x^i} + \frac{\partial f}{\partial p}\frac{dp}{dt} + \mathcal{O}(\epsilon^2). \qquad (5.114)$$

We find dp/dt from the geodesic equation. After some algebraic legwork, we find at linear order:

$$\frac{dp}{dt} = -p \left[H + \dot{\Phi} + \frac{\hat{p}^i}{a} \frac{\partial \Psi}{\partial x^i} \right]. \tag{5.115}$$

Now we need to define our perturbed distribution function in terms of useful cosmological fields. We use the *temperature anisotropy*:

$$\Theta = \delta T / T; \quad \Theta_0(\vec{x}, t) \equiv \frac{1}{4\pi} \int d\Omega \, \Theta(\hat{p}, \vec{x}, t), \tag{5.116}$$

where the second equation defines the *monopole* as the integral of the temperature anisotropy over the direction of the 3-momentum. In terms of the temperature anisotropy, the perturbed distribution function is

$$f(\vec{x}, p, \hat{p}, t) = \left[\exp \left(\frac{p}{T(t)[1 + \Theta(\vec{x}, \hat{p}, t)]} \right) - 1 \right]^{-1}. \tag{5.117}$$

The relevant collision term is *Compton scattering*:

$$e^-(\vec{p}_1) + \gamma(\vec{p}_2) \longleftrightarrow e^-(\vec{p}_3) + \gamma(\vec{p}_4), \tag{5.118}$$

which is expressed as a Feynman diagram shown in Fig. 5.6. The matrix element is

$$|\mathcal{M}|^2 \approx 8\pi \sigma_T m_e^2, \tag{5.119}$$

where σ_T is the *Thomson cross section*. Terms beyond '$\approx$' have to do with polarisation, which we neglect here.

Using these results and our distribution function Eq. (5.117), the collisional Boltzmann equation for photons turns out to be

$$\boxed{\dot{\Theta} + i\frac{\hat{p}^i}{a} k_i \Theta + \dot{\Phi} + i\frac{\hat{p}^i}{a} k_i \Psi = n_e \sigma_T \left[\Theta_0 - \Theta + \hat{p} \cdot \vec{v}_b \right],} \tag{5.120}$$

where we are using the *free electron density* n_e as above (related to the baryon density by the ionization fraction) and work in the *tight coupling limit* such that the electron distribution follows the baryon distribution, in particular the baryon velocity, $\vec{v}_b$. The baryon velocity is assumed to be parallel to $\vec{k}$, that is, an irrotational flow. We make a few more notational simplifications. First, we define $\hat{p}^i k_i = k \cos \theta = k\mu$ (with θ the angle between the wavevector defining the direction to the perturbation in our coordinate system and the photon momentum at that location). Next, we introduce the *Thomson optical depth*:

$$\tau = \int_t^{t_0} dt \, n_e \sigma_T, \quad \text{(i.e. } n_e \sigma_T = \dot{\tau}). \tag{5.121}$$

Finally we switch to conformal time η with $a d\eta = dt$ and use primes to denote derivatives w.r.t. conformal time:

$$\Theta' + ik\mu\Theta + \Phi' + ik\mu\Psi = -\tau' \left[\Theta_0 - \Theta + \mu v_b \right]. \tag{5.122}$$

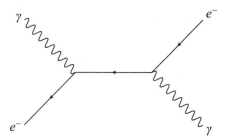

Figure 5.6 Compton scattering.

Note that Θ depends on $\hat{p}$ direction, and so to fully solve these equations we need a *moment hierarchy* beyond Θ_0, of which more later. Eq. (5.122) with no collision term holds also for massless neutrinos: the differences between the Bose–Einstein and Fermi–Dirac distributions do not appear in this limit.

5.2.2.2 Boltzmann Equation for CDM

We need to play the same game to derive equations of motion (e.o.m.'s) for CDM, but we now assume the on-shell condition for a massive, collisionless particle:

$$g_{\mu\nu}P^{\mu}P^{\nu} = -m^2, \quad C[f] = 0. \tag{5.123}$$

We do not need to assume a form for f_c, the CDM distribution function; we instead look for equations of motion for the moments with respect to $\hat{p}^i$:

$$n_c = \int \frac{d^3p}{(2\pi)^3} f_c \quad \text{(zeroth moment)}, \tag{5.124}$$

$$v_c^i = \frac{1}{n_c} \int \frac{d^3p}{(2\pi)^3} f_c \frac{p\hat{p}^i}{E} \quad \text{(first moment)}, \tag{5.125}$$

where v_c is the CDM 'peculiar velocity'. We perturb as

$$n_c = \bar{n}_c(1 + \delta_c), \tag{5.126}$$

where here and in the following we use overbars for background quantities in order to be careful about our perturbation definition, and δ_c is the *overdensity*. The background number density is related to the energy density as $\bar{\rho}_c = m\bar{n}_c$ which obeys the conservation equation with $w = 0$ discussed already:

$$\dot{\bar{\rho}}_c = -3H\bar{\rho}_c. \tag{5.127}$$

Taking moments of the Boltzmann equation, we can derive equations of motion for the moments giving the *Boltzmann hierarchy*. In general, the e.o.m. for the ℓth moment depends on the $(\ell + 1)$th moment. However, for CDM, we work in the limit $p/E \ll 1$ and neglect terms $\mathcal{O}(p/E)^2$. In this case, the Boltzmann hierarchy closes at the first moment, with no dependence on the higher moments. Furthermore, the

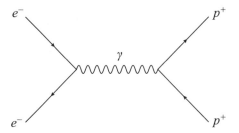

Figure 5.7 Coulomb scattering.

CDM mass, m, drops out: the mass of the CDM particle plays no role in cosmological perturbation theory.[4] The algebra can be found in Ref. [22]. The resulting equations for CDM are

$$\dot{\delta}_c + \frac{ik}{a}v_c + 3\dot{\Phi} = 0. \tag{5.128a}$$

$$\dot{v}_c + Hv_c + \frac{ik}{a}\Psi = 0. \tag{5.128b}$$

Note that CDM is coupled to the photons by the gravitational potentials.

5.2.2.3 Boltzmann Equation for Baryons

Coulomb scattering, shown in Fig. 5.7, keeps e^- and p^+ (i.e. hydrogen ions) tightly coupled at early times. Thus the overdensities and velocities follow one another:

$$\frac{\rho_e - \bar{\rho}_e}{\bar{\rho}_e} = \frac{\rho_p - \bar{\rho}_p}{\bar{\rho}_p} \equiv \delta_b, \tag{5.129}$$

$$\vec{v}_e = \vec{v}_p \equiv \vec{v}_b, \tag{5.130}$$

where subscript 'b' refers to total baryons. Cosmological perturbation theory is concerned with times after e^+e^- annihilation. Neither Coulomb scattering nor Compton scattering changes the electron number. Therefore, *the collision term vanishes for the zeroth moment of the Boltzmann equation*, and the δ_b e.o.m. must match the CDM e.o.m.:

$$\dot{\delta}_b + + \frac{ik}{a}v_b + 3\dot{\Phi} = 0. \tag{5.131}$$

Compton scattering leads to an exchange of momentum between photons and electrons, and so couples the first moments of the baryon distribution, that is, $\vec{v}_b$, to the

---

[4] These simplifications do not occur for hot or warm DM, which have $p/E \sim 1$ at some time. In these cases we have to solve more moments of the Boltzmann equation, and the particle mass appears explicitly. If we were being very careful, we could do this for CDM too, but by definition the CDM momentum does not affect cosmological length scales.

photons. It can be shown that this leads to the e.o.m.:

$$\dot{v}_b + Hv_b + \frac{ik}{a}\Psi = \dot{\tau}\frac{4}{3}\frac{\bar{\rho}_\gamma}{\bar{\rho}_b}[3i\Theta_1 + v_b].$$

(5.132)

We have used the first Legendre moment of the photon distribution:

$$\Theta_1(k,t) \equiv i\int_{-1}^{1}\frac{d\mu}{2}\,\mu\Theta(\mu,k,t).$$

(5.133)

The first Legendre moment of the photon distribution is a temperature dipole, and is related to photon 'velocity field perturbation'.

5.2.3 The Energy-Momentum Tensor

The last thing we need to do is relate the quantities in the Boltzmann equations to the components of the energy-momentum tensor on the RHS of the Einstein equations. We have that

$$T^0_{\ 0} = -\rho,$$

(5.134)

which for CDM and baryons is trivially

$$\rho_b = \bar{\rho}_b(1+\delta_b), \quad \rho_c = \bar{\rho}_c(1+\delta_c).$$

(5.135)

For photons and neutrinos, we defined Θ as the temperature fluctuation. Since $\rho \propto T^4$ we find $\delta\rho/\bar{\rho} = 4\Theta_0$ from the monopole. Thus:

$$\rho_\gamma = \bar{\rho}_\gamma(1+4\Theta_{0,\gamma}), \quad \rho_\nu = \bar{\rho}_\nu(1+4\Theta_{0,\nu}).$$

(5.136)

The source term for the longitudinal traceless component of $G^i_{\ j}$ is trickier. We use the general definition of $T_{\mu\nu}$ from the distribution function:

$$T_{\mu\nu} = \int\frac{d^3p}{(2\pi)^3}\frac{1}{\sqrt{-g}}\frac{P^\mu P^\nu}{P_0}f.$$

(5.137)

We define the ℓth Legendre moment from the ℓth Legendre polynomial P_ℓ:

$$\Theta_\ell = (-i)^{-\ell}\int_{-1}^{1}\frac{d\mu}{2}\,P_\ell(\mu)\Theta.$$

(5.138)

One can then show that the longitudinal traceless projection

$$\left[\hat{k}_i\hat{k}_j - \frac{1}{3}\delta^i_j\right]T^i_{\ j}$$

(5.139)

depends on $P_2(\mu) = 3\mu^2/2 - 2$. Since this is clearly second order in p/E, it vanishes for non-relativistic matter and is only sourced by photons and neutrinos. The source in this case is known as 'anisotropic stress' and depends on the quadrupole moments of the photon and neutrino distribution. The relevant Einstein equation is

$$k^2(\Phi + \Psi) = -32\pi G_N a^2 \left(\bar{\rho}_\gamma \Theta_{2,\gamma} + \bar{\rho}_\nu \Theta_{2,\nu}\right).$$ (5.140)

Thus the photon and neutrino hierarchies need to be expanded to at least second order.

Eq. (5.140) shows that, in the absence of relativistic matter and quadrupole moments, we have $\Phi = -\Psi$. In practice, it turns out that Compton scattering leads to a vanishingly small $\Theta_{2,\gamma}$. Thus only $\Theta_{2,\nu}$ neutrino anisotropic stress is relevant. The background densities in neutrinos and radiation redshift as a^{-4}, and this term is only relevant during the radiation-dominated epoch.

Another useful component of the Einstein equation is

$$k^2 \left(\dot{\Phi} - \frac{\ddot{a}}{a}\Psi\right) = 4\pi G_N a^2 i k^j \delta T^0{}_j,$$ (5.141)

where $\delta T^0{}_j$ is the velocity perturbation of the energy-momentum tensor. From Eq. (5.137) we notice that the dipole of the temperature field, $\Theta_{r,1}$, plays the role of a velocity field for the photon fluid.

We now have equations of motion for the metric perturbations and how these couple to the photons, DM, baryons and neutrinos. The baryons and photons are coupled in addition by the Thomson scattering term. For neutrinos and photons we in principle have an infinite Boltzmann hierarchy to solve for the momentum dependence, but in practice, we will be able to truncate this. Now our task is to solve these equations of motion, given some initial conditions, and relate the solutions to observables. Our initial conditions will be Gaussian and scale invariant. The observables we will consider are two point correlation functions for the photon temperature and the matter density field. Before we turn to solutions and observables from cosmological perturbation theory, there is one more item required in our cosmology toolkit.

5.3 INFLATION

Inflation provides an explanation for why the CMB is very uniform and why the initial conditions for cosmological perturbations take the form they do (the 'primordial power spectrum'). This not only explains the CMB we see but also provides the link between the observed CMB and the cosmological evidence for DM, which we discuss in chapter 6. We also need to discuss the theory of inflation because we can use it to produce the relic density of one of our DM candidates, primordial black holes, which we discuss in chapter 11.

5.3.1 The Horizon Problem

Cosmological inflation refers to a hypothetical period preceding the thermal epoch we studied in section 5.1. Inflation solves three problems that arise if we extrapolate the thermal Universe back to arbitrary times: the *relics*, *flatness* and *horizon* problems. We will only be concerned with the horizon problem here.

We begin by recalling Eq. (5.4). A light ray has $ds^2 = 0$, and so we can divide by the factor of a (away from the singularity when $a = 0$) and integrate to find the separation $\Delta\chi = \pm\Delta\eta$. Thus, light rays in the (χ, η) coordinates move on $45°$ angles, just like lightcones in special relativity. Now recall the solution to Friedmann's equation in a radiation-dominated Universe, $a \propto \eta$, which implies that the Big Bang singularity occurred at $\eta = 0$. The *particle horizon* is defined as the distance that light can travel between some initial time η_i and some future time η:

$$\chi_p(\eta) = \eta - \eta_i = \int_{t_i}^{t} \frac{dt}{a(t)}. \tag{5.142}$$

If we draw a lightcone back from an observer at some time η, then the particle horizon is the size of one-half of the lightcone at the time η_i. This defines the size of the region at η_i that is in causal contact with the observer (see the upper half of Fig. 5.8 defining the past lightcone of the points p_1 and p_2 at recombination).

More generally, by using our solutions to Friedmann's equation already derived for a Universe dominated by any fluid with an equation of state w (Eq. (5.30)), you can integrate Eq. (5.4) to show that the particle horizon is given by

$$\chi_p = \frac{2}{3(1+w)}(aH)^{-1}, \tag{5.143}$$

so long as $w > -1/3$, such that $\eta_i = 0$ in general (see Problems). In Eq. (5.143) we recognise $(aH)^{-1}$ as the size of the comoving Hubble radius.

We know when the CMB was formed, so we can compute the size of the particle horizon at recombination and compare it to the size of the particle horizon today (which is essentially the size of the visible Universe). The ratio of the particle horizon at recombination to the particle horizon today will be roughly given by the ratio of the sizes of the Hubble radius. Using $H \sim t^{-1}$ and the fact that recombination happened at $t \sim 300,000$ years and the Universe today is roughly 14 billion years old, we can see that the ratio $\chi_p(t_{rec})/\chi_p(t_0)$ is very small. This implies that the visible Universe today is made up of many, many tiny patches that were in causal contact at recombination.

The fact that the particle horizon today is so much larger than the particle horizon at recombination is a problem because we observe the CMB to be almost completely uniform. The temperature of the CMB at any point in the sky is $T(\theta,\phi) = T_0 + \Delta T$ with $\Delta T/T_0 \approx 10^{-5}$. But if each region is not in causal contact with its neighbour, we should expect $\Delta T/T_0 \approx 1$. This problem arose from assuming that the early Universe was radiation dominated by a fluid with $w > -1/3$.

The theory of inflation solves this problem by positing a period in the history of the Universe with $w < -1/3$, which corresponds to an *accelerated rate of expansion*, $\ddot{a} > 0$ (see Eq. (5.19)). In such a case two important things happen:

1. The comoving Hubble horizon shrinks:

$$\frac{d}{dt}(aH)^{-1} < 0. \tag{5.144}$$

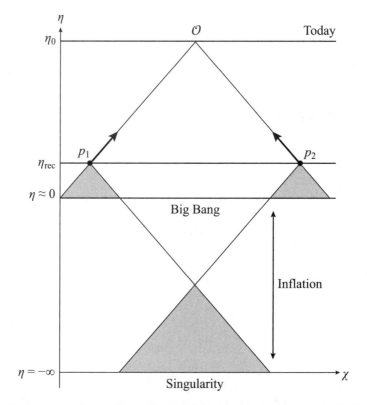

Figure 5.8 Spacetime diagram illustrating the horizon problem and its solution by inflation. An observer at $\mathcal{O}$ today, at conformal time η_0, can see their past lightcone, which encompasses two points p_1 and p_2 on the CMB, which were emitted at conformal time η_{rec}. If the Universe was radiation dominated at early times ('the hot Big Bang') then the initial singularity would occur at $\eta = 0$. The particle horizon of p_1 is given by the size of the base of the past lightcone and similarly for p_2. These lightcones don't cross, indicating that, in the Big Bang cosmology, there is a horizon problem if p_1 and p_2 have correlated temperatures. Inflation extends conformal time further back, and the initial singularity occurs at $\eta = -\infty$. Extending the lightcones of p_1 and p_2 backwards, they overlap at some time in the past. This solves the horizon problem, since p_1 and p_2 were in the same causal horizon during inflation, and can thus be correlated with one another.

2. The initial conformal time (the time at which the Big Bang, $a = 0$, occurred) is at $\eta_i = -\infty$.

The second of these points solves the horizon problem. We can keep drawing our lightcones at recombination back past the particle horizon defined by $\eta = 0$ and see that they meet again as $\eta_i \to -\infty$ (see Fig. 5.8).

The first point defines the history of what happens to perturbations during the evolution of the Universe. Take a perturbation of any fixed wavelength initially smaller than the Hubble horizon (as $a \to 0$ in the past, this will always be true at some time): we call this a 'mode'. At early times, regions within the wavelength of the mode are in causal contact with each other. During inflation if $w \approx -1$ then a grows rapidly as $a \sim e^{Ht}$; however, despite this, thanks to Eq. (5.144), the Hubble horizon actually shrinks. Thus, the wavelength of the mode will become larger than the horizon at some point, and we say 'the mode exits the horizon'. Regions separated by more than one Hubble horizon are no longer considered to be in causal contact, and so the regions contained within the mode wavelength are no longer in causal contact. We should expect, therefore, that when the mode exits the horizon it stops evolving, and indeed this occurs for so-called adiabatic perturbations. Inflation ends when $w > -1/3$ and the expansion of the Universe is no longer accelerating. At this point, the Hubble horizon starts to grow again and at some point will again become larger than the wavelength of our mode, and we say 'the mode re-enters the horizon'. When the mode re-enters the horizon, we are able to observe its causal effects.

5.3.2 Scalar Fields and Inflation

A classical scalar field, ϕ, that dominates the energy density of the Universe provides a model for inflation. We call the scalar field that drives inflation the *inflaton*. The Higgs field is one example of a scalar field, which we will discuss in chapter 7. The axion is another example, which we will discuss in chapter 10. Neither the QCD axion nor the Higgs of the Standard Model work quite right to explain inflation, but small modifications to either theory can make them work, and there is a whole zoo of other inflation models, reviewed in the *Encyclopaedia Inflationaris* [24].

If our scalar field depends only on time, $\phi = \phi(t)$, then it takes the same value everywhere in space, that is, it is homogeneous. A homogeneous classical scalar field obeys the Klein–Gordon equation (this equation is derived in section 10.1.1):

$$\ddot{\phi} + 3H\dot{\phi} + \frac{\partial V}{\partial \phi} = 0. \tag{5.145}$$

The scalar field energy density is

$$\rho = \frac{1}{2}\dot{\phi}^2 + V. \tag{5.146}$$

If the scalar field dominates the energy density of the Universe then the Friedmann equation is

$$3H^2 M_{\mathrm{Pl}}^2 = \frac{1}{2}\dot{\phi}^2 + V(\phi). \tag{5.147}$$

Eqs. (5.145, 5.147) define a coupled set of non-linear ordinary differential equations that determine the evolution of ϕ and a given a potential V and some initial conditions. In general, they cannot be solved analytically, but are relatively easy to solve numerically since, thanks to our assumption that ϕ depends only on time, we are only dealing with ordinary differential equations.

Successful inflation requires that the Hubble radius shrinks, which in turn implies that

$$\epsilon = -\frac{\dot{H}}{H^2} < 1. \tag{5.148}$$

This parameter, ϵ, is known as the first *slow-roll* parameter. During inflation, while $\epsilon \ll 1$, the Hubble parameter and scale factor satisfy

$$H \sim \text{constant},$$
$$a \sim e^{Ht}. \tag{5.149}$$

The spacetime described by these parameters is known as *de Sitter spacetime*. However, since this is not exactly true during inflation and $\epsilon > 0$, we say that inflationary spacetime is quasi-de Sitter.

Inflation should also last sufficiently long to put all of the modes of the CMB into causal contact, and this length of time happens to be between 50 and 60 e-folding times (i.e., a has to grow by a factor e^{50-60}). The length of inflation is governed by the following parameter:

$$\eta \equiv \frac{\dot{\epsilon}}{H\epsilon} < 1, \tag{5.150}$$

where this new η, the second slow-roll parameter, should not be confused with conformal time η. Using the Friedmann, Raychaudhuri and Klein–Gordon equations, one can show that the slow-roll conditions are satisfied by a scalar field if

$$\dot{\phi}^2 \ll V(\phi),$$
$$\ddot{\phi} \ll H\dot{\phi}. \tag{5.151}$$

Under these slow-roll conditions, it is possible to find analytic solutions to the coupled Klein–Gordon and Friedmann equations.

A scalar field is an elegant and simple model for inflation since the slow-roll conditions can be satisfied as long as we have a sufficiently flat potential V. The field begins slowly rolling somewhere on the potential where it is flat (how it got there is not a question we will answer here). However, eventually, the Hubble parameter becomes smaller and smaller and, at some point, the field will roll down to a steeper part of the potential, and oscillate in some local or global minimum. This naturally ends inflation: this is known as *graceful exit*. The end of inflation is known as *reheating*. These phases are illustrated in Fig. 5.9. Successful reheating requires that the inflaton should decay into Standard Model particles and begin the hot phase of the Universe we discussed earlier in this chapter. The theory of reheating is discussed in Ref. [25].

5.3.3 Quantum Fluctuations and the Primordial Power Spectrum

Quantum field theory replaces fields, ϕ, with operators, $\hat{\phi}$, and is concerned with the correlation functions/expectation values of these operators. We can get a feeling for this by considering the field operator expanded in terms of creation and annihilation operators, $\hat{a}^\dagger$ and $\hat{a}$. These operators create plane wave perturbations of the field. The field we quantise in inflation is $u = a\phi$ (where this a is the scale factor, not an annihilation operator). We carry out a Fourier expansion and include creation and annihilation operators for each mode, k, corresponding to particles with

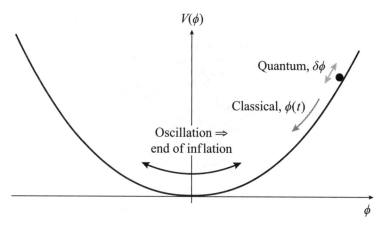

Figure 5.9 Motion of the inflaton field in its potential.

momentum k:

$$\hat{u}(\eta,\vec{x}) = \frac{1}{(2\pi)^{3/2}} \int d^3k \left[\hat{a}_k u_k(\eta) e^{i\vec{k}\cdot\vec{x}} + \hat{a}_k^\dagger u_k^*(\eta) e^{-i\vec{k}\cdot\vec{x}} \right]. \qquad (5.152)$$

We next consider perturbations of the metric and of the Klein–Gordon equation. The perturbed Klein–Gordon equation leads to an equation of motion for the 'mode functions' $u_k(\eta)$. The energy-momentum tensor of the scalar field acts as the source for metric perturbations in Einstein's equations.

We are familiar with the idea of the zero point energy of a harmonic oscillator in quantum mechanics. Similarly, there is a zero point energy for a quantum field, $\langle \hat{\phi}_k \hat{\phi}_{k'} \rangle$. Using Einstein's equations, we can re-express this zero point energy of the inflaton field in terms of the *curvature perturbation*, ζ, and we can express the expectation value in terms of the power spectrum we have met already in this chapter. In cosmology, spacetime is not static, and we do not arrive at a constant solution for the zero point energy but rather one that depends on time. However, in the slow-roll approximation, this can be solved exactly.

We are not concerned with the technical derivation of such results here, so let's just state the results and try to understand the physics intuitively. The exponentially fast expansion of the Universe during inflation and the shrinking Hubble radius lead to the existence of a causal horizon: a region of the Universe that an observer present during inflation could never access, even given infinite time. We can imagine such a situation as being like existing inside a black hole (but one with a singularity in its past, rather than at its centre). If quantum pair production happens right on the edge of the horizon, then one particle may fall inwards, towards us, and the other one moves out of the horizon and is forever lost to our view (see Fig. 5.10). The horizon is one-way during the exponential expansion: pairs can never re-annihilate.

The observer inside the horizon would see this flux of particles as if the horizon had an effective temperature. This temperature is known as the *Gibbons–Hawking temperature*, $T_{\mathrm{GH}} \sim H_I$, where H_I is the Hubble scale during inflation and leads

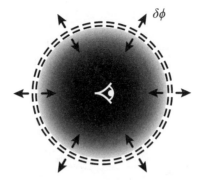

Figure 5.10 Production of inflaton fluctuations, $\delta\phi$, at the horizon. The observer is at the origin of coordinates, and only the incoming fluctuations can be observed.

to fluctuations in the inflaton field $\delta\phi \sim T_{\text{GH}}$. The phenomenon is an almost exact analogy of the famous Hawking radiation of a black hole, which we will meet in chapter 16. Because the Hubble expansion is not exactly uniform during slow-roll inflation, different modes pick up different amounts of vacuum perturbations. This is because H_I changes slowly during inflation, and different modes thus see different values of T_{GH} as they leave the horizon.

The fluctuations in ϕ generate curvature,

$$\zeta \sim \frac{H^2}{\dot{\phi}} \qquad (5.153)$$

and the power spectrum is (we define the power spectrum for a general random variable in section 6.2)

$$P_\zeta = A_s \left(\frac{k}{k_0}\right)^{n_s-1}, \qquad (5.154)$$

where

$$A_s = \frac{1}{2\epsilon}\left(\frac{H_*}{2\pi M_{\text{Pl}}}\right)^2 \qquad a_* H(a_*) = k_0 \qquad n_s - 1 = -6\epsilon + 2\eta, \qquad (5.155)$$

with k_0 being a normalisation scale known as the pivot scale, a_* being the scale factor and $H(a_*) \equiv H_I$ being the Hubble rate when k_0 left the horizon during inflation.

At the end of inflation, the inflaton decays into radiation, and the perturbations of the radiation follow the curvature perturbations laid down from the inflaton zero-point energy. Thus, the theory of inflation not only solves the horizon problem but also predicts the initial conditions of the Universe and gives a fundamental origin for them: *the initial conditions of the Universe arise from quantum fluctuations of the inflaton field.* As we will see in the next chapter, given a certain form of the initial conditions, and if the Universe is dominated by DM at late times, then these initial fluctuations grow under gravity and eventually form galaxies and ultimately us. Both inflation and DM are essential to our understanding of the cosmic origins of galaxies, and the theories stand up quantitatively to observational data.

From observations of the CMB $A_s \approx 2 \times 10^{-9}$, $n_s = 0.96$ in the range near $k_0 = 0.05\text{Mpc}^{-1}$. These two measurements can be used to fix two of the three quantities ϵ, η and H_I. Leaving H_I as the unfixed variable, this observation can be used to fix the inflaton potential $V(\phi)$ in some range. Currently, we can exclude the possibility that the CMB perturbations were caused by a potential close to $V(\phi) \propto \phi^2$ (a simple harmonic potential), and observations appear to favour convex 'hilltop' shapes. Inflation also generates fluctuations in the metric of tensor type, leading to primordial gravitational waves, which imprint a particular type of polarisation into the CMB known as a B-mode. The amplitude of these fluctuations, if measured, could be used to determine H_I. Currently, null results on the 'tensor-to-scalar ratio' imply $H_I \lesssim 10^{14}$ GeV. A lower bound on H_I can be found by demanding that reheating gives a temperature at least as high as that required for Big Bang nucleosynthesis at around 5 MeV, which from the Friedmann equation gives $3H_I^2 M_{\text{Pl}}^2 \gtrsim T_{\text{BBN}}^4 \Rightarrow H_I \gtrsim 10^{-15}$ eV. There is a lot of room for inflationary model building within these constraints.

QUIZ

i. How many scalar metric degrees of freedom are there in the Newtonian gauge?
   a. None, the metric is a tensor.
   b. One.
   c. Two.

ii. Why is Fourier wavenumber k always paired with an inverse power of the scale factor?
   a. Momentum increases as the Universe expands.
   b. Momentum decreases as the Universe expands.
   c. Momentum space is non-Euclidean.

iii. The longitudinal traceless component of the perturbed Einstein tensor (select all that apply):
   a. Is proportional to the dark matter overdensity.
   b. Is proportional to the photon temperature dipole.
   c. Is proportional to the photon and neutrino temperature quadrupoles.
   d. Vanishes in the non-relativistic limit.

iv. The temperature anisotropy multipole moments, Θ_i, serve what purpose when solving the perturbed Boltzmann equation?
   a. They are time independent.
   b. They have no spatial dependence.
   c. They factorise the spatial and momentum dependence.
   d. They have no momentum dependence.

v. Why does the cold dark matter Boltzmann equation moment hierarchy close (i.e. involve only the overdensity and velocity, no quadrupole)?
   a. Cold dark matter is collisionless.
   b. We don't know what cold dark matter is.
   c. Due to the non-relativistic limit, $p/E \ll 1$.
   d. Cold dark matter is a fermion.

vi. Baryons and (on-shell) photons are coupled by:
   a. The strong nuclear force.
   b. Coulomb scattering.
   c. The weak nuclear force.
   d. Bremsstrahlung.
   e. Compton scattering.

Answers

i. c; ii. b; iii. c,d; iv. c; v. c; vi. e

Chapter Six

Cosmological Evidence for Dark Matter

Cosmology provides some of the most precise and incontrovertible evidence for DM. In the early Universe, baryons and photons form a tightly coupled plasma, which supports large-amplitude acoustic oscillations, observed in exquisite detail in the CMB anisotropies. These same acoustic oscillations are not observed in the clustering of galaxies. The existence of a new, very cold and almost pressureless degree of freedom is required to explain the detailed shape of the CMB anisotropies and to achieve consistency with the observed distribution of galaxies. Cosmological measurements are extremely statistically accurate and measure the required density of DM to within 1%.

6.1 CONCORDANCE, FLATNESS AND THE BARYON BUDGET

We begin by recalling the Friedmann equation in its most general form, Eq. (5.18). In the present-day Universe, at redshifts long after matter-radiation equality, the dominant components making up the right-hand side of Eq. (5.18) could include anything with $w \leq 0$. Out of the standard components, this includes baryons, CDM (both with $w = 0$), the cosmological constant (or any 'dark energy', with $w \approx -1$) and curvature ($w = -1/3$). We can write Friedmann's equation including curvature but ignoring radiation as

$$H^2 = H_0^2[(\Omega_b + \Omega_c)(1+z)^3 + \Omega_k(1+z)^2 + \Omega_\Lambda]. \qquad (6.1)$$

Since baryons and CDM redshift in the same way, when we are considering quantities related to the FRW metric (such as cosmological distances and the expansion rate of the Universe), we tend to lump them together as 'matter', Ω_m. If we can measure $H(z)$, for example by determining distances to objects of known redshift, then we can fit this model and determine the unknowns, Ω_m, Ω_Λ, Ω_k and H_0. At redshift zero, we have the constraint $1 = \Omega_m + \Omega_\Lambda + \Omega_k$.

Since 1998 what is known as the 'concordance cosmological model' has emerged, which constrains these different components observationally in a self-consistent manner. In 1998, and until relatively recently, a variety of measurements were needed to pin down the components. For example, supernova measurements of the expansion rate of the Universe pinned down one combination of Ω_m and Ω_Λ, while the CMB pinned down an orthogonal combination. Thus together the two measurements determine both Ω_Λ and Ω_m, and from the constraint determine Ω_k (which can be positive or negative).

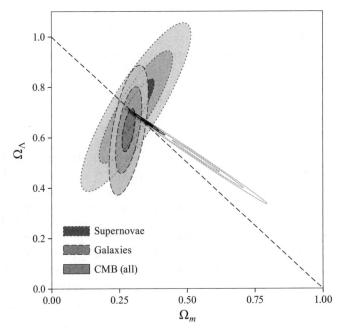

Figure 6.1 The 'concordance' standard cosmological model: ΛCDM. The dotted line has $\Omega_\Lambda = 1 - \Omega_m$, representing a flat Universe. The ellipses represent the statistical confidence, at one, two, and three standard deviations, for Ω_m and Ω_Λ to lie in the indicated region, as determined by supernovae, galaxies, and the CMB. The open ellipse shows the CMB result without including gravitational lensing. SNe and BAO results are taken from the Sloan Digital Sky Survey legacy archive, https://svn.sdss.org/public /data/eboss/DR16cosmo/tags/v1_0_1/, and CMB results from the Planck legacy archive, https://pla.esac.esa.int/#home. See colour insert.

This picture of using multiple, complementary observables to determine the contents of the Universe is sketched in Fig. 6.1. The ellipses represent the statistical confidence, at one, two and three standard deviations, for Ω_m and Ω_Λ to lie in the indicated region. Each observable-supernova distance measurements, galaxy measurements of the baryon acoustic oscillation (BAO) scale and the CMB-independently constrains Ω_m and Ω_Λ. The accuracies of different observables on the different parameters are also different, leading to stretched ellipses in different directions. In addition, for each observable, the determination of one parameter is correlated with the determination of the other due to *statistical degeneracies*, leading to tilted ellipses; for example, Ω_m and Ω_Λ have a positive correlation in the supernova distances but a negative correlation in the CMB. This complementarity, ellipses oriented in different directions, means that when we combine two different observables, say CMB and supernovae, the ellipses overlap in a small region, due to their being nearly orthogonal along their long directions. Therefore, we say that combining the observables *breaks the degeneracies* and leads to a consistent (the ellipses overlap) and well-constrained (small overlap region) model.

These days, the CMB alone is accurate enough to determine all of these key cosmological parameters, and many more (we will meet the details of the CMB measurements and theory later in this chapter), although combining the CMB with other data is still important to test the consistency of the model and improve precision, particularly the use of galaxies to measure the BAO to pin down $\Omega_k = 0$ and to test models beyond the standard 'ΛCDM' paradigm.[1] The measurements are consistent with $\Omega_k \approx 0$ (with an error of about 10^{-2} on the absolute value): *we live in a Universe that is very close to being flat*.

Furthermore, again in 1998, it became clear that cosmological measurements require an accelerating Universe. From the Raychaudhuri equation, Eq. (5.19), we see that acceleration $\ddot{a} > 0$ requires $(\rho + 3P) < 0$ or, equivalently, $w < -1/3$, that is, in terms of our standard components $\Omega_\Lambda > 0$. *An accelerating Universe requires dark energy*. The concordance model has an amount of dark energy roughly in the range of $\Omega_\Lambda = 0.7$.

What does all of this have to do with dark matter? Well, we can rewrite Friedmann's equation at $z = 0$ one more time:

$$\Omega_c = 1 - \Omega_b - \Omega_\Lambda - \Omega_k. \tag{6.2}$$

If the Universe is flat, and observation of its acceleration fixes the value of the cosmological constant, then if we also determine Ω_b somehow we can infer how large Ω_c has to be to ensure the consistency of Eq. (6.2), that is, to ensure flatness. Put another way, if we know how many baryons there are in the Universe, how much dark energy there is and that the Universe is flat, then we can determine if there is any additional missing matter, that is, we can determine the DM relic density.

The total baryon budget of the Universe is very accurately determined by observations of the light element abundances in the Universe, using a theory known as *Big Bang nucleosynthesis* (BBN). BBN is a theory that emerged in the early twentieth century and explains how the hot, dense plasma in the early Universe leads to the nuclear fusion of primordial hydrogen into other light elements such as deuterium, lithium and helium. BBN involves solving a series of coupled collisional Boltzmann equations for the number densities of the different elements using the experimentally measured valued of $\langle \sigma v \rangle$ for nuclear interactions. Detailed calculations can be found in Mukhanov's book, Ref. [36], and up-to-date numbers in the Particle Data Group (PDG) [37].

The results of BBN calculations depend on the *baryon-to-photon ratio*, η_b. Normalising to a nominal value of $\eta_b = 10^{-10}$ we define

$$\eta_{10} = 10^{10} \frac{n_b}{n_\gamma}, \tag{6.3}$$

---

[1] As of the time of writing, one area where this is of particular importance is around the apparent discrepancy between the value of H_0 determined from the CMB compared to using supernovae calibrated with Cepheids. This discrepancy could point to problems in the measurements, including systematics, or to physics beyond ΛCDM. An independent and precise determination of H_0 using gravitational wave 'standard sirens' could resolve the issue definitively one way or another.

where n_b here is the number of baryons leftover after annihilation with anti-baryons (see footnote in section 5.1.5: η_b is a measure of the baryon asymmetry). We define the abundance of an element, i, with atomic number A by weight as

$$X_i = A \frac{n_i}{n_b}. \tag{6.4}$$

BBN calculations then give the abundance of deuterium, D, and helium, He, as

$$X_D = 4.5 \times 10^{-5} \exp(-0.1\eta_{10}), \quad X_{He} = 0.23 + 0.012(N_{eff} - 3) + 0.005 \ln \eta_{10}, \tag{6.5}$$

where we have defined the 'effective number of neutrinos', N_{eff}, which is simply a parameterisation of the energy density in relativistic particles. What we have called X_{He} here is also known as Y_p in the literature.

The deuterium abundance is particularly sensitive to η_b and is also a key piece of evidence for BBN, since deuterium is not produced in normal stellar burning, and hence estimates for its abundance from observation are clean probes of BBN physics. The observed deuterium abundance is fit by

$$5.8 \leq \eta_{10} \leq 6.5 \Rightarrow 0.021 \leq \Omega_b h^2 \leq 0.024, \tag{6.6}$$

where the conversion to $\Omega_b h^2$ is done by fixing the photon number density n_γ from the CMB temperature, $T_{CMB} = 2.7255$ K (see the Problems). The CMB can also independently measure $\Omega_b h^2$, giving a value consistent with, and more accurate than, the one derived from BBN. For any reasonable value of $h \approx 0.7$ this implies $\Omega_b \ll 1$, and we therefore see that the observed flatness of the Universe requires a significant amount of DM. Take for example the overlap point in Fig. 6.1 with $\Omega_m = 0.3$, $\Omega_\Lambda = 0.7 \rightarrow \Omega_k = 0$. If we now use the BBN value $\Omega_b h^2 = 0.022$ from Eq. (6.6) and $h = 0.7$, then the flatness condition Eq. (6.2) gives $\Omega_c = 0.26$ and $\Omega_c h^2 = 0.12$. A more precise determination of the relic density of DM requires studying the growth of structure, galaxy clustering and the CMB, a study that we now embark on.

6.2 THE GROWTH OF STRUCTURE

In this section, we will look at solutions to the equations of cosmological perturbation theory we derived in the last chapter and see how cosmology provides very strong evidence for the existence of DM. But first, let's look at the qualitative, big picture, which goes under the name *the gravitational instability paradigm*:

1. We begin with small initial seed perturbations. These lead to acoustic oscillations in the baryon-photon plasma.[2]

2. Gravity causes these fluctuations to grow. Baryons and DM behave very differently. Baryons are coupled to photons, and for $z > 1100$, the baryon perturbations

---

[2] This is in contrast to models of 'active perturbations' seeded by cosmic strings or other topological defects, which are now known to be at most a sub-dominant effect in the CMB.

oscillate and don't grow. DM, on the other hand, is free to gravitationally cluster at times prior to recombination.

3. The shape of the acoustic features in the CMB and the consistency between the CMB (seen as an initial state) and galaxy clustering (seen as a final state) imply that DM must have been present.

This may seem relatively complicated; however:

- The whole story only relies on linear perturbation theory. There is thus a lot of theoretical control and rigour to the calculations.
- The data can be confronted with a statistically rigorous analysis. Thus the evidence can be quantified and different models systematically compared.

With these points in mind, let's begin at the beginning.

6.2.1 Initial Conditions

Throughout this section we are concerned with statistics of density fields. We define the *power spectrum*, P_Φ, of the curvature perturbation Φ as

$$\underbrace{\left\langle \Phi(\vec{k})\Phi(\vec{k}') \right\rangle}_{\text{ensemble average}} = (2\pi)^3 \underbrace{\delta^{(3)}(\vec{k}-\vec{k}')}_{\text{'momentum conservation'}} P_\Phi(\vec{k}) . \qquad (6.7)$$

Units are, as ever, important to bear in mind:

$$[\Phi(\vec{x})] = 1 \quad \text{(the metric is dimensionless)}, \qquad (6.8)$$

$$[\Phi(\vec{k})] = L^3 \quad \left(\text{recall: } \Phi(\vec{x}) = \int \frac{d^3k}{(2\pi)^3} e^{i\vec{k}\cdot\vec{x}} \Phi(\vec{k}) \right), \qquad (6.9)$$

$$[\delta^{(3)}(\vec{k}-\vec{k}')] = L^3 \quad \left(\int d^3k \, \delta^{(3)}(\vec{k}-\vec{k}')f(\vec{k}) = f(\vec{k}') \Rightarrow [\delta] = [k]^{-3} \right), \qquad (6.10)$$

$$\Rightarrow [P(\vec{k})] = L^3 = [k]^{-3}. \qquad (6.11)$$

Consider the variance of the fluctuations, $\sigma^2 = \sqrt{\langle x \rangle^2 - \langle x^2 \rangle}$, where we assume zero mean, $\langle x \rangle = 0$. From the power spectrum we find the variance of the curvature perturbations:

$$\sigma^2 = \underbrace{\int \frac{d^3k}{(2\pi)^3} \frac{d^3k'}{(2\pi)^3} \left\langle \Phi(\vec{k})\Phi(\vec{k}') \right\rangle}_{\text{average over all space}}, \qquad (6.12)$$

$$= \int \frac{d^3k}{(2\pi)^3} \underbrace{P_\Phi(\vec{k})}_{\text{using } \delta^{(3)}} . \qquad (6.13)$$

Isotropy implies $P_\Phi(\vec{k}) = P_\Phi(k)$; thus we use spherical polars in k space, $d^3k = k^2 \sin\theta\, dk\, d\theta\, d\phi$ and

$$\int \sin\theta\, d\theta\, d\phi = 4\pi, \tag{6.14}$$

which reduces our integral:

$$\sigma^2 = \int \underbrace{\frac{dk}{k}}_{d\ln k} \frac{k^3 P_\Phi(k)}{2\pi^2}. \tag{6.15}$$

We now define the *dimensionless power*, $\Delta_\Phi(k)$:

$$\Delta_\Phi(k) \equiv \frac{k^3 P_\Phi(k)}{2\pi^2}, \tag{6.16}$$

which gives the contribution to the variance per log-interval in k. *Scale-invariant perturbations* have $\Delta_\Phi(k)$ constant. We parameterise the initial power spectrum as

$$\boxed{\Delta_\Phi(k) = A_s \left(\frac{k}{k_0}\right)^{n_s - 1}}, \tag{6.17}$$

where

$$A_s = \text{scalar amplitude}, \tag{6.18}$$

$$n_s = \text{scalar spectral index}, \tag{6.19}$$

$$k_0 = \text{some arbitrary reference scale}. \tag{6.20}$$

As we saw in section 5.3, a power spectrum of the form Eq. (6.17) is provided by the theory of inflation.[3] However, historically, such a spectrum was just guessed and used for simplicity/symmetry reasons. When $n_s = 1$ we have scale invariance and this spectrum is known as a Harrison-Zel'dovich-Peebles spectrum after its historical parents. Modern precision CMB observations are consistent with $n_s \approx 0.96$ and $A_s \approx 2 \times 10^{-9}$ measured at $k_0 = 0.05$ Mpc^{-1}. That such a simple form for the initial conditions stands up to precision measurement is somewhat remarkable, and shows that the early Universe was an especially symmetric place.

Consider a matter overdensity, $\delta = \delta\rho/\bar{\rho}$, sourced by Φ. The *matter power spectrum* P_δ is our main concern in this section, as it can be related to observables. How does it scale with k initially? From the definition of P_Φ we have

$$P_\Phi \sim \Phi^2 \sim k^{-3 + (n-1)}. \tag{6.21}$$

The Poisson equation implies

$$k^2 \Phi \sim \delta, \tag{6.22}$$

---

[3] This is true assuming a slowly rolling inflaton field: we will revise this assumption when we construct a theory for primordial black holes in section 11.3.

and so we have

$$\delta \sim k^2 k^{-3/2+(n-1)/2} \sim k^{n_s/2}.$$ (6.23)

Thus the *primordial matter power spectrum* scales as

$$P_\delta \sim \delta^2 \sim k^{n_s}.$$ (6.24)

We will now show that this behaviour is preserved on the largest scales where Φ does not evolve and derive how this changes on small scales.

6.2.2 Solutions of the Cosmological Perturbations

This section closely follows Ref. [22]. We are interested in correlation functions of random fields. Because we have a linear system, we can factor out the initial random field. Consider any field $\xi(\vec{k}, a)$, where we use the scale factor as a time variable. We can factorise it as

$$\xi(\vec{k}, a) = \xi_i(\vec{k}) \times T_\xi(\vec{k}) \times D_\xi(a).$$ (6.25)

The initial conditions are specified by $\xi_i(\vec{k})$ which is a random field (Gaussian for simplicity, although this is also predicted by inflation). The initial conditions encode all of the correlation information in the sense that for a linear system we can write any correlation function as

$$\langle \xi_1 \cdots \xi_n \rangle = \langle \xi_{i,1} \cdots \xi_{i,n} \rangle \times [\text{deterministic evolution}].$$ (6.26)

We have factorised the deterministic evolution given some prior knowledge of the form of the solution:

$$T(k) = transfer\ function,$$ (6.27)

$$D(a) = growth\ function,$$ (6.28)

where $T(k)$ gives the early time scale-dependent evolution, and $D(a)$ captures the late time scale-independent evolution.[4]

We assume the initial conditions are sourced entirely by curvature, Φ, known as the *adiabatic mode*. Historically this is also assumed for simplicity but again turns out to be a prediction of the theory of single-field inflation. Thus we factor the initial curvature power spectrum Eq. (6.17) out of the matter power spectrum and see that it is given by

$$P_\delta \propto A_s \left(\frac{k}{k_0}\right)^{(n_s-1)} k^{-3} T_\delta^2(k) \left[\frac{D(a)}{D(a=1)}\right]^2.$$ (6.29)

---

[4] Late time evolution is only scale independent in the standard ΛCDM model, and even here the exact scale invariance of $D(a)$ is broken by massive neutrinos. Stronger scale dependence can be induced by modified theories of gravity, by dark energy and in theories where there is a component of DM that is not cold, collisionless and pressureless. Scale dependence is strongly broken on small scales when fluctuations become non-linear.

Note that here our notation differs from Ref. [22], which defines the transfer function as the curvature transfer function, $T \equiv T_\Phi = T_\delta / k^2$.

We do not have space here for a complete treatment of cosmological perturbation theory. To get a rough solution for the shape of the matter power spectrum, we can solve a simplified system of equations neglecting the baryons (at early times the baryons are strongly coupled to the photons, while at late times the baryon perturbations behave in the same way as CDM). Using primes to denote derivatives with respect to conformal time, our simplified system of radiation and CDM perturbations coupled with gravity is

$$\Theta'_{r,0} + k\Theta_{r,1} = -\Phi', \tag{6.30}$$

$$\Theta'_{r,1} - \frac{k}{3}\Theta_{r,0} = -\frac{k}{3}\Phi, \tag{6.31}$$

$$\delta'_c + ikv_c = -3\dot{\Phi}, \tag{6.32}$$

$$v'_c + \left(\frac{a'}{a}\right)v_c = ik\Phi, \tag{6.33}$$

$$k^2\Phi + 3\left(\frac{a'}{a}\right)\left[\Phi' + \left(\frac{a'}{a}\right)\Phi\right] = 4\pi Ga^2\left[\bar{\rho}_c\delta_c + 4\bar{\rho}_r\Theta_{r,0}\right]. \tag{6.34}$$

Note that a'/a is the *conformal Hubble rate*, $\mathcal{H} = aH$.

Taking a combination of Einstein equations, we can also arrive at a modified Poisson equation:

$$k^2\Phi = 4\pi Ga^2\left[\sum_i \delta\rho_i + i\frac{3\mathcal{H}}{k}\sum_i \rho_i v_i\right], \tag{6.35}$$

where the first term comes from the Poisson equation, and the second term represents the gravitation of the fluid velocity field.

6.2.2.1 Large Scales

Typical time scales imply that $d/d\eta \sim 1/\eta$. The comoving horizon size is $1/aH = 1/\mathcal{H} \sim \eta$, and k is the comoving wavevector with wavelength $L \sim 1/k$. Thus modes are 'superhorizon' when $k\eta \ll 1$, our definition of 'large scale'. For such modes we can drop $k\Theta_1$ in Eq. (6.30), kv_c in Eq. (6.32) and $k^2\Phi$ in Eq. (6.34) relative to the other terms. This leaves

$$\Theta'_{r,0} = -\Phi' = \delta'_c \Rightarrow \delta_c - 3\Theta_{r,0} = \text{const.} = 0, \tag{6.36}$$

where the last equality is fixed by the initial conditions, the *adiabatic mode*.

We now change time variables relative to matter-radiation equality in the following way. The background densities evolve as

$$\rho_c = \frac{\rho_c^{(0)}}{a^3}; \quad \rho_r = \frac{\rho_r^{(0)}}{a^4}, \tag{6.37}$$

which defines equality at a_{eq}:

$$\rho_c(a_{eq}) = \rho_r(a_{eq}) \Rightarrow y \equiv \frac{a}{a_{eq}} = \frac{\rho_c}{\rho_r}. \tag{6.38}$$

We can rewrite Eq. (6.34) in the super-horizon limit as

$$3\left(\frac{a'}{a}\right)\left[\Phi' + \left(\frac{a'}{a}\right)\Phi\right] = 4\pi Ga^2\rho_c\delta_c\left[1 + \frac{4}{3y}\right]. \tag{6.39}$$

Let's assume pure CDM and radiation (neglecting the cosmological constant), so that Friedmann's equation is

$$\frac{8\pi G}{3}(\rho_c + \rho_r) = \frac{8\pi G}{3}\rho_c\frac{(1+y)}{y} = H^2. \tag{6.40}$$

Eq. (6.34) now becomes a second-order equation of motion for Φ:

$$\frac{d^2\Phi}{dy^2} + \frac{21y^2 + 54y + 32}{2y(y+1)(3y+4)}\frac{d\Phi}{dy} + \frac{\Phi}{y(y+1)(3y+4)} = 0. \tag{6.41}$$

Remarkably, this has an analytic power law solution:

$$\Phi = \frac{\Phi(0)}{10}\frac{1}{y^3}\left[16\sqrt{1+y} + 9y^3 + 2y^2 - 8y - 16\right], \tag{6.42}$$

where $\Phi(0)$ is the initial condition from the primordial power spectrum. At early times, $y \ll 1$, a Taylor expansion shows that the leading behaviour of Φ is constant. At late times, large $y \gg 1$, we find

$$\Phi = \frac{9}{10}\Phi(0) = \text{const.} \tag{6.43}$$

Thus the super-horizon evolution of Φ is k-independent and thus does not change the k-dependence of the power spectrum and only changes the normalisation by a factor of $9/10$. We can plug this solution for Φ into the right-hand side of Eqs. (6.32, 6.33) to find the solution for δ_c, which at leading order in the super-horizon limit remains constant.

It turns out that Φ also remains constant in the matter era even after horizon re-entry, as you are invited to prove. With this in mind, once modes re-enter the horizon during matter domination ($k\eta \gg 1$, $y \gg 1$), Eq. (6.34) reduces to

$$k^2\Phi \sim \delta_c \Rightarrow \delta_c \sim k^2. \tag{6.44}$$

Then using Eq. (6.29) we find that the large-scale modes deep in the matter-dominated era have matter power spectrum:

$$P_\delta(k) \propto k^{n_s}. \tag{6.45}$$

This solution Eq. (6.45) holds for all 'small k', defined as all modes where our approximations thus far hold, that is, all modes that are, or were, super horizon with $y > 1$, that is, for modes which enter the horizon during matter domination.

6.2.2.2 Small Scales

Next we are interested in modes that cross the horizon *before matter-radiation equality*. We work in the radiation era when $\rho_r \gg \rho_c$ and so drop the CDM. The RHS of Eq. (6.35) becomes

$$16\pi G a^2 \left[\Theta_{r,0} + \frac{3\mathcal{H}}{k} \Theta_{r,1} \right] \rho_r = \frac{6\mathcal{H}}{k^2} \left[\Theta_{r,0} + \frac{3\mathcal{H}}{k} \Theta_{r,1} \right], \qquad (6.46)$$

where we used the Friedmann equation $H^2 = 8\pi G \rho_r / 3$ before equality.

Eq. (6.35) is a constraint equation that we can use to eliminate $\Theta_{r,1}$ in favour of Φ and substitute in Eq. (6.30) and Eq. (6.31), thus arriving at a second-order equation of motion for Φ:

$$\Phi'' + \frac{4}{\eta} \Phi' + \frac{k^2}{3} \Phi = 0. \qquad (6.47)$$

This has the form of a *damped harmonic oscillator* such that each k mode acts like a different mass, and the *Hubble term provides time-dependent friction*. We will encounter such equations again, especially when studying axion DM. Eq. (6.47) has an exact solution given by a first-order spherical Bessel function. The solution is depicted in Fig. 6.2.

We can approximate the solution for the curvature perturbation as a source for the CDM perturbations. Combining Eqs. (6.32) and (6.33) yields a second-order equation for δ_c:

$$\delta_c'' + \frac{1}{\eta} \delta_c' = \underbrace{-3\Phi'' - \frac{3}{\eta} \Phi' + k^2 \Phi}_{\Phi \text{ sourced by } \delta_r \,=\, \text{source for } \delta_c}. \qquad (6.48)$$

Using the fact that Φ decays allows for an approximate solution for δ_c:

$$\delta_c \propto \ln(C_i k \eta), \qquad (6.49)$$

where C_i is a constant of integration. Thus we learn that *CDM perturbations have scale-dependent logarithmic growth during the radiation era*. The resulting power spectrum in this regime is

$$P_\delta \sim k^{-3} \ln \left(\frac{k}{k_{\mathrm{eq}}} \right) k^{n_s - 1}, \qquad (6.50)$$

where the solution of Eq. (6.49) is normalised at the horizon size wavevector at equality, $k_{\mathrm{eq}} = a_{\mathrm{eq}} H(a_{\mathrm{eq}})$.

The total power spectrum interpolates between the small-scale (high k) solution, Eq. (6.50), and the large-scale (low k) solution, Eq. (6.45), with the crossover

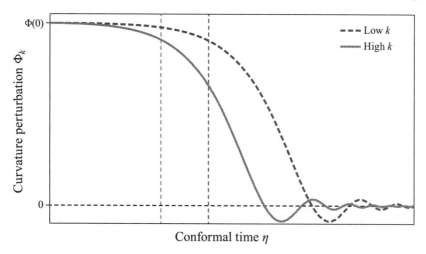

Figure 6.2 Solution for the curvature perturbation, Φ, inside the horizon during radiation domination, Eq. (6.47). Two k modes are shown. The vertical dashed lines mark the time when the mode enters the horizon, $k\eta = 1$. After entering the horizon, the solution oscillates around zero (horizontal dashed line) and decays in amplitude. The smaller scale mode has a larger k and so enters the horizon sooner and decays more. The η-axis scale is arbitrary and log-scaled.

implying that *the power spectrum peaks and turns over around* $k = k_{\text{eq}}$. The solution is sketched in Fig. 6.3.

We will show shortly why this solution provides the cosmological evidence for DM, because:

- The solution with no DM and only baryons has damped oscillations instead of the behaviour in Eq. (6.49).
- The CMB gives the normalisation of P_δ at small k, implying that non-linear structure formation (galaxies, us!), which is defined by large variance fluctuations, $\sigma^2 > 1$, requires CDM.

Historically, Peebles [38] did this the other way a round. He observed that galaxy formation requires $\sigma^2 = 1$ on some scale and used this to predict the amplitude of temperature anisotropies in the CMB from the assumption that DM provides the required potential wells for non-linear structure. The prediction, $\delta T/T \approx 10^{-5}$, was subsequently observed, confirming the cosmological DM hypothesis and the gravitational instability paradigm. We require two more ingredients to make this leap. The first concerns the growth of the matter power spectrum during the matter-dominated era.

6.2.2.3 Growing Mode

In the previous sections we essentially estimated $T(k)$. Now we consider the late time behaviour of CDM perturbations for *sub-horizon modes in the matter-dominated era* (the reason to consider sub-horizon modes is to avoid any issues of

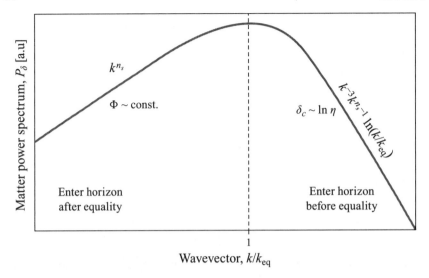

Figure 6.3 Sketch of the matter power spectrum. Axes are log-scaled to show the power law behaviour.

gauge dependence). In the sub horizon limit, $k\eta \gg 1$, the Einstein equations reduce to the Poisson equation. The equations of relevance are

$$\frac{d\delta_c}{dy} + \frac{ik}{aHy}v_c = -3\frac{d\Phi}{dy} \qquad \text{(conservation)}, \qquad (6.51)$$

$$\frac{dv_c}{dy} + \frac{v_c}{y} = \frac{ik}{aHy}\Phi \qquad \text{(Euler)}, \qquad (6.52)$$

$$k^2\Phi = \frac{3y}{y+1}a^2H^2\delta_c \qquad \text{(Poisson)}. \qquad (6.53)$$

These equations can be combined into a single second-order equation for δ_c:

$$\frac{d^2\delta_c}{dy^2} + \frac{2+3y}{2y(y+1)}\frac{d\delta_c}{dy} - \frac{3}{2y(y+1)}\delta_c = 0. \qquad (6.54)$$

This is another damped oscillator-type equation, but now the 'mass squared' term (coefficient of δ_c) is negative, implying growing and decaying modes. Furthermore, note that k does not appear in Eq. (6.54), and thus the solutions are scale independent, as promised. As can easily be verified by substitution, the following power laws solve the Eq. (6.54):

$$\delta_c = c_+ a + c_- a^{-3/2}, \qquad (6.55)$$

where the first term is the growing mode and the second term is the decaying mode. The decaying mode is conventionally set to zero at late times (since it decays), and so we have the simple behaviour that matter perturbations grow proportional to the scale factor, that is,

$$D(a) \propto a \quad \text{(matter-dominated era growing mode)}. \qquad (6.56)$$

Since the solution Eq. (6.55) is scale independent, the shape of P_δ does not change during the matter-dominated era; it simply grows proportional to $D(a)^2$:

$$P_\delta(a_1) = \left(\frac{a_1}{a_2}\right)^2 P_\delta(a_2). \qquad (6.57)$$

6.2.3 Baryons, Photons and the CMB

For $z < 1100$ (recall, this is when *recombination* happens; see section 5.1.6) the baryons behave just like CDM: they are decoupled from the photons and the baryons themselves are non-relativistic. Therefore, at late times, baryon perturbations grow with $D(a)$. However, for $z > 1100$, the baryons are *tightly coupled to the photons*. Thus, sub-horizon modes for $z > 1100$ behave very differently for baryons.

Consider the baryon equation of motion:

$$\delta_b' = -kv_b - \Phi'. \qquad (6.58)$$

Tight coupling between baryons and photons means $v_b = v_\gamma$, that is, they share a common velocity field, and recall that Φ decays on sub-horizon scales. Thus, δ_b will follow the temperature perturbations $\Theta_{r,0}$. Without going into the derivation (see Ref. [39], section 3.6) the full photon equation of motion is

$$\underbrace{c_s^2 \frac{d}{d\eta}\left(c_s^{-2}\Theta_{r,0}'\right)}_{\sim\Theta''} + \overbrace{\frac{k^2 c_s^2}{\dot{\tau}}\left[\frac{16}{15} + \frac{R^2}{1+R}\right]\Theta_{r,0}'}^{\text{damping}} + \underbrace{c_s^2 k^2}_{\text{pressure}}\Theta_{r,0} = \overbrace{-\frac{k^2}{3}\Psi - c_s^2 \frac{d}{d\eta}\left(c_s^{-2}\Phi'\right)}^{\text{gravitational forcing}},$$

$$(6.59)$$

where

$$R = \frac{P_b + \rho_b}{P_\gamma + \rho_\gamma} \approx \frac{1 + z_{\text{dec}}}{1 + z} 30\Omega_b h^2, \qquad (6.60)$$

and the sound speed is:[5]

$$c_s^2 = \frac{1}{3(1+R)}. \qquad (6.61)$$

As should now be familiar, Eq. (6.59) is a damped oscillator. The sub-horizon modes oscillate, and these oscillations in the baryon-photon plasma are known as *baryon acoustic oscillations (BAO)*. The damping term is known as *Silk damping, or diffusion damping* and arises from the Thomson scattering term (appearing here through $\dot{\tau}$) and the non-zero photon quadrupole, $\Theta_{r,2} \propto \Theta_{r,1}/\dot{\tau}$.

Sound waves in Eq. (6.59) propagate distances of order the horizon size when c_s^2 is large, that is, when $R \approx 0$ at large $z \gg 1 + z_{\text{dec}} \approx 1100$. When $z < z_{\text{dec}}$, R grows, and c_s^2 goes to zero. The distance sound waves can travel prior to z_{dec} is known as

---

[5] The speed of sound in a pure photon gas is $c_s^2 = 1/3$, which can be derived from the equipartition theorem. Here, the sound speed is reduced by the presence of the now-decoupled baryons at $z < z_{\text{dec}}$.

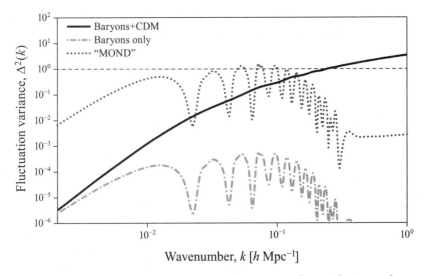

Figure 6.4 The matter power spectrum fluctuation variance, $\Delta^2(k) = k^3 P(k)/(2\pi^2)$. The spectra are computed numerically with CAMB [40]. The case with baryons only is normalised to have the same large-scale amplitude as the baryons+CDM case, broadly consistent with the large-scale CMB temperature fluctuations $(\delta T/T)^2 \sim 10^{-9}$. The 'MOND' model is inspired by the one discussed by Ref. [41], where the large-scale (small k) amplitude is simply increased in order that non-linear structure formation can take place, that is, reaching $\Delta^2 = 1$ (dotted horizontal line), without introducing more degrees of freedom or considering a relativistic completion of MOND.

the *sound horizon* and is given by

$$r_s = \int c_s \, d\eta = \int_0^{\eta_{\text{dec}}} \frac{d\eta}{\sqrt{3(1+R)}}. \tag{6.62}$$

The baryon perturbations for $k > k_{\text{eq}}$ do not grow like our CDM solution but oscillate, and r_s sets the scale of the oscillations of δ_b for different wavenumbers k. Although we haven't derived it rigorously, hopefully you can now appreciate how sound waves arise in δ_b and how a baryon-only Universe does not grow structure as efficiently as a Universe with CDM.

The matter fluctuation variance is given by $\Delta^2(k) = k^3 P(k)/(2\pi^2)$. The variance with baryons only is shown in Fig. 6.4, compared to the case with baryons plus CDM. Notice that, while the case with baryons plus CDM and the case with baryons only can be normalised to share the same large-scale amplitude, they differ dramatically on small scales (larger wavenumbers). This is because, on the scales $k < k_{\text{eq}}$, the baryon perturbations have effectively $c_s^2 = 0$ and thus follow our CDM solutions. However, for $k > k_{\text{eq}}$, the baryon perturbations oscillate instead of growing. CDM solutions during the radiation era have logarithmic growth compared to their primordial amplitude, while the baryon fluctuations retain their small primordial amplitude. It is the small-scale modes that give rise to galaxy formation (galaxies are typically separated by around 1 Mpc, while k_{eq} is around 10^{-2} Mpc^{-1}), and it is the logarithmic growth of CDM perturbations that makes CDM necessary for galaxy formation.

Now we need to discuss the CMB observables. The CMB anisotropies arise from the solution for Θ_r evolution and angular dependence. The full observable is the temperature autocorrelation, and a gauge invariant description requires the Doppler shift and gravitational redshift, which we don't cover here. The solution for Θ gives the temperature fluctuation, which is decomposed as

$$\frac{\delta T}{T} = \sum_{\ell,m} a_{\ell m} Y_{\ell m}(\theta, \phi), \tag{6.63}$$

where $Y_{\ell m}$ are the Legendre polynomials for sky coordinates θ, ϕ. This is a multipole expansion of the temperature, with multipole moment ℓ. Large ℓ corresponds to small angular scales, while small ℓ corresponds to large angular scales. The observable is the two-point correlation function, C_ℓ, defined as

$$\langle a_{\ell m} a_{\ell' m'} \rangle = \delta_{\ell \ell'} \delta_{m m'} C_\ell. \tag{6.64}$$

C_ℓ measures the variance of the components $a_{\ell m}$ of the multipole expansion of the temperature. The $a_{\ell m}$ have zero mean and are Gaussian distributed with width $\sqrt{C_\ell}$. The power spectrum measures the amount of correlation in the temperature on angular scales given by ℓ, as encoded in the θ dependence of the Legendre polynomials, Y.

The solution for C_ℓ is found numerically and is displayed in Fig. 6.5. Notice that C_ℓ has acoustic oscillations on small scales (large ℓ): these arise from the sound speed term in Eq. (6.59). The large first peak at $\ell \sim 200$ corresponds to the angular size of the sound horizon, Eq. (6.62), at $z = z_{\text{dec}}$, corresponding to the furthest distance the sound waves could travel before c_s^2 dropped to zero and the sound waves stalled. This solution in first-order perturbation theory depends on the cosmological parameters: of particular interest to us, it depends on the DM density. In the absence of DM, the acoustic oscillations have very large amplitude, while including DM brings the amplitude down by providing gravitational potential wells for the baryon photon fluid. By fitting this solution to the C_ℓ observed on the sky, we are able to *measure the DM density to percent-level accuracy.*

The CMB provides the most precise evidence for a new, cold, collisionless component of the Universe: cold dark matter. The density is measured extremely precisely and is given by $\Omega_c h^2 = 0.120 \pm 0.001$, compared to the baryon density which is $\Omega_b h^2 = 0.0224 \pm 0.0001$ [43]. The fits are found by varying a set of cosmological parameters to find the best fit and confidence regions, which is done numerically, typically using a Bayesian statistical method like Markov chain Monte Carlo.

For galaxy clustering observations, a key parameter is $\sigma_8 = \sigma(R = 8 \text{ Mpc})$, the standard deviation of the matter fluctuations on 8 Mpc scales ($\sigma(R)^2$ can be found theoretically by integrating $P(k)$ with an appropriate window function $W(k, R)$; see section 6.3). The scale 8 Mpc is chosen as approximately the scale of galaxy clusters, that is, where galaxies become bound to each other in a non-linear way, so that perturbations on scales larger than this are in the linear regime. If this separation of linear and non-linear regimes is correct, then σ_8 should be of order unity.

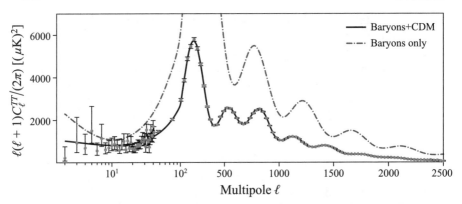

Figure 6.5 The CMB power spectrum. The spectrum is computed numerically with CAMB [40], and the data are taken from *Planck* [42]. The case 'baryons+CDM' has CDM density $\Omega_c h^2 = 0.120$ compared to baryon density $\Omega_b h^2 = 0.0224$, while the case 'baryons only' sets $\Omega_c h^2 = 0$, leaving all other cosmological parameters the same. Removing CDM, the model is still a reasonable fit at low multipoles, but the higher acoustic peaks are badly off, and the amplitude of acoustic oscillations is much larger without the CDM to provide gravitational potential wells. The equivalent to the 'MOND' model in Fig. 6.4 is not visible on this plot, the amplitude being around 1000 times larger than the baryons+CDM model. See colour insert.

Using this fact, and assuming a scale-invariant initial power spectrum, a critical Universe with curvature $\Omega_k = 0$, *which requires CDM to make up the total density*, Peebles [38] found $\delta T / T_0 \sim \sqrt{C_2} T_0 \sim 10^{-5}$ (where T_0 is the mean temperature of the CMB, around 3 K), which was subsequently measured by the COBE satellite in 1992.

Cosmological parameter fits in the COBE era were rough and allowed for a range of possibilities for the Hubble parameter, H_0, spectral slope, n_s, and cosmological constant density, Ω_Λ. All of the fits, however, involved a total matter density far in excess of the baryon density required by nucleosynthesis [44], that is, *all of the fits required CDM*. Nowadays, one of the key parameters is the primordial power spectrum amplitude, A_s, which normalises the large-scale CMB fluctuations (the 'low ℓ plateau' in the log-scale region of Fig. 6.5). Such a cosmology is then found to be consistent, via the predicted $P(k)$ from the gravitational instability paradigm, *with no additional fitting of parameters* with the slightly less well determined σ_8 derived from observations of large-scale structure, such as galaxy clustering and lensing. In fact, the consistency extends beyond σ_8 and is visible in the whole shape of $P(k)$ measured by many different tracers of the large-scale structure of the Universe, as we discuss shortly.

On the other hand, if the galaxy power is kept fixed in a Universe with no CDM, the primordial amplitude has to be increased dramatically to reach a fluctuation variance of order unity required for non-linear galaxy formation (see Fig. 6.4). The required value of A_s to achieve $\Delta^2 \approx 1$ without CDM is large and leads to a low ℓ plateau in C_ℓ that is much too high compared to the measurements (see Fig. 6.5). As we have seen in Fig. 6.4, the resulting matter fluctuations with no CDM (galaxy $P(k)$

and other observables) contain dramatic oscillations, which are also not observed. In the context of general relativity and the gravitational instability paradigm, there is no way around the introduction of CDM to reconcile observations of the CMB and the matter power spectrum (see Ref. [41] and appendix A for more details). *A Universe with only baryons cannot reconcile the theory of structure growth from perturbations (the gravitational instability paradigm) with the observed fluctuation amplitude in the CMB.*

In Fig. 6.4 the model with only baryons but with increased amplitude is inspired by Ref. [41] as a proxy for a minimal MOND model with no additional degrees of freedom. Relativistic completions of MOND with more (dark!) degrees of freedom are required to better fix this problem and obtain a $P(k)$ with no dramatic acoustic oscillations and the correct amplitude.

One such theory is known as 'TeVeS', for 'Tensor Vector Scalar gravity' due to Beckenstein. TeVeS takes ordinary GR and adds two new fields, A_μ (vector) and ϕ (scalar). These fields are thought of as part of the gravitational sector, in the sense that they mix with the metric $\tilde{g}_{\mu\nu}$ (the 'Beckenstein metric') to form a 'universally coupled metric', $g_{\mu\nu}$, that determines the motion of ordinary matter particles. The universally coupled metric is $g_{\mu\nu} = e^{-2\phi}\tilde{g}_{\mu\nu} - 2\sinh\phi A_\mu A_\nu$, and it is this metric that determines, for example, the geodesic equation for a test particle. The dynamics of TeVeS are determined from an action principle. For ordinary GR the action that leads to Einstein's equations is called the 'Einstein-Hilbert action' and looks like

$$S = \int d^4x \sqrt{-g}R, \qquad (6.65)$$

where R is the Ricci scalar, a two-derivative function of the metric. TeVeS writes this action instead for the Beckenstein metric, $\tilde{g}_{\mu\nu}$, and supplements it with a quadratic action for A_μ (which resembles the Maxwell action) and an action for ϕ (we will study similar scalar field Lagrangians and equations of motion in chapter 10). In addition, there are a number of free functions, which, if chosen correctly, modify Poisson's equation for the matter fields coupled to $g_{\mu\nu}$ such that a MONDian force law is recovered.

The new fields ϕ and A_μ are new degrees of freedom and when quantised will introduce new particles into the Standard Model. The coupling of ϕ and A_μ to ordinary matter via $g_{\mu\nu}$ is of gravitational strength; it is thus very weak compared to the other forces, and this new ϕ and A_μ matter is rather dark. It is thus not surprising that ϕ and A_μ are candidates for DM like the axion (a scalar field) or a 'dark photon'. However, the structure of TeVeS is rather arcane and restricted.

As described in detail in Ref. [45], TeVeS can match the observed matter fluctuations given by $P(k)$ but only the first two acoustic peaks in the CMB power spectrum (note in Fig. 6.5 the presence of more than seven damped acoustic peaks well fit by CDM). Furthermore, TeVeS has recently been falsified by gravitational wave observations of event GW170817, a binary neutron star merger with associated counterparts across the electromagnetic spectrum, which measures the ratio of the speed of gravitational waves to the speed of light and finds it to be consistent with unity to high accuracy. In TeVeS this ratio is not unity, thus falsifying the theory. Further MONDian variants of TeVeS have been developed to avoid these problems, yet all continue to feature new dark degrees of freedom. Theories of 'MOND' that work on cosmological scales are all really theories of DM.

6.3 DARK MATTER HALOS

The average dark matter density across the Universe is estimated to be around $1 \, \text{keVcm}^{-3}$. However, as we know, dark matter is not distributed smoothly. Instead, much of the dark matter has collapsed into clumps starting from very small perturbations in the early Universe. We call these clumps of dark matter 'halos'.

As mentioned in section 3.4.2, the Milky Way has its own dark matter halo. In fact, most *but not all* galaxies are observed to require halos. The 'but not all' caveat is very important. If DM is a substance, it can be in some galaxies and not in others, thanks to accidents of their formation and growth. If DM were explained instead by modifying gravity, this should be universal, and there would be no way one galaxy could have modified gravity and another could not.

We have discussed how on average at the solar radius the dark matter density is roughly 0.3–$0.4 \, \text{GeVcm}^{-3}$. This is around 10^5 times more dense than the Universe average. The Sun is relatively close to the centre of the Milky Way, where the dark matter density is higher than average. The average dark matter density in a dark matter halo is 200 times denser than the cosmological average, where the numerical value comes from the virial overdensity, which we computed for a pure CDM Universe in Eq. (2.43).

By establishing connections between the dark matter halos and observable galaxy properties, such as stellar mass, luminosity and star formation rate, we can explore the relationship between dark matter and visible matter, furthering our understanding of galactic evolution. But first, we need to know how to compute the properties of DM halos.

6.3.1 Halo Mass Function

The halo mass function (HMF), dn/dM, describes the number density, n, of dark matter halos as a function of their mass, M. Expressed in this way it has units solar mass per cubic Mpc, $M_\odot \, \text{Mpc}^{-3}$. Often we use $dn/d \ln M = M dn/dM$, which gives the number of halos in a logarithmic mass window centred on mass M and has units Mpc^{-3}. Despite halos being non-linear objects, it is possible to compute (an estimate of) the HMF using analytical methods.

The Press–Schechter (PS) formalism, developed by Press and Schechter in the 1970s [26], provides a statistical approach to model the process of non-linear structure formation. By assuming that the linear density fluctuations follow a Gaussian distribution and applying a simple criterion for halo formation, the PS formalism provides a rough approximation of the HMF. Due to its simplicity, the PS formalism is widely used in particular for exploring the effects of the underlying cosmological model and the model of DM on the statistics of halos.

From the spherical collapse model, we know how long it takes a region to collapse. From perturbation theory, we can only find the linear evolution of the Fourier modes of the DM overdensity field, $\delta(k, z)$. The first question PS asks is: what value does the linear overdensity have to take before non-linear spherical collapse happens? A first rough guess would be $\delta = 1$ since this is where linear perturbation theory breaks down. We can actually get a better estimate by calculating the 'linearly extrapolated overdensity' of an initially small perturbation at the time when the same perturbation collapses in the full non-linear spherical collapse problem (section 2.2). Performing this calculation (see Problem I.10) we find that the linear threshold for collapse is $\delta_c = 1.686$.

Therefore, we expect that regions of a linearly evolved overdensity field above the critical overdensity, $\delta_c = 1.686$, will have collapsed to form virialised objects. This can be written as the criterion that, locally, if a peak in the overdensity field satisfies $\delta(\mathbf{x}, z) > \delta_c(z)$, then a DM halo should form. How massive will the halo be? The DM density field is described by the power spectrum, which is a sum of Fourier modes and will not give a smooth density field of spherically collapsing peaks of well-defined mass and radius. To assign mass and radius to a peak region we consider the smoothed density field given by

$$\delta_s(\mathbf{x}; R) \equiv \int \delta(\mathbf{x}')W(\mathbf{x} + \mathbf{x}'; R)\mathrm{d}^3\mathbf{x}', \qquad (6.66)$$

where $W(\mathbf{x}; R)$ is a window/filter function of characteristic radius R. For the simplest case of a spherical tophat filter in real space, the mass contained within the filter function is $M = \frac{4}{3}\pi\bar{\rho}R^3$. The PS formalism works by assuming that the probability of $\delta_s > \delta_c(z)$ is equal to the fraction of mass contained within halos of mass M.

Assuming a Gaussian density field with zero mean and variance σ, the probability of $\delta_s > \delta_c$ is found to be

$$\mathscr{P}[> \delta_c(t)] = \frac{1}{2}\mathrm{erfc}\left[\frac{\delta_c(t)}{\sqrt{2}\sigma(M)}\right]. \qquad (6.67)$$

The variance $\sigma(M)$ is the mass variance of the smoothed linear overdensity field, which can be computed from the power spectrum:

$$\sigma^2(M) = \langle\delta_s^2(\mathbf{x}; R)\rangle = \frac{1}{2\pi^2}\int_0^\infty P(k)\tilde{W}^2(\mathbf{k}R)k^2\mathrm{d}k, \qquad (6.68)$$

where $\tilde{W}^2(\mathbf{k}R)$ is the Fourier transform of the real space filter that defines δ_s.

Thus, if we have the linear power spectrum and specify the filter function, we can find $\sigma(M)$ and therefore the probability. For a spherical tophat, the window function in Fourier space is

$$\tilde{W}_{\mathrm{th}}(kR) = \frac{3(\sin kR - kR\cos kR)}{(kR)^3}. \qquad (6.69)$$

However, underdense regions can be enclosed within a larger overdense region, causing them to be included in the larger collapsed object. This 'cloud-in-cloud' problem is accounted for by introducing a famous 'fudge factor' of 2 (which can be derived rigorously in the 'extended PS' or 'excursion set' model [27]). This modifies the original ansatz to give the following:

$$F(> M) = 2\mathscr{P}[> \delta_c(t)]. \tag{6.70}$$

Using the above, the HMF is given by

$$\frac{dn}{d\ln M} = M\frac{\rho_0}{M^2}f(\sigma)\left|\frac{d\sigma}{d\ln M}\right|, \tag{6.71}$$

where ρ_0 is the mean DM density and $f(\sigma)$ is the multiplicity function, which for a Gaussian field undergoing spherical collapse is given by

$$f_{PS}(\sigma) = \sqrt{\frac{2}{\pi}}\frac{\delta_c}{\sigma}\exp\left(\frac{-\delta_c^2}{2\sigma^2}\right), \tag{6.72}$$

where δ_c is the critical overdensity. As discussed, for CDM $\delta_c \approx 1.686$. This applies at any fixed time. However, because the late time growth is in general scale independent (see Eq. 6.56), we can use the linear power spectrum $P(k)$ today, at $z = 0$, and consider the barrier to depend on z as $\delta_c = 1.686/D(z)$, where $D(z)$ is the linear growth function. Our derivation of PS is somewhat vague and motivated by very simplifying assumptions. The whole thing can be derived rigorously for a random field using statistical mechanics, as was done in 1985 in a famous paper by Bardeen, Bond, Kaiser and Szalay (BBKS) [28], although the problem can only be solved analytically for a Gaussian field with a 'sharp-k' filter function that leaves the mass assignment somewhat ambiguous.

Going beyond the spherical collapse approximation leads to different multiplicity functions, which are often found to better match simulation data than PS. One of the most popular alternatives is the Sheth–Tormen multiplicity function, which fits ellipsoidal collapse:

$$f_{ST}(\sigma) = A\sqrt{\frac{2a}{\pi}}\left[1 + \left(\frac{\sigma^2}{a\delta_c^2}\right)^p\right]\frac{\delta_c}{\sigma}\exp\left(\frac{-a\delta_c^2}{2\sigma^2}\right), \tag{6.73}$$

where $A = 0.3222$, $a = 0.707$ and $p = 0.3$ [29].

Applying the outlined formalism on the linear power spectrum gives the HMF as a function of redshift and halo mass. The cosmological parameters and the model of DM enter through the computation of the linear power spectrum. For the PS spherical collapse multiplicity function and some reference cosmological parameters, the HMF is shown in Fig. 6.6. We observe that low-mass halos are the most numerous: this is expected since the dimensionless power spectrum $\Delta^2(k) = k^3 P(k)$ rises continuously at large k, meaning there is the most power on small scales and that low mass halos form first chronologically: this is known as *hierarchical structure formation*. Secondly, the HMF has a strong cut-off at high mass: on large scales the power is small and regions with $\delta > \delta_c$ are rare, which translates into a very small number of the most massive halos.

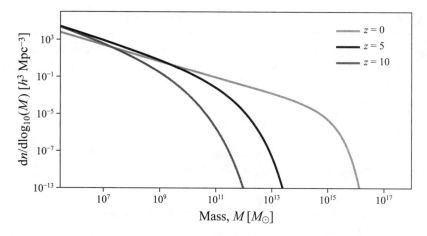

Figure 6.6 Typical HMF calculated using the Press–Schechter formalism from the linear matter power spectrum using the standard multiplicity function.

6.3.2 NFW Profile

We expect NFW profile for DM halos, Eq. (3.30), if they are formed via hierarchical structure formation, that is, the merger of many smaller already collapsed objects [30]. This is essentially an empirical observation arising from numerical simulations. After many hierarchical mergers, the NFW profile, or something very similar to it, appears to be a universal attractor. Recall that the concentration parameter is then defined by $c = R_{\text{vir}}/r_s$ where R_{vir} is the radius that encloses a region of mean density equal to $\rho_{\text{vir}} = 200\rho_{\text{crit}}$ [31].

N-body simulations have shown that it is possible to relate the concentration parameter of a halo to its collapse redshift. The collapse redshift, sometimes also called the formation redshift, is defined to be the redshift at which, for some halo, half of its final mass M_{final} is contained within 'progenitors' (smaller halos that merge to form the final halo) of a mass larger than $0.01M_{\text{final}}$. This can be estimated using the PS formalism, giving us an analytical way to estimate the 'concentration-mass relation', $c(M,z)$. Using the definition just given, the collapse redshift is estimated for a Gaussian field as

$$\text{erfc}\{X(z_{\text{col}}) - X(z)\} = \frac{1}{2}, \tag{6.74}$$

where

$$X(z) = \frac{\delta_c(z)}{\sqrt{2[\sigma^2(0.01M,z) - \sigma^2(M,z)]}}, \tag{6.75}$$

and $\sigma^2(M,z)$ is the variance of the linear power spectrum at the redshift z of interest. Given $\sigma(M,z)$, we can numerically find the root of the equation giving z_{col}. We then assume that the scale density of the halo, ρ_s, is proportional to the density of the Universe at the collapse redshift with some constant of proportionality, κ_{NFW}.

Since we have calculated the characteristic density, the concentration parameter can then be calculated using Eq. (3.32).

Once the concentration parameter curve is calculated, the free parameter κ_{NFW} can be fitted to results from N-body simulations. For CDM it is found that the approximate formula:

$$c(M,z) = 4 \left(\frac{1 + z_{\mathrm{col}}(M,z)}{1+z} \right), \tag{6.76}$$

is a reasonable fit. So if we can use PS to solve for z_{col}, we have $c(M)$, and the collapse redshift is another useful way to parameterise halos.

6.3.3 The Halo Model for the Non-Linear Power Spectrum

The halo model builds upon the PS formalism to provide a more realistic framework for describing the distribution of dark matter, from which we can calculate the matter power spectrum. It does this by accounting for halo 'bias' (which we explain below) and the internal density profile of dark matter halos.

The matter power spectrum is approximated to be the sum of two terms

$$P(k) = P^{1\mathrm{h}}(k) + P^{2\mathrm{h}}(k), \tag{6.77}$$

where $P^{1\mathrm{h}}(k)$ is the one-halo term which gives the correlation of the density within a single halo, and $P^{2\mathrm{h}}(k)$ is the two-halo term which gives the correlation between two different halos.

The one- and two-halo terms are given by

$$P^{1\mathrm{h}}(k) = \frac{1}{\bar{\rho}^2} \int \mathrm{d}M M^2 n(M) |\tilde{u}(k,M)|^2, \tag{6.78}$$

and

$$P^{2\mathrm{h}}(k) = P^{\mathrm{L}}(k) \left[\frac{1}{\bar{\rho}} \int \mathrm{d}M M^2 n(M) b(M) |\tilde{u}(k,M)| \right]^2, \tag{6.79}$$

where $b(M)$ is the halo bias, $\tilde{u}(k,M)$ is the Fourier transform of the halo density profile, $P^{\mathrm{L}}(k)$ is the linear matter power spectrum and $n(M)$ is the halo mass function which can be calculated using PS as shown above. Since halos form only at the peaks of the density field, they will be 'biased tracers' of it, that is, they are only in dense places, not underdense ones: the halo bias parameter, $b(M)$, accounts for this statistically in the power spectrum.

For ellipsoidal collapse with a Sheth–Tormen multiplicity function given by Eq. (6.73), the halo bias can be shown to be

$$b(M,z) = 1 + \frac{1}{\delta_{\mathrm{c}}} \left(a\nu^2 - 1 + \frac{2p}{1 + (\sqrt{a}\nu)^{2p}} \right), \tag{6.80}$$

where a and p are the Sheth–Tormen fitting parameters given previously and $\nu = \delta_{\mathrm{c}}(z)/\sigma(M,z)$.

We will consider a normalised radial density profile $u(r, M, z) = \rho(r, M, z)/M$ truncated at the virial radius R_{vir}. Its Fourier transform is given by

$$\tilde{u}(k, M, z) = 4\pi \int_0^{R_{vir}} u(r, M, z) \frac{\sin(kr)}{kr} r^2 dr. \tag{6.81}$$

If we assume that the densities of our halos are given by the NFW profile discussed previously (Eq. (3.30)), the Fourier transform is given by

$$\tilde{u}(k, M, z) \bigg(\cos(b)(\text{Ci}(b + kR_{vir}) - \text{Ci}(b))$$

$$+ \sin(b)(\text{Si}(b + kR_{vir}) - \text{Si}(b)) - \frac{\sin(kR_{vir})}{b, +kR_{vir}} \bigg), \tag{6.82}$$

where $\text{Si}(x)$ and $\text{Ci}(x)$ are the sine and cosine integrals, and $b = kR_{vir}/c$ in which c is the NFW concentration parameter. Therefore using the halo model, using the HMF and $c(M)$ (or any halo density profile model) which we can compute from the linear theory using PS, we can go from the linear power spectrum to the non-linear power spectrum.

The halo model gives a simple and intuitive way to model the non-linear power spectrum, and the result qualitatively reproduces the non-linear power spectrum in simulations and in the real Universe. The advantage of the halo model and PS is that they are physically motivated and can be adapted to different theories of DM. However, an accurate model of the non-linear power spectrum, which is necessary to statistically compare theory to data and learn about cosmology and DM in more detail, requires additional tuning parameters fitted to grids of simulations. Higher orders in perturbation theory, on the other hand, have the advantage that accuracy is under full theoretical and analytical control, but the disadvantage of being strictly fixed to regimes with $\delta < 1$, before halos form. This limits the range of wavenumbers, k, where perturbation theory can be applied to roughly less than 0.1 Mpc^{-1}. Simulations, to which we now turn, have the advantage of retaining all non-linearities. However, since they require a large amount of development, they are less commonly available for exotic theories of cosmology and DM.

6.3.4 Simulations

In section 5.2, we discussed how the evolution of matter density in the very early Universe could be modelled using linear theory due to the small size of the initial perturbations. However, as these perturbations collapse and grow in size, our linear equations can no longer be used. Additionally, while semi-analytical approaches such as the Press–Schechter method discussed above can be used to estimate very useful statistics such as the halo mass function, they are unable to shed light on the finer details such as the internal structure of dark matter halos or the non-linear clustering that produces features such as the cosmic web.

Instead, a full non-linear treatment is needed. In this section, we discuss so-called N-body methods, which aim to simulate the DM phase space distribution function by discretising it into individual particles. These particles are not fundamental

DM particles, but a kind of quasiparticle, usually of mass around $10^6 \, M_\odot$ or so in a cosmological simulation. These particles interact gravitationally like CDM, and their dynamics produce a coarse-grained view of structure formation.[6]

Consider a system of N particles of equal mass $m = 1$ in some units, with positions and velocities $(\mathbf{r}_i, \mathbf{u}_i)$ where $i = 1, 2, ..., N$. In cosmological simulations, the background space is uniformly expanding. Therefore, it is convenient to choose coordinates that expand with the background. The comoving positions and velocities are given by

$$\mathbf{x}_i = \mathbf{r}_i / a, \tag{6.83}$$

and

$$\mathbf{v}_i = \frac{d\mathbf{x}_i}{dt}, \tag{6.84}$$

respectively where a is the scale factor. The comoving acceleration is given by

$$\frac{d\mathbf{v}_i}{dt} + 2H(t)\mathbf{v}_i = -\frac{1}{a^2}\nabla_x \Phi|_x, \tag{6.85}$$

where $H(t) = \dot{a}/a$ is the Hubble parameter and Φ is the gravitational potential due to density *perturbations* given by

$$\nabla_x \Phi|_x = 4\pi G a^2 \left[\rho(\mathbf{x}, t) - \bar{\rho}(t)\right], \tag{6.86}$$

where $\bar{\rho}(t)$ is the average background density.

While the exact approach may vary, the evolution of a system of particles is simulated by:

1. Calculating the gravitation field based on the positions of the particles at some time t_n.
2. Hence calculating the new speed and position of each particle:

$$v_i(t_n) = v_i(t_{n-1}) + \frac{dv_i}{dt}\Delta t, \tag{6.87}$$

$$x_i(t_n) = x_i(t_{n-1}) + v_i(t_n)\Delta t. \tag{6.88}$$

3. Recalculating the gravitation force due to the particles at their new positions.

Typically, to improve the precision and stability of the solution, an additional half-step is added such that the acceleration on each particle is calculated from its intermediate position at $t_{n+1/2}$ instead of its initial position at t_n. This is known as the 'leap-frog scheme'.

The simplest way to calculate the motion of the particles is to directly sum the gravitation force between each pair of particles as given by

$$F_i = \sum_{j \neq i} Gm \frac{r_i - r_j}{|r_i - r_j|^3}. \tag{6.89}$$

---

[6] Section 10.3 discusses briefly how this method can be modified to describe very light bosonic dark matter like axions.

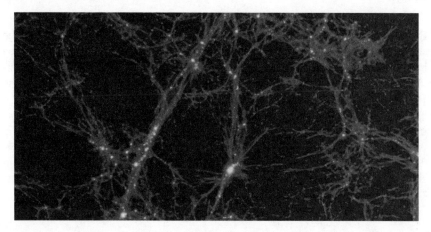

Figure 6.7 Cosmological simulation performed using ENZO, shared by the ENZO project [35]. See colour insert.

Since this does not introduce any approximations, using a direct sum offers the highest level of accuracy. However, it is also the most computationally expensive as $\mathcal{O}(N^2)$ operations need to be performed for each time-step [32].

One way of reducing the computation time is to use the tree code method. By subdividing the space into a hierarchy of cells, this method enables the force contribution of many distant particles to be approximated together. Doing so reduces the computation time to $\mathcal{O}(N \log N)$. This can be sped up further by considering the fact that the effect of distant particles will be very similar for all particles in a close group. By doing this, the Fast Multipole Method further reduces the calculation time to just $\mathcal{O}(N)$ [33]. Another method used to simplify and speed up the calculations is by using a particle-mesh (PM) algorithm. Here, instead of calculating the force directly between each pair of particles, the gravitational field is calculated by solving the Poisson equation for the mean density field.

Additionally, in order to avoid nonphysical behaviour at small radii such as the formation of binary systems of the particles (which, recall, are not physical particles), N-body simulations typically employ a 'softening length' below which the forces between particles are reduced.

It is also possible to develop N-body simulations that incorporate baryonic physics. Doing so enables us to study the impacts of gas, star formation and so on (often collectively called 'baryonic effects' or 'feedback') on structure formation, which are often neglected by analytical methods.

There is a large variety of publicly available N-body simulation codes, each with its own strengths. Some simulations make use of adaptive mesh refinement to probe greater levels of detail without significantly increasing the computational requirements to do so. Much of the detail in an N-body simulation is highly localised in dark matter halos with vast areas of near-empty space between them. Adaptive mesh refinement enables us to 'zoom in' on the areas of greatest detail while spending less computation resources on the space between halos. One popular publicly available code that uses this technique is ENZO [34].

An example of a cosmological simulation performed using ENZO is shown in Fig. 6.7. Here we can see the bright dark matter halos, joined together on large scales by filaments, making up the grand cosmic web. The cosmic web is observed by galaxy surveys like SDSS and has a striking similarity to the one seen in simulations: further evidence for the need for DM to drive structure formation in the Universe.

While there is not enough space to discuss simulations in any more detail here, it is important to note that in the 1980s evidence for the stability of galaxies in the presence of DM halos, which came from simulations, provided strong evidence to many for the existence of DM (see Ref. [1] and the historical note in chapter 1).

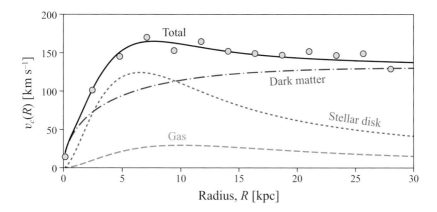

PLATE 3.6. Sketch of the total rotation curve for a galaxy containing gas, disk, bulge, and DM.

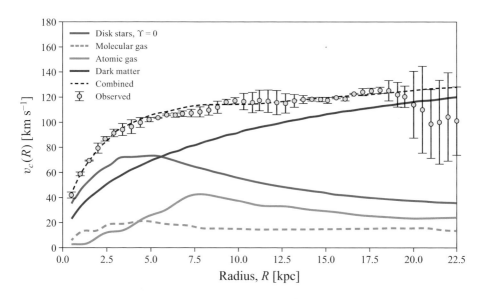

PLATE 3.7. Rotation curve of M33. Data and models are taken from Ref. [7]. The model shown takes the mass-to-light ratio as a free parameter. Best-fit NFW parameters are $c = 4.55 \pm 0.01$ and $v_{vir} = 139.9 \pm 0.5$, where v_{vir} is the computed from the virial mass defined in Eq. (2.44) with $\Delta_{vir} = 200$.

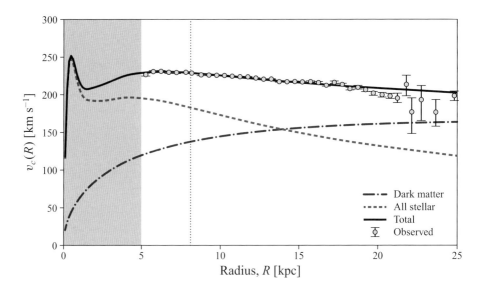

PLATE 3.8. Milky Way rotation curve. Data and models are taken from Ref. [13]. Best-fit NFW parameters are $c = 12.8 \pm 0.3$, and $M_{vir} = 7.25 \pm 0.25 \times 10^{11} M_\odot$, which imply $\rho_0 = 1.06 \pm 0.09 \times 10^7 M_\odot$ kpc^{-3}, $R_{vir} = 189.3 \pm 2.2$ kpc, and $R_s = 14.8 \pm 0.4$ kpc. The shaded region denotes the inner galaxy, which is not used in the fit. The vertical dotted line denotes the radial location of the Sun.

PLATE 4.5. *The Bullet Cluster*, composite public domain image from NASA. Red colours indicate the hot gas, measured using X-rays. Blue colours indicate the mass distribution inferred from the gravitational lensing κ map, measured from the ellipticities of the visible galaxies.

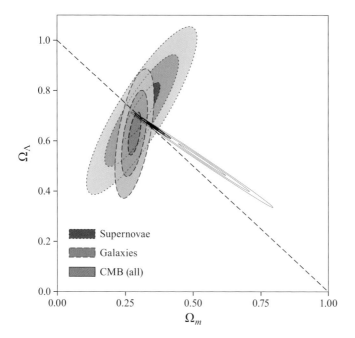

PLATE 6.1. The 'concordance' standard cosmological model: ΛCDM. The dotted line has $\Omega_\Lambda = 1 - \Omega_m$, representing a flat Universe. The ellipses represent the statistical confidence, at one, two, and three standard deviations, for Ω_m and Ω_Λ to lie in the indicated region, as determined by supernovae, galaxies, and the CMB. The open ellipse shows the CMB result without including gravitational lensing. SNe and BAO results are taken from the Sloan Digital Sky Survey legacy archive, https://svn.sdss.org/public/data/eboss/DR16cosmo/tags/v1_0_1/, and CMB results from the Planck legacy archive, https://pla.esac.esa.int/#home.

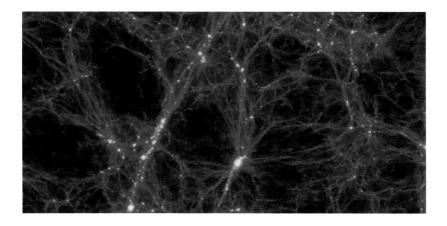

PLATE 6.7. Cosmological simulation performed using ENZO, shared by the ENZO project [35].

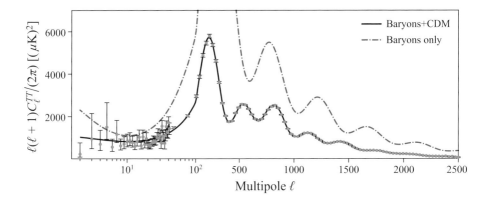

PLATE 6.5. The CMB power spectrum. The spectrum is computed numerically with CAMB [40], and the data are taken from Planck [42]. The case 'baryons+CDM' has CDM density $\Omega_c h^2 = 0.120$ compared to baryon density $\Omega_b h^2 = 0.0224$, while the case 'baryons only' sets $\Omega_c h^2 = 0$, leaving all other cosmological parameters the same. Removing CDM, the model is still a reasonable fit at low multipoles, but the higher acoustic peaks are badly off, and the amplitude of acoustic oscillations is much larger without the CDM to provide gravitational potential wells. The equivalent to the 'MOND' model in Fig. 6.4 is not visible on this plot, the amplitude being around 1000 times larger than the baryons+CDM model.

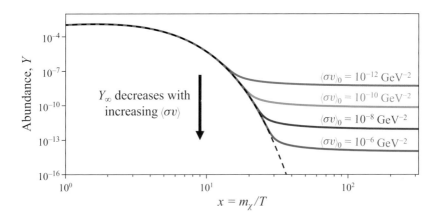

PLATE 9.2. Freeze out of a massive particle. Calculated for a WIMP with temperature-independent annihilation cross section, for four different values of the cross section. The result is found by numerically solving the Boltzmann equation as given in Eq. (9.3) with equilibrium initial conditions. We use the full temperature dependence of number density, Hubble rate, entropy and relativistic degrees of freedom $g_*(T)$ and $g_{*s}(T)$. The black dashed line shows the equilibrium value for the abundance.

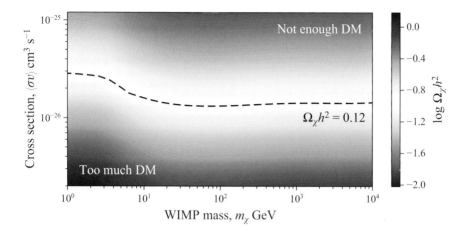

PLATE 9.3. WIMP relic density—calculated for a WIMP with temperature-independent annihilation cross section. The result is found by numerically solving the Boltzmann equation as given in Eq. (9.3) with equilibrium initial conditions. We use the full temperature dependence of number density, Hubble rate, entropy and relativistic degrees of freedom $g_{\star,r}(T)$ and $g_{\star,s}(T)$. The black dashed line shows the $(m_\chi, \langle\sigma v\rangle)$ combinations which give the observed relic density, $\Omega_\chi h^2 = 0.12$.

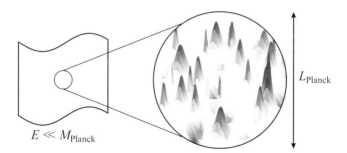

PLATE 9.12. At large scales, spacetime is homogenous; however, at the Planck scale, L_{Pl}, quantum effects like potential barriers and wells become visible. This is often called the *spacetime foam*.

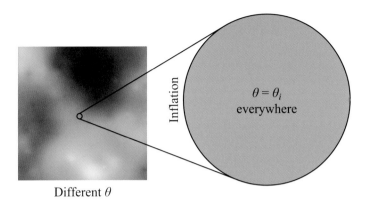

Different θ

PLATE 10.5. If SSB occurs before inflation, θ takes a random value in different patches. One of these patches, that is, a fixed value of θ, is then inflated and $\theta = \theta i$ everywhere.

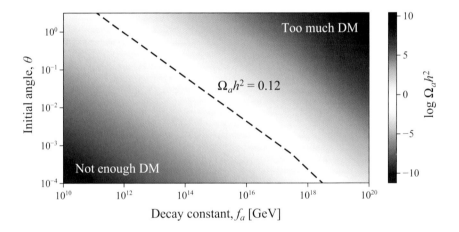

PLATE 10.11. Axion relic density for $ma(T)$ of the QCD axion. The solution here contains no anharmonic corrections.

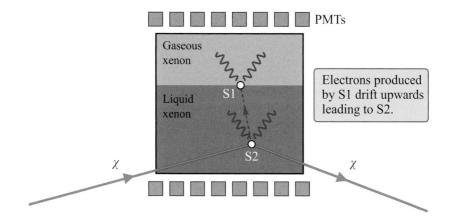

PLATE 12.7. XENON1T–DM particles collide with 2 tonnes of liquid xenon causing a scintillation event. This produces an electron which drifts upwards due to an applied field causing an additional scintillation event. Both events produce photons which are then detected using photomultiplier tubes.

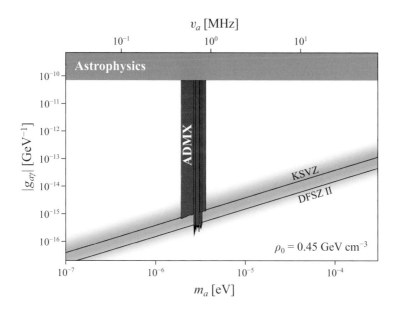

PLATE 14.6. ADMX constraints—constraints on the coupling of the QCD axion coupling to photons from the ADMX haloscope experiment. Generated using AxionLimits [85].

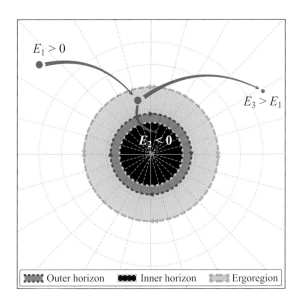

PLATE 15.2. Schematic of the Penrose process. A particle with energy E_1 enters the Kerr ergoregion where it is forced to co-rotate with the BH. It then splits in two, with one particle falling into the event horizon with $E_2 < 0$ and another escaping to infinity with $E_3 > E_1$, with the balance of energy paid by a reduction in the BH spin and a shrinking of the ergoregion. If the released particle is confined by a 'mirror', then this process repeats over and over, spinning the BH down. Reproduced from "Superradiance in string theory" by Viraf M. Mehta et al. (*Journal of Cosmology and Astroparticle Physics*, Vol. 2021) ©2021 IOP Publishing Ltd and Sissa Medialab.

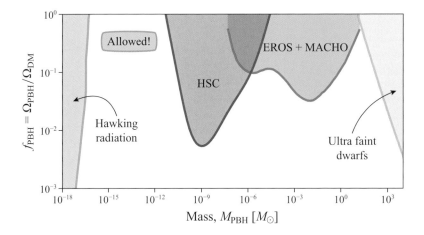

PLATE 16.7. Compilation of key constraints on the fraction of DM composed of PBHs including microlensing (HSC and EROS), γ and X-ray measurements and the survival of ultra faint dwarf galaxies. Data collated by B. Kavanagh as part of the PBHbounds project, https://github.com/bradkav/PBHbounds (Ref. [181]).

QUIZ

i. What are the units of the power spectrum $P(k)$?
   a. k^{-1}.
   b. k^{-2}.
   c. k^{-3}.
   d. k^{-4}.

ii. What is the value of the scalar spectral index for scale invariance?
   a. $n_s = 0$.
   b. $n_s = 1$.
   c. $n_s = 0.96$.
   d. $n_s = 3$.

iii. What defines a 'superhorizon mode'?
   a. $k\eta \gg 1$.
   b. $k\eta \ll 1$.
   c. $k \ll 1$.
   d. $k \gg 1$.

iv. Under what conditions is the gravitational potential approximately constant in time? (Select all that apply.)
   a. Superhorizon, matter era.
   b. Superhorizon, radiation era.
   c. Superhorizon, transition from matter to radiation.
   d. Sub-horizon, matter era.
   e. Sub-horizon, radiation era.

v. The general oscillator equation is $\ddot{x} + \Gamma(t)\dot{x} + m(t)^2 x$. Select the correct properties this equation has.
   a. Damped oscillations when $m^2 > 0$, $m > \Gamma$.
   b. Always has an analytical solution.
   c. Growing solutions when $m^2 > 0$.
   d. Growing solutions when $m^2 < 0$.

vi. The baryons cannot cluster as strongly as the dark matter because (select all that apply):
   a. Protons weigh less than DM.
   b. The plasma pressure (sound speed) leads to acoustic oscillations.
   c. The baryons are relativistic at early times.
   d. The baryons and photons are tightly coupled.
   e. There is a new force that acts between DM particles.
   f. The baryon Boltzmann equations have no quadrupole moment.

vii. Without DM, the theoretical power spectrum of density perturbations in matter (normalised to the CMB quadrupole on large scales) is:

a. Too large compared to the observed power spectrum.

b. Too small compared to the observed power spectrum.

c. Strongly non-linear.

d. Contains no acoustic oscillations.

Answers

i. c; ii. b; iii. b; iv. b; v. a,b,d; vi. a,d; vii. b,d; viii. b

Problems on Evidence for Dark Matter

Problem I.1. Derive an expression for the baryon density in terms of η_{10} and T_{CMB} and thus verify Eq. (6.6). Convert this into the implied value of Ω_c, and see what range of values Ω_c can take if you vary $|\Omega_k| \approx 10^{-2}$, $\Omega_\Lambda \approx 0.7$ and $h \approx 0.7$ within some reasonable limits. Try randomly numerically sampling all of the input parameters with Gaussian errors, and also from uniform priors, and thus produce histograms of Ω_c and of the relic density $\Omega_c h^2$ for these choices.

Problem I.2. Recall the virial theorem:

$$2\langle T \rangle = -\langle U \rangle. \tag{6.90}$$

Astronomers studying a galaxy have measured its brightness and calculated that the mass of visible matter within a radius of 2.2×10^{21} m is 4.1×10^{40} kg. Stars orbiting at this radius have been measured to have a velocity dispersion of 1.5×10^5 ms^{-1}. Is the total mass of the stars enough to explain their velocity dispersion? If not, what percentage of the mass is 'missing'?

Hint: assume a spherical distribution of matter.

Problem I.3. We will now re-derive the virial theorem using Hamiltonian dynamics. For this we will make use of Hamilton's equations of motion

$$\dot{p} = -\frac{\partial H}{\partial q}; \quad \dot{q} = \frac{\partial H}{\partial p}, \tag{6.91}$$

and the definition of the Poisson bracket

$$\{A, B\} = \frac{\partial A}{\partial q}\frac{\partial B}{\partial p} - \frac{\partial A}{\partial p}\frac{\partial B}{\partial q}. \tag{6.92}$$

a) Compute $\frac{df}{dt}$ for $f = f(\mathbf{q}, \mathbf{p}, t)$ in terms of partial derivatives and show that

$$\frac{df}{dt} = \{f, H\} + \frac{\partial f}{\partial t}. \tag{6.93}$$

b) Consider 'the virial':

$$Q = \sum_i^N \mathbf{p} \cdot \mathbf{q}. \tag{6.94}$$

Find an expression for $\frac{dQ}{dt}$.

c) Following the argument in chapter 2, show that

$$\left\langle u_\alpha \frac{\partial H}{\partial u_\alpha} \right\rangle = k_B T, \tag{6.95}$$

where T is temperature, for $u_\alpha = p, q$.

d) For $H = K(p) + U(q)$, with $K(p) = p^2/2m$, show that

$$\left\langle \frac{dQ}{dt} \right\rangle = 0 \tag{6.96}$$

gives rise to the virial theorem.

e) By considering the meaning of the $\langle ... \rangle$ time average, argue that

$$\left\langle \frac{dQ}{dt} \right\rangle = 0, \tag{6.97}$$

as $t \to \infty$ for bound systems.

Problem I.4. Given that

$$\frac{d^2 r}{dt^2} = -\frac{GM}{r^2} \tag{6.98}$$

has parametric solutions

$$r = A(1 - \cos\theta), \tag{6.99}$$

$$t = B(\theta - \sin\theta), \tag{6.100}$$

show that

$$A^3 = GMB^2. \tag{6.101}$$

Problem I.5. a) The light from two stars, A and B, within the Milky Way has flux densities of 4.00×10^{-15} W/m² and 1.72×10^{-15} W/m² respectively. Using parallax, astronomers measure their distances from Earth as 4.1 kpc and 7.2 kpc. Which star is brighter?
Flux and intrinsic luminosity are related by $F = L/4\pi d^2$ for a source at distance d.

b) Stars A and B emit 21 cm neutral hydrogen lines with wavelengths of 21.1030 cm and 21.1042 cm respectively. What are their velocities in the direction of the Earth's line of sight? What are their velocities with respect to the Earth? *Hint: the 21 cm neutral hydrogen line has a vacuum wavelength of 21.106114 cm.*

c) In galactic coordinates, star A has a longitude of 105° and a latitude of 2.1° while star B has a longitude of −12° and a latitude of 5.2°. What are the stars' orbital radii? What are their velocities relative to the centre of the galaxy?

d) Assuming circular motion, estimate the amount of mass contained within the orbital radius of each star. *Hint: use the fact that at early times $\theta \ll 1$.*

Problem I.6. The density of dark matter in our local region of the galaxy is $\rho_{DM} \approx$ 0.3 GeV cm^3. Estimate the number of DM particles passing through the Earth every minute if the DM is composed of:

- WIMPs of mass ≈ 100 GeV.
- QCD axions of mass $\approx 10^{-6}$ eV.
- Primordial black holes of mass $\approx 10^{-12} M_\odot$.

Problem I.7. Derive the fluid equation

$$\dot{\rho} + 3H(\rho + p) = 0 \qquad (6.102)$$

from the acceleration and Friedmann equations

$$\ddot{a} = \frac{-4\pi G}{3}(\rho + 3p)a, \qquad (6.103)$$

$$H^2 = \frac{\dot{a}^2}{a^2} = \frac{8\pi G}{3}\rho - \frac{K}{a^2}. \qquad (6.104)$$

Problem I.8. a) The deceleration parameter is defined as

$$q_0 = -\frac{\ddot{a}(t_0)a(t_0)}{\dot{a}(t_0)^2}. \qquad (6.105)$$

Show that for a Universe containing matter (m), radiation (r) and a cosmological constant (Λ):

$$q_0 = \frac{1}{2}(\Omega_m + 2\Omega_r - 2\Omega_\Lambda). \qquad (6.106)$$

Problem I.9. *The Age of the Universe*
  Friedmann's equation is one of Einstein's equations that applies to an expanding Universe with 'scale factor' a. It reads:

$$\frac{da}{dt} = aH_0\sqrt{\frac{\Omega_m}{a^3} + \Omega_\Lambda}, \qquad (6.107)$$

where Ω_m and Ω_Λ are the fractions of the Universe in matter (dark and ordinary) and the cosmological constant. The scale factor has value $a = 0$ at $t = 0$: it is zero at the Big Bang. It is also normalised to be equal to one today: $a = 1$ at $t = t_u$, where t_u is the age of the Universe.
  Show the following:

$$t_u = \frac{1}{H_0}\int_0^1 \frac{da\, a}{\sqrt{a\Omega_m + a^4\Omega_\Lambda}} \qquad (6.108)$$

$$= \frac{1}{H_0}\frac{2}{3}\frac{1}{\sqrt{\Omega_\Lambda}}\sinh^{-1}\sqrt{\frac{\Omega_\Lambda}{\Omega_m}}. \qquad (6.109)$$

The values $\Omega_\Lambda = 0.68$, $\Omega_m = 0.32$ are measured by the *Planck* satellite from the cosmic microwave background. Use this to verify that the factor multiplying the

Hubble time, $1/H_0$, is approximately one. Therefore, $t_u \approx 1/H_0$, and the estimate that the Hubble time is approximately the age of the Universe, which you used in Problem 2, is a good one.

b) For relatively nearby objects, the apparent luminosity of a source can be written as

$$\ell = \frac{L}{4\pi d^2},$$
(6.110)

where L is the absolute luminosity and d is the distance to the object. This however must be modified for larger distances to account for:

- At the time t_0 that the light reaches the Earth, the proper area of the sphere into which light is emitted is $4\pi r_1^2 a^2(t_0)$ where r_1 is the coordinate distance.
- Due to the expansion of space (i.e. redshifting of frequency and time), the power of each detected photon is reduced by a factor $a(t_1)/a(t_0) = 1/(1+z)^2$.

Modify equation 6.110 accordingly and hence define the 'luminosity distance' d_L in terms of $a(t_0)$, r_1 and z such that

$$\ell = \frac{L}{4\pi d_L^2}.$$
(6.111)

c) By expanding for small z show that, to second order, the luminosity distance is given by

$$d_L = \frac{1}{H_0}\left[z + \frac{1}{2}(1 - q_0)z^2\right].$$
(6.112)

Hence, by measuring the apparent brightness of Type 1a supernovae as a function of redshift, it is possible to infer the deceleration parameter q_0. Given that $q_0 \sim -0.6$, calculate Ω_m.

Problem I.10. *Linear collapse threshold*

Consider a small overdensity $\delta_i \ll 1$ with shell radius r_i and velocity v_i in the matter-dominated epoch defined such that $\rho(t_i) = \bar{\rho}_m(t_i)(1 + \delta_i)$.

a) Show that

$$3GM = r_i^3 \Omega_{m,i} H_i^2 (1 + \delta_i).$$
(6.113)

b) By equating the total energy of this shell at its maximum radius R_* to the initial time, hence show that

$$\frac{r_i}{R_*} = 1 - \frac{(v_i/(r_i H_i))^2}{\Omega_{m,i}(1 + \delta_i)}.$$
(6.114)

c) Due to the conservation of mass,

$$r_i^3 \bar{\rho}_m (1 + \delta_i) = \text{constant},$$
(6.115)

and its time derivative is therefore zero. Using the chain rule and the conservation equation for the background, $d\bar{\rho}_m/dt_i = -3\bar{\rho}_m H_i$, show that

$$v_i = \frac{dr_i}{dt_i} = H_i r_i \left(1 - \frac{1}{3H_i t_i} \frac{\delta_i}{1 + \delta_i} \frac{d \ln \delta_i}{d \ln t_i} \right). \tag{6.116}$$

d) For a matter-only Universe $H_i t_i = 2/3$ and $\delta_i \propto t_i^{2/3}$ for small t_i. Hence, for small δ_i, show that

$$v_i = H_i r_i \left(1 - \frac{\delta_i}{3} \right). \tag{6.117}$$

e) Hence use your answer from part **a** to show that

$$\frac{r_i}{R_*} = 1 - \left(1 - \frac{5}{3} \delta_i \right) \Omega_{m,i}, \tag{6.118}$$

for small δ_i.

f) Recall from section 2.2 that

$$R_*^3 = \frac{GM}{\pi^2} t_c^2, \tag{6.119}$$

where t_c is the time at which a non-linear overdensity collapses to a point in non-linear spherical collapse. Hence show that

$$t_c^2 = \left(\frac{3\pi}{2} \right)^2 \frac{t_i^2}{\left(1 - [1 - \frac{5}{3}\delta_i] \Omega_{m,i}^{-1} \right)^3} \frac{1}{\Omega_{m,i}(1 + \delta_i)}. \tag{6.120}$$

g) Finally, using the fact that $\Omega_{m,i} = 1$ in a flat matter-only Universe, and that $\delta_i \propto t_i^{2/3}$, show that the linearly extrapolated overdensity at the time of collapse is $\delta(t_c) = 1.686$.

PART II

Theories of Dark Matter

Chapter Seven

Particle Physics and the Standard Model

The Standard Model of particle physics (SM) details our understanding of the dynamics governing the interactions of elementary particles at the shortest distance scales currently accessible.

The treatment of this topic will be very brief, attempting to only introduce the most important concepts or those that we will need later. For more details on this vast and extremely interesting subject, see the reading list.

In particle physics, we express the matter content and interactions via a Lagrangian density, $\mathcal{L}$.[1] The Lagrangian for the Standard Model of particle physics reads

$$\mathcal{L} = -\frac{1}{4}\overbrace{F_{\mu\nu}F^{\mu\nu}}^{\text{gauge fields}} + \underbrace{\overline{\psi}\left(i\slashed{D}\psi\right)}_{\text{fermions}} + \overbrace{\left(D_\mu\phi\right)^\dagger\left(D^\mu\phi\right) - V\left(\phi\right)}^{\text{Higgs}}$$

$$+ \underbrace{Y_{ij}\phi\overline{\psi}_i\psi_j + \text{Hermitian conjugate}}_{\text{Yukawa interactions}}. \qquad (7.1)$$

This is the 'compact' version of the Lagrangian.[2] When writing down this Lagrangian, we have collected the various symmetries and fields involved.

- The SM is *Lorentz invariant*, and its mathematical underpinning—quantum field theory—is a combination of quantum mechanics and special relativity.
- The gauge symmetry of the SM is $SU(3)_C \times SU(2)_L \times U(1)_Y$ and the matter content of the SM, for the most part, transforms non-trivially under these symmetry groups.
- The matter content of the SM is made up of spin-$\frac{1}{2}$ fermions, that is, leptons and quarks, and integer-spin bosons, that is, gauge fields like gluons and photons and the Higgs field (spin-0) which breaks electroweak symmetry to electromagnetism.

---

[1] Normally, the particle physics literature tends to use the Minkowski metric $\eta = \text{diag}(+1, -1, -1, -1)$, which has the opposite 'signature' to the flat-space metric used in GR. For the remainder of this chapter, when not talking about gravity, we will use this convention for the metric. The metric convention affects the sign of the kinetic terms appearing in the Lagrangian. You can deduce the sign from the physical requirement that kinetic energy is positive semi-definite, that is, $\text{KE} \geq 0$.

[2] You can see the full gory details here https://www.symmetrymagazine.org/article/the-deconstructed-standard-model-equation.

The mathematical underpinnings of the Standard Model are described using quantum field theory (QFT). The interested reader should consult the reading list for references. For our purposes, it is useful to think of a Lagrangian simply as a way of writing down the particle content of a model, that is, what fields appear. QFT tells us we can write down a Feynman diagram, a shorthand for the calculation of an interaction, for every term in the Lagrangian, with a line representing every type of field. The terms in the Lagrangian that are products of more than two fields tell us what interactions our model has: the Feynman diagram for the product of two fields is just a straight line representing a particle moving from one place to another, while for a product of more than two fields we draw vertices joining the lines, which represent interactions between the particles. Lastly, a Lagrangian encodes symmetries. The Lagrangian is a scalar, so all allowed field redefinitions must leave it as such. If they leave it looking the same, with the same products of fields, we say the theory has a symmetry under the change of variables.

7.1 GLOBAL AND 'GAUGE' SYMMETRIES

Here we will define the kinetic terms of the gauge fields and fermions and discuss aspects of the Higgs field and its potential. First, we consider a free complex scalar field, ϕ, with mass, m. Its Lagrangian reads

$$\mathcal{L} = \left(\partial_\mu \phi\right)^* \left(\partial^\mu \phi\right) - m^2 \phi^* \phi, \tag{7.2}$$

and we notice that, under field rotation in the complex plane

$$\phi \to e^{i\alpha} \phi \tag{7.3}$$

$$\phi^* \to e^{-i\alpha} \phi^*, \tag{7.4}$$

Eq. (7.2) is unchanged. This is due to $\alpha \neq f(\mathbf{x})$, that is, α has no local coordinate dependence. Thus, this is known as a *global symmetry*.

The rotation, $e^{i\alpha}$, is a member of the *group* U(1): the group of *unitary*, 1×1, *complex matrices*. A matrix is unitary if it obeys

$$M^\dagger M = M M^\dagger = \mathbf{1}, \tag{7.5}$$

where the †—the 'Hermitian conjugate'—is simply the transpose of the complex conjugate of the matrix. So an element of U(1) is just a complex number with unit length.

One of the key concepts of symmetries is the notion of *conserved quantities*. This was famously derived by Emmy Noether.

Noether's theorem: for every (continuous, differentiable) symmetry of a system, there is a conserved quantity.

More specifically, we can define the conserved current,

$$\partial_\mu \underbrace{j^\mu}_{\text{conserved 4-current}} = 0, \tag{7.6}$$

and its corresponding conserved charge,

$$Q = \int d^3x j^0. \tag{7.7}$$

In the Lagrangian in Eq. (7.2) the conserved current is

$$j^\mu = -i[\phi^* \partial^\mu \phi - (\partial^\mu \phi^*)\phi]. \tag{7.8}$$

You are invited to prove (1) that j^μ can be derived considering the infinitesimal U(1) transformations and (2) using the Euler–Lagrange equations, that the corresponding conserved charge is equivalent to the number density of ϕ particles, minus the number density of ϕ^* particles, that is, we interpret the two degrees of freedom ϕ and ϕ^* as having charges +1 and −1 respectively.

Now, if we promote our global symmetry to a *local* or *gauge* symmetry, that is, $\alpha = q\alpha(x)$, instead,[3]

$$\phi \to e^{iq\alpha(x)}\phi, \tag{7.9}$$

the Lagrangian is no longer invariant, due to the appearance of additional $\partial_\mu \alpha$ terms in $\mathcal{L}$ after applying the transformation. Is there a way for these terms to be absorbed and the *local* U(1) symmetry to be preserved, that is, a way to make $\mathcal{L}$ *invariant*?

In fact, this 'gauging procedure' is done by introducing another field into the theory: a so-called *gauge field*. Let's introduce the vector field A_μ with the Maxwell Lagrangian

$$\mathcal{L} = -\frac{1}{4}F_{\mu\nu}F^{\mu\nu}, \tag{7.10}$$

where $F_{\mu\nu} := \partial_\mu A_\nu - \partial_\nu A_\mu$. The Euler–Lagrange equations arising from this Lagrangian are Maxwell's equations of electromagnetism. Eq. (7.10) is invariant under the U(1) gauge transformation

$$A_\mu \to A_\mu + \partial_\mu \alpha(x). \tag{7.11}$$

By construction:

- $F_{\mu\nu} = -F_{\nu\mu}$: *field strength* is antisymmetric.
- Field strength is gauge invariant since $\partial_\mu \partial_\nu \alpha = \partial_\nu \partial_\mu \alpha$: partial derivatives commute.

Note that there is **no mass term** for A_μ (that is, $m^2 A_\mu A^\mu$): it is forbidden by invariance under the gauge transformation, Eq. (7.11). How does this gauge field help us with promoting our complex scalar field theory to a U(1) gauge symmetric theory?

Returning to our complex scalar field in Eq. (7.2), we introduce the *covariant derivative*:

$$\partial_\mu \to D_\mu \equiv \partial_\mu - iqA_\mu \tag{7.12}$$

---

[3] With this choice, we say that ϕ now has charge q under the U(1).

and notice that the kinetic term in the Lagrangian

$$\mathcal{L} = \left(D_\mu \phi\right)^* \left(D^\mu \phi\right) \tag{7.13}$$

is now invariant under the combination of the local U(1) transformation, Eq. (7.9), and the gauge transformation, Eq. (7.11): we thus interpret the two symmetries as two aspects of the same U(1) *gauge symmetry*.[4] Plugging D into the Lagrangian Eq. (7.13), we notice that the covariant derivative for matter fields results in interactions with the gauge field, that is,

$$\mathcal{L} \supset \overbrace{q\left(\partial_\mu \phi^*\right) A^\mu \phi + \text{h.c.}}^{\text{trilinear interactions}} + \underbrace{q^2 \phi^2 A_\mu A^\mu}_{\text{quartic interactions}}. \tag{7.14}$$

Thus, imposing gauge symmetries on matter fields automatically introduces interactions with the gauge fields. We can generalise the definition of the field strength tensor of a U(1) field by using the commutator of the covariant derivatives, that is, defining

$$F_{\mu\nu}\Psi \propto \left[D_\mu, D_\nu\right] \underbrace{\Psi}_{\text{arbitrary field}} \tag{7.15}$$

implies **no A_μ interactions** arising from the U(1) symmetry,

$$\left[D_\mu, D_\nu\right] = \left[\partial_\mu, \partial_\nu\right]. \tag{7.16}$$

This general definition, however, leads to a different result for non-Abelian groups, as we will see.

For the scalar potential $V(\phi)$ in Eq. (7.1), we note that:

- It must be invariant under a U(1) gauge symmetry.
- It must be stable, that is, it must diverge to positive infinity or go to a constant at $\phi \gg 1$, such that the global minimum is at finite ϕ.

Thus, $V(\phi)$ can take two possible forms, including terms up to ϕ^4:[5]

$$V(\phi) = \pm\frac{1}{2}m^2\phi^*\phi + \lambda\left(\phi^*\phi\right)^2 \xrightarrow[\text{potential}]{\text{minimise}} \phi_{\text{min}} \equiv \langle\phi\rangle \begin{cases} +\text{ sign} \rightarrow \langle\phi\rangle = 0 \\ -\text{ sign} \rightarrow \langle\phi\rangle = v \end{cases}. \tag{7.17}$$

You should be able to work out v in terms of m and λ. Note that linear or cubic terms in ϕ would not be invariant under a gauge transformation, and the positive sign for λ is fixed by demanding that the potential is stable at large ϕ.

Finally, fluctuations around the vacuum expectation value (vev), of course, may be present. Accounting for these as $\delta\phi$, we write

$$\mathcal{L}(\phi) \rightarrow \mathcal{L}(\langle\phi\rangle + \delta\phi). \tag{7.18}$$

---

[4] In group theory language, the gauge field, A_μ, transforms in the *adjoint representation* of U(1).

[5] In quantum field theory, these are all the 'renormalisable' interactions that lead to a finite theory at all energies. Non-renormalisable interactions are allowed in *effective field theories*, that is, ones that are only defined up to some maximum cut-off energy and are derived from more fundamental theories.

The quartic interaction term now reads

$$\phi^2 A_\mu A^\mu \rightarrow \underbrace{\langle\phi\rangle^2 A_\mu A^\mu}_{\langle\phi\rangle^2 \equiv \text{const.} \rightarrow m_\gamma^2} + \mathcal{O}(\delta\phi), \tag{7.19}$$

that is, an effective photon mass has arisen! Therefore, the non-zero vev has induced a mass term for the gauge field and *spontaneously broken* the U(1) symmetry: without including the $\delta\phi$ fluctuations, the vacuum term alone violates the gauge symmetry. It is the derivatives of $\partial_\mu \delta\phi$ that keep the symmetry preserved dynamically: you should write this out and convince yourself.

This model is known as the **Abelian Higgs** model, as U(1) is an Abelian (that is, commutative) group,

$$\forall\, (g_1, g_2) \in \mathrm{U}(1), g_1 g_2 = g_2 g_1. \tag{7.20}$$

This is **not** the Standard Model, as the SM does not have a mass term for the photon, that is, the Higgs boson, ϕ, does not directly interact with the photon, and the U(1) gauge symmetry of electromagnetism is unbroken. Demonstrated above, however, is a very important mechanism in theoretical physics, that of **spontaneous symmetry breaking** (SSB), occurring in a variety of settings in nature:

A falling pencil – breaks rotational symmetry;
Galaxy formation – breaks translational symmetry;
Crystal formation – also breaks translational symmetry;

to name a few. As we will see later, this mechanism induces a mass for three of the four vector bosons of $\mathrm{SU}(2)_L \times \mathrm{U}(1)_Y$ and also for the fermions.

7.2 FERMIONS, CHIRALITY AND NON-ABELIAN SYMMETRIES

7.2.1 Fermions and Chirality

We will now define the spinor notation, used in the 'fermions' term in Eq. (7.1). The Lagrangian for Dirac fermions is

$$\mathcal{L}_D = \overline{\psi}\left(i\partial_\mu \gamma^\mu - m\right)\psi, \tag{7.21}$$

where ψ is a 4-component *spinor*, $\overline{\psi}$ its conjugate—known as the *adjoint*—and γ^μ are the *Dirac matrices*. These are generators of the Clifford algebra,

$$\{\gamma^\mu, \gamma^\nu\} = \gamma^\mu \gamma^\nu + \gamma^\nu \gamma^\mu = 2\eta^{\mu\nu}, \tag{7.22}$$

where

$$\left(\gamma^0\right)^2 = \mathbb{1};$$

$$\left(\gamma^i\right)^2 = -\mathbb{1}; \tag{7.23}$$

$$\left(\gamma^\mu\right)^\dagger = \gamma^0 \gamma^\mu \gamma^0.$$

The γs are four-by-four matrices. We introduce the shorthand 'slashed' notation when contracting γ matrices with any 4-vectors,

$$\gamma_\mu X^\mu = \gamma^\mu X_\mu \equiv \slashed{X};$$
$$\partial_\mu \gamma^\mu = \slashed{\partial}; \tag{7.24}$$
$$D_\mu \gamma^\mu = \slashed{D}.$$

The adjoint of ψ is

$$\overline{\psi} = \psi^\dagger \gamma^0, \tag{7.25}$$

and we can also define a fifth γ-matrix

$$\gamma_5 = -i\gamma^0 \gamma^1 \gamma^2 \gamma^3, \tag{7.26}$$

where $\gamma_5^2 = 1$. As the γ-matrices are generators of an algebra, their representation is basis dependent, that is, when we choose different coordinates to represent ψ on field space, the form of γ^μ changes. A useful representation is the **chiral representation**,

$$\gamma^0 = -\begin{pmatrix} 0 & 1_2 \\ 1_2 & 0 \end{pmatrix};$$
$$\gamma_5 = \begin{pmatrix} 1_2 & 0 \\ 0 & -1_2 \end{pmatrix}, \tag{7.27}$$

where the fermions are then written as

$$\psi = \begin{pmatrix} \psi_L \\ \psi_R \end{pmatrix} \tag{7.28}$$

with $\psi_{L,R}$ two-component **Weyl fermions**. The projection operators

$$P_{L,R} = \frac{1}{2}(1 \pm \gamma_5) \tag{7.29}$$

extract the L,R Weyl components of the Dirac fermions,

$$P_L \psi = \begin{pmatrix} \psi_L \\ 0 \end{pmatrix}; \qquad P_R \psi = \begin{pmatrix} 0 \\ \psi_R \end{pmatrix}. \tag{7.30}$$

Using γ_0 and the definition of $\overline{\psi}$ we find that in this representation the Dirac mass term, $m\overline{\psi}\psi$, is

$$m\overline{\psi}\psi = m\,(\psi_L\psi_R + \psi_R\psi_L), \tag{7.31}$$

that is, fermion mass terms mix left- and right-handed chiral states.

7.2.2 Symmetries of the Dirac Lagrangian

As we saw earlier with the scalar field theory, we notice that the Dirac Lagrangian in Eq. (7.21) is also invariant under a $U(1)$ global symmetry:

$$\psi \to e^{i\alpha}\psi$$
$$\psi^\dagger \to e^{-i\alpha}\psi^\dagger. \tag{7.32}$$

However, gauging the U(1), that is, $\alpha \to \alpha(x)$, results in an additional term that does not cancel. Again, using the covariant derivative given in Eq. (7.12), that is,

$$\overline{\psi}\partial_\mu\gamma^\mu\psi \to \overline{\psi}D_\mu\gamma^\mu\psi, \tag{7.33}$$

makes the Dirac Lagrangian U(1) gauge invariant. It also introduces an interaction term between the photons and the fermions,

$$\boxed{q\overline{\psi}A\!\!\!/\,\psi.} \tag{7.34}$$

Note: an A^2 is absent (since the fermion kinetic term is linear in D), and thus no photon mass can be introduced by giving $\overline{\psi}\psi$ a vev, unlike the case of the Higgs in the last section.

This simple U(1) gauge invariant Dirac + Maxwell theory allows for an electron mass,

$$\mathcal{L}_{\text{mass}} = m_e\overline{\psi}\psi, \tag{7.35}$$

that is, the term $\mathcal{L}_{\text{mass}}$ is invariant under local U(1) transformations. The theory is known as **quantum electrodynamics** (QED) for which Feynman, Schwinger and Tomonaga were awarded the Nobel Prize in Physics in 1965.

Let's consider a different type of global U(1) symmetry, namely **chiral symmetry**,

$$\psi \to e^{i\alpha_{L,R}P_{L,R}}\psi \tag{7.36}$$

with

$$\psi = \begin{pmatrix} \psi_L \\ \psi_R \end{pmatrix} \to \begin{pmatrix} e^{i\alpha_L}\psi_L \\ e^{i\alpha_R}\psi_R \end{pmatrix} \tag{7.37}$$

under $U(1)_{L(R)}$ (chiral comes from the Greek word for hand, hence acting only on the left- or right-handed Weyl spinors).

The kinetic term, $i\overline{\psi}\partial\!\!\!/\,\psi$, is invariant under the chiral symmetries $U(1)_L$ and $U(1)_R$,

$$i\overline{\psi}\partial\!\!\!/\,\psi = i\overline{\psi_L}\partial\!\!\!/\,\psi_L + i\overline{\psi_R}\partial\!\!\!/\,\psi_R, \tag{7.38}$$

but the mass term is not, that is, *Chiral symmetry forbids fermion masses.*[6]

However, with the introduction of a scalar, we can again get around this and have chiral symmetries and mass terms. We introduce the **Yukawa interaction** term,

$$Y\left(\overline{\psi}_L \phi \psi_R + \overline{\psi}_R \phi^* \psi_L\right), \tag{7.39}$$

which is invariant under global chiral U(1) transformations *if ϕ also transforms chirally*, that is, ϕ picks up a $U(1)_{L,R}$ rotation. What charge does ϕ need in each case?

For a global chiral symmetry, the vev of the scalar field, $\langle \phi \rangle$, gives the fermion its mass. Following our logic in this section, you can also gauge chiral symmetries, that is, make $\alpha_{L,R} \to \alpha_{L,R}(x)$ if the gauge transformation on A_μ is chiral (we say that A_μ 'knows' about the fermion chirality). If we only gauged $U(1)_L$, then only left-handed fermions would interact with the gauge field.

7.2.3 Non-Abelian Symmetries and the Standard Model Gauge Group

Let's take N 4-component spinors, that is, in the Dirac basis, and construct a $4N$-component 'vector' (i.e. a vector in some abstract vector space, not a spacetime vector):

$$\vec{\psi} = (\psi_1, \ldots, \psi_N)^T. \tag{7.40}$$

Defining an $N \times N$ transformation matrix:

$$\vec{\psi} \to \mathbf{M}\vec{\psi}, \tag{7.41}$$

we notice that, if $\mathbf{M}$ is unitary, the Dirac kinetic term remains invariant, that is, $\mathbf{M} \in U(N)$ – the unitary group of $N \times N$ complex matrices—with

$$\mathbf{M}^\dagger \mathbf{M} = \mathbf{M}\mathbf{M}^\dagger = \mathbb{1}_{N \times N}. \tag{7.42}$$

The determinant $\det \mathbf{M} = \pm 1$ for $\mathbf{M} \in U(N)$. A subset of unitary symmetries are those with

$$\det \mathbf{M} = 1. \tag{7.43}$$

The groups of such matrices are known as **special unitary** groups, written as SU(N). Since matrices do not in general commute, $\mathbf{M}_1 \mathbf{M}_2 \neq \mathbf{M}_2 \mathbf{M}_1$, we call such groups 'non-Abelian'.

In the Standard Model, two such symmetry groups determine the dynamics of the strong, SU(3), and weak, SU(2), interactions, with only quarks charged under SU(3). In the *fundamental* representation (for SU(3), the three-dimensional representation),

$$\psi = \begin{pmatrix} \psi_r \\ \psi_b \\ \psi_g \end{pmatrix}, \tag{7.44}$$

---

[6] This can also be phrased as 'massless fermions are chiral'.

the quarks are in a triplet of Dirac fermions, where the indices represent 'colours': red, blue and green. A general element of SU(3) is written (with summation over the index a) as

$$\mathbf{M} = e^{i\alpha(x)_a T^a} \qquad (7.45)$$

where T^a are $(N^2 - 1) = 8$ *generators of* SU(3) satisfying a **Lie algebra**,

$$[T^a, T^b] = i \underbrace{f^{ab}_{c}}_{\text{structure constants}} T^c. \qquad (7.46)$$

There are different representations for the T^a matrices that generate different representations of SU(3) (i.e. matrices that satisfy the same commutation relations and group properties but which need not be 3×3 matrices themselves). For instance, one particular representation of the generators of SU(3) is commonly known as the Gell-Mann matrices. The theory of quarks and strong interactions is called **quantum chromodynamics** (QCD). Gauging SU(3) means that we use $\alpha \equiv \alpha(x)$ for a matrix Eq. (7.45) and transform $\psi \to \mathbf{M}\psi$ for the triplet Eq. (7.44) in the Dirac Lagrangian. Gauge invariance requires introducing the covariant derivative

$$\partial_\mu \to D_\mu = \partial_\mu \mathbb{1}_3 + ig A_\mu^a T^a. \qquad (7.47)$$

As there are eight generators, T, there must be *eight gauge boson fields*, which we call **gluons**. The gluons transform as

$$A_\mu^a \to A_\mu^a + \partial_\mu \alpha^a. \qquad (7.48)$$

We can generalise the dynamics of the photon given by the field strength introduced in Eq. (7.15):

$$[D_\mu, D_\nu]\Psi = F_{\mu\nu}^a T^a \Psi \equiv F_{\mu\nu}\Psi, \qquad (7.49)$$

such that the field strength is a matrix valued object. Unlike in the case of electromagnetism, the generators, T^a, of non-Abelian groups, for example, SU(3) do not commute in general, since they are matrices. Thus, non-linear terms in the gluon fields, A_μ^a, are introduced in $F_{\mu\nu}^a$,

$$\boxed{F_{\mu\nu}^a = \partial_\mu A_\nu^a - \partial_\nu A_\mu^a + g f^a_{bc} A_\mu^b A_\nu^c,} \qquad (7.50)$$

which in turn introduces cubic and quartic gluon self-interactions in the SU(3) invariant Lagrangian:

$$\mathcal{L} = -\frac{1}{4} \text{Tr}\left[F_{\mu\nu} F^{\mu\nu}\right]. \qquad (7.51)$$

Quadratic Lagrangians correspond to free-field theories, but the non-linearities in $F_{\mu\nu}$ introduce non-quadratic terms in $\mathcal{L}$, which lead to interactions between the gauge bosons. **Thus, non-Abelian gauge theories are self-interacting and their gauge bosons carry 'charge'.** We call the charge of SU(3) 'colour charge'.

The overall gauge group of the Standard Model is

$$G = \underbrace{SU(3)_c}_{\text{colour}} \times \overbrace{SU(2)_L}^{\text{left-handed 'weak isospin'}} \times \underbrace{U(1)_Y}_{\text{hypercharge}}. \tag{7.52}$$

$SU(2)_L$ isospin is *chiral* and is generated by the Pauli matrices, σ^a, as

$$\mathbf{M} \in SU(2)_L = e^{i\alpha(x)_a \sigma^a P_L}. \tag{7.53}$$

In the SM, all fermions are charged under $SU(2)_L$, that is, they are all put together in 2-component 'vectors' that $\mathbf{M}$ in Eq. (7.53) acts on. Thus, as a fermion mass term violates $SU(2)_L$, fermions must have mass generated by the Higgs mechanism. In the fundamental two-dimensional representation of $SU(2)$ used in the SM, fermions are 'paired' up in doublets:

$$\begin{aligned} \begin{pmatrix} \nu_L \\ e_L \end{pmatrix} &\rightarrow \text{Lepton doublets,} \\[2mm] \begin{pmatrix} d_L \\ u_L \end{pmatrix} &\rightarrow \text{Quark doublets,} \end{aligned} \tag{7.54}$$

that is, neutrinos pair up with electrons, and up- and down-type quarks pair up. The same is true for all three generations of the SM. The right-handed particles do not transform under $SU(2)$.

7.3 PARTICLE CONTENT OF THE STANDARD MODEL[7]

7.3.1 Constructing the Lagrangian

We can now write down the SM by returning to the Lagrangian density in Eq. (7.1). To do so, let us first recollect the symmetries and fields involved.

- As mentioned, the SM respects Lorentz invariance, so we need to write down a *Lorentz-invariant* action.
- The SM has a $SU(3)_C \times SU(2)_L \times U(1)_Y$ gauge symmetry. This says that the Lagrangian density will be invariant under a $SU(3)$ transformation, a $SU(2)$ transformation and a $U(1)_Y$ transformation, and all fields in principle transform non-trivially under these three transformations. In reality, not all fields will have non-trivial transformation laws under all three factors of the gauge group. For example, out of all the fermions only quarks transform non-trivially under $SU(3)$ whereas leptons are 'scalars' under $SU(3)$ and do not transform at all.

  Note the subscript Y on $U(1)_Y$. The reason is that we want to distinguish this $U(1)$ symmetry from the $U(1)$ symmetry associated with electromagnetism, which is *not* identical with $U(1)_Y$. The $U(1)_Y$ symmetry of the SM is called the

---

[7] The following two subsections are adapted from particle physics lectures given by V. M. Mehta and L. T. Witkowski at Birzeit University, Occupied Palestinian Territories, in 2019.

hypercharge U(1), and the charge under it is called, well, hypercharge. We will later see how electromagnetism arises from the SU(2)×U(1)$_Y$ part of the SM, known as the *electroweak* sector. At lower energies this symmetry is in fact violated and broken down to SU(2)×U(1)$_Y$ →U(1)$_{EM}$. What is left over from the electroweak symmetry is a U(1)$_{EM}$ factor describing electromagnetism (QED).

Having specified the gauge group of the SM, we can now count the number of gauge bosons (= number of vector fields) in the SM. This is just given by the number of generators for the corresponding group. For a gauge group SU(3)$_C$ × SU(2)$_L$ × U(1)$_Y$ we have

$$\text{SU(3):} \quad 3^2 - 1 = 8 \quad \text{gluon fields } G_\mu^a,$$

$$\text{SU(2):} \quad 2^2 - 1 = 3 \quad \text{gauge fields } W_\mu^1, W_\mu^2, W_\mu^3,$$

$$\text{U(1)}_Y: \quad 1 \quad \text{gauge field denoted by } B_\mu.$$

After electroweak symmetry breaking, a combination of W^3 and B will become the photon field A, while another combination will become the Z boson field. The W^1 and W^2 bosons can be combined into the W^- and W^+ fields. We will do this analysis explicitly later.

- Let us now turn to the matter sector. For one, we have fermions, which we will describe by Dirac spinor fields $\psi_i(x)$, which we will often break down into L- and R-chiral spinor fields. In addition, there will be a complex scalar $\Phi(x)$, which will ultimately give rise to the Higgs boson.

Under a SU(3)$_C$ × SU(2)$_L$ × U(1)$_Y$ gauge transformation the matter fields transform as

$$\psi_i(x) \rightarrow \exp\left(i\alpha_1(x)g_1\frac{Y}{2} + ig_2\alpha_2^l(x)\frac{\sigma_l}{2}P_L + ig_3\alpha_3^a(x)\frac{\lambda_a}{2}\right)\psi_i(x), \quad (7.55)$$

where $\alpha_{1,2,3}(x)$ are the transformation parameters. We also introduced the coupling constants g_1, g_2, g_3 associated with the three factors of the SM gauge group. The combination $g_1\frac{Y}{2}$ plays the role of the charge q for the U(1) symmetry. We also inserted the appropriate generators $\frac{\sigma_l}{2}$ for SU(2) and $\frac{\lambda_a}{2}$ for SU(3) transformations. For fields that do not transform under SU(3), like leptons, just set $\lambda_a = 0$ in the above. One crucially important ingredient of the above is the appearance of the projection operator $P_L = \frac{1}{2}(\mathbb{1} - \gamma^5)$. This ensures that SU(2) transformations only act on L-chiral fermions, while leaving R-chiral fermions unchanged. The fact that the SM treats L-chiral and R-chiral fermions differently is an important though somewhat peculiar property of the SM.

For the Higgs field, just replace ψ_i by Φ in the above. The Higgs does not transform under SU(3), so we just set $\lambda_a = 0$. There is no notion of chirality for scalars, so the P_L is not needed. Thus we obtain

$$\Phi(x) \rightarrow \exp\left(i\alpha_1(x)g_1\frac{Y}{2} + ig_2\alpha_2^l(x)\frac{\sigma_l}{2}\right)\Phi(x). \quad (7.56)$$

We are now in a position to write down the Lagrangian density of the SM. It is given by

$$\mathcal{L}_{\text{SM}} = -\frac{1}{4}G^a_{\mu\nu}G^{\mu\nu,a} - \frac{1}{4}W^l_{\mu\nu}W^{\mu\nu,l} - \frac{1}{4}F_{\mu\nu}F^{\mu\nu} \tag{7.57}$$

$$+ \sum_i \overline{\psi}_i i\gamma^\mu D_\mu \psi_i$$

$$+ (D_\mu\Phi)^\dagger D_\mu\Phi - V(\Phi^\dagger\Phi) - \sum_{i,j}\left(y_{ij}\,\Phi\,\overline{\psi}_i\psi_j + \text{h.c.}\right).$$

That's it! This is a surprisingly compact expression for a theory governing everything there is on the level of elementary particles. Let us explain some of the terms appearing in Eq. (7.57). The first line contains the kinetic terms for the gauge bosons, and we defined

$$G^a_{\mu\nu} \equiv \partial_\mu G^a_\nu(x) - \partial_\nu G^a_\mu(x) - g_3 f^a{}_{bc}\left[G^b_\mu(x), G^c_\nu(x)\right], \tag{7.58}$$

$$W^a_{\mu\nu} \equiv \partial_\mu W^l_\nu(x) - \partial_\nu W^l_\mu(x) - g_2 \epsilon^l{}_{mn}\left[W^m_\mu(x), W^n_\nu(x)\right], \tag{7.59}$$

$$F_{\mu\nu} \equiv \partial_\mu B_\nu(x) - \partial_\nu B_\mu(x), \tag{7.60}$$

where $f^a{}_{bc}$ are the structure constants for SU(3) introduced in Eq. (7.46), and $\epsilon^l{}_{mn}$ are the SU(2) structure constants, which are just the usual totally anti-symmetric Levi-Civita symbol. These are just the kinetic terms for SU(3), SU(2) and U(1) gauge bosons as described in the previous section.

On the second line we have a Dirac Lagrangian describing the SM fermions. To make sure this is invariant under $SU(3)_C \times SU(2)_L \times U(1)_Y$ gauge transformation, the derivative has been replaced by the appropriate covariant derivative. It is given by

$$D_\mu = \partial_\mu - ig_1\frac{Y}{2}B_\mu(x) - ig_2 W^l_\mu(x)\frac{\sigma_l}{2}P_L - ig_3 G^a_\mu(x)\frac{\lambda_a}{2}. \tag{7.61}$$

This then gives rise to all interaction terms between fermions and the gauge bosons. Last, note that we did not include a mass term $\sim m_i\overline{\psi}_i\psi_i$ in the SM Lagrangian density (Eq. (7.57)). As we have already discussed, since this term mixes left- and right-handed components of ψ, such a term is inconsistent with the $SU(2)_L$ symmetry of the SM.

The third line of (Eq. (7.57)) is the so-called Higgs sector. The first term is a kinetic term. The covariant derivative will give rise to an interaction between the Higgs and gauge bosons. The term $V(\Phi^\dagger\Phi)$ is called the potential and contains all of the self-interactions of the Higgs field with itself. We will later see that the potential plays an important role in the breaking of $SU(2)_L \times U(1)_Y \to U(1)_{\text{EM}}$. The last term is an interaction between the Higgs field and the SM fermions. Interaction terms between scalars and spinors of the form $\sim \Phi\overline{\psi}\psi$ are called *Yukawa* couplings. The parameters y_{ij} are the coupling constants.

7.3.2 Fermionic Spectrum and a Game of Indices

Let us have a closer look at the fermionic matter content of the SM. As the SM treats L- and R-chiral spinors differently, let us write down the matter content in terms of L- and R-chiral spinors. Thus we could write down the spectrum in terms of L- and R-chiral 2-component Weyl spinors. Following tradition, we will however write the fermionic spectrum in terms of L- and R-chiral 4-component Dirac spinors. The reason is simplicity and brevity. When working with L- and R-chiral Weyl spinors, we need to use two sets of matrices $\bar{\sigma}^\mu$ and σ^μ to write down a Lorentz-invariant Lagrangian. When working with Dirac spinors, we can write everything more compactly by just using one set of γ^μ matrices.

What do we then mean by an L-chiral Dirac spinor (or a R-chiral Dirac spinor)? Typically, Dirac spinors contain both L- and R-chiral parts. An L-chiral Dirac spinor is then just a Dirac spinor that only contains the L-chiral part, while the R-chiral part is absent (and vice-versa for R-chiral Dirac spinors).[8]

In the following, we will see that we can organise the spectrum according to the transformation properties under $SU(3)_C \times SU(2)_L \times U(1)_Y$. One striking property of the SM is that the fermionic spectrum comes in three *generations*, that is, spinors with the same transformation properties under the SM gauge group come in three copies. For every particle in one generation there are two partner particles in the other two generations with exactly the same properties (charge, interactions) except the mass. For example, the electron, the muon and the tau have exactly the same physical properties except for their rest masses.

Within one generation, we will use the transformation properties under the SM gauge group $SU(3)_C \times SU(2)_L \times U(1)_Y$ to distinguish spinors and their corresponding particles further.

- Spinors that transform under the $SU(3)_C$ part will carry an index $\alpha = 1, 2, 3$ on which $SU(3)_C$ matrices can act. The three components $\alpha = 1, 2, 3$ are typically called colours and are referred to as red (r), green (g) and blue (b). Spinors that transform non-trivially under $SU(3)_C$ describe quarks.
- All L-handed spinors also transform under $SU(2)_L$ and thus carry an index $i = 1, 2$. The two components form a so-called *electroweak doublet*. R-chiral spinors do not carry an $SU(2)_L$ index and are thus *singlets* under $SU(2)_L$. One peculiar property of our nomenclature for particles is that we refer to the two components of an electroweak doublet as different particles.[9] This is best illustrated by an example. The L-chiral lepton doublet in the first-generation $L_{L,1}$ is

---

[8] More precisely, an L-chiral (R-chiral) Dirac spinor will be a 4-component object that can be entirely built out of an L-chiral 2-component Weyl spinor (an R-chiral Weyl spinor, respectively). Thus, while the Dirac spinor will have four components, only two of these will be free.

[9] This is partly for historical reasons but can be understood from the fact that the $SU(2)_L$ part of the SM gauge group is broken at low energies.

given by

$$L_{L,1} = \begin{pmatrix} v_{e,L} \\ e_L \end{pmatrix}, \tag{7.62}$$

that is, one component is called the electron neutrino and the other component is the electron. This implies that at high energies, where the SU(2) symmetry of the SM is intact, the electron neutrino and the electron are just two realisations of the same physical object, just like a red and a green quark are two realisations of the notion of a quark.

- Finally, we also assign a hypercharge Y to the various spinors that determines their behaviour under U(1)$_Y$ transformations. In the following, we will give the hypercharge as a subscript, that is, ψ_Y.

Then, the first generation of fermions of the SM is given by the following set of Dirac spinors.

First Generation:

quarks:
$$Q_{L,1} = \overbrace{\begin{pmatrix} u_L^r \ u_L^g \ u_L^b \\ d_L^r \ d_L^g \ d_L^b \end{pmatrix}}^{\text{SU(3) index } i} {}_{\frac{1}{3}} \left.\right\} \text{SU(2) index } i,$$

$$U_{R,1} = \begin{pmatrix} u_R^r \ u_R^g \ u_R^b \end{pmatrix}_{\frac{4}{3}},$$

$$D_{R,1} = \begin{pmatrix} d_R^r \ d_R^g \ d_R^b \end{pmatrix}_{-\frac{2}{3}},$$

leptons:
$$L_{L,1} = \begin{pmatrix} v_{e,L} \\ e_L \end{pmatrix}_{-1} \left.\right\} \text{SU(2) index } j.$$

$$L_{R,1} = \begin{pmatrix} e_R \end{pmatrix}_{-2}$$

As usual, the subscripts L and R distinguish between L- and R-chiral spinors. As observed before, L-chiral spinors come as electroweak doublets (two components) while R-chiral spinors are electroweak singlets (one component). L-chiral up (u) and down (d) quarks form an electroweak doublet, while R-chiral u and d quarks come as two distinct singlets. Also recall that every object in the above table is a 4-component Dirac spinor, that is,

$$d_R^g = \begin{pmatrix} (d_R^g)_1 \\ (d_R^g)_2 \\ (d_R^g)_3 \\ (d_R^g)_4 \end{pmatrix}, \quad \text{etc.} \tag{7.63}$$

The remaining two generations are as follows.

Second Generation:

SU(3) index i

$$Q_{L,2} = \left. \begin{pmatrix} c_L^r & c_L^g & c_L^b \\ s_L^r & s_L^g & s_L^b \end{pmatrix}_{\frac{1}{3}} \right\} \text{SU(2) index } j,$$

quarks:

$$U_{R,2} = \left(c_R^r \ c_R^g \ c_R^b \right)_{\frac{4}{3}},$$

$$D_{R,2} = \left(s_R^r \ s_R^g \ s_R^b \right)_{-\frac{2}{3}},$$

leptons:

$$L_{L,2} = \left. \begin{pmatrix} \nu_{\mu,L} \\ \mu_L \end{pmatrix}_{-1} \right\} \text{SU(2) index } j.$$

$$L_{R,2} = \left(\mu_R \right)_{-2}$$

Third Generation:

SU(3) index i

$$Q_{L,3} = \left. \begin{pmatrix} t_L^r & t_L^g & t_L^b \\ b_L^r & b_L^g & b_L^b \end{pmatrix}_{\frac{1}{3}} \right\} \text{SU(2) index } j,$$

quarks:

$$U_{R,3} = \left(t_R^r \ t_R^g \ t_R^b \right)_{\frac{4}{3}},$$

$$D_{R,3} = \left(b_R^r \ b_R^g \ b_R^b \right)_{-\frac{2}{3}},$$

leptons:

$$L_{L,3} = \left. \begin{pmatrix} \nu_{\tau,L} \\ \tau_L \end{pmatrix}_{-1} \right\} \text{SU(2) index } j.$$

$$L_{R,3} = \left(\tau_R \right)_{-2}$$

As you can see above, we did not list any R-chiral neutrinos. The reason is that we do not know whether they exist. If they existed they would not transform under SU(3), as they are leptons, and they would not transform under SU(2), as they are R-chiral. In addition, they would have $Y = 0$ and thus not transform under the hypercharge U(1). As a result, R-chiral neutrinos would not interact with gauge bosons at all; in fact, they would not interact with any known particles. Because of this, we cannot really test whether they exist.

7.3.3 No Dirac Masses in the SM

When writing down the SM Lagrangian density in Eq. (7.57) we did not include Dirac mass terms of the form $m_i \overline{\psi}_i \psi_i$ for the SM fermions. We are now in a position to understand this. The reason is the gauge symmetry of the SM and, in particular, the fact that the SM treats L-chiral and R-chiral fermions differently.

Let us write a Dirac spinor ψ as a sum of its L-chiral and R-chiral parts as $\psi = \psi_L + \psi_R$. We can always extract the L-chiral and R-chiral parts using the projection operators $P_{L,R}$, that is,[10]

$$\psi_L = P_L \psi, \quad \text{with} \quad P_L = \frac{1}{2}\left(\mathbb{1} - \gamma^5\right), \tag{7.65}$$

$$\psi_R = P_R \psi, \quad \text{with} \quad P_R = \frac{1}{2}\left(\mathbb{1} + \gamma^5\right). \tag{7.66}$$

We showed above that the Dirac mass term can be written in terms of the L- and R-chiral components as

$$m\overline{\psi}\psi = m(\overline{\psi}_L \psi_R + \overline{\psi}_R \psi_L). \tag{7.67}$$

Thus, for the electron, a Dirac mass term would be given by

$$m_e(\overline{e}_L e_R + \overline{e}_R e_L). \tag{7.68}$$

Now let us act on this with a $U(1)_Y$ gauge transformation. Under this the electron mass term becomes

$$\rightarrow m_e \left(e^{i\alpha_1(x)\frac{g_1}{2}(Y_{e_R} - Y_{e_L})} \overline{e}_L e_R + e^{-i\alpha_1(x)\frac{g_1}{2}(Y_{e_R} - Y_{e_L})} \overline{e}_R e_L \right). \tag{7.69}$$

However, in the SM we have $Y_{e_L} = -1$ and $Y_{e_R} = -2$. Thus $Y_{e_R} - Y_{e_L} = -1 \neq 0$, and hence an electron mass term is not invariant under a $U(1)_Y$ gauge transformation. As a result, an electron mass term is not allowed by the SM gauge symmetry.

You can repeat this analysis for any SM fermion. As the L- and R-chiral fermions transform differently under $U(1)_Y$ (and $SU(2)$), no Dirac mass terms are allowed in the SM. At the same time we know that fermions such as electrons, muons, taus and quarks do have a mass. Thus there must be a way of endowing fermions with mass, consistent with the gauge symmetry of the SM. This is the Higgs mechanism which we will now discuss.

7.3.4 The Higgs Mechanism

Let's recap what we discussed above about the Higgs mechanism and go into slightly more detail.

7.3.4.1 Higgs Mechanism in an Abelian Gauge Theory

In particular, to understand the role of the Higgs field in the SM better, it will be useful to first discuss the Higgs mechanism in a simpler gauge theory. Thus let us set the SM with its complicated gauge group $SU(3)_C \times SU(2)_L \times U(1)_Y$ aside and consider an Abelian $U(1)$ gauge theory for the moment. We then also include a

---

[10] In the chiral basis we could write this more explicitly as

$$\psi = \psi_L + \psi_R, \quad \text{with} \quad \psi_L = \begin{pmatrix} \chi_L \\ 0 \end{pmatrix}, \quad \text{and} \quad \psi_R = \begin{pmatrix} 0 \\ \chi_R \end{pmatrix}. \tag{7.64}$$

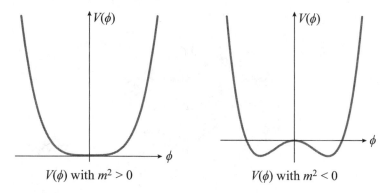

$V(\phi)$ with $m^2 > 0$ $V(\phi)$ with $m^2 < 0$

Figure 7.1 Higgs potential with differing signs of m^2.

complex scalar charged under this U(1) gauge group, which will be the Higgs field in this theory. The corresponding Lagrangian density is

$$\mathcal{L} = -\frac{1}{4}F_{\mu\nu}F^{\mu\nu} + (D_\mu\Phi)^\dagger(D^\mu\Phi) - V(\Phi^\dagger\Phi), \qquad (7.70)$$

with $F_{\mu\nu} = \partial_\mu A_\nu - \partial_\nu A_\mu$,

and $D_\mu = \partial_\mu - iqA_\mu$,

with A_μ the gauge field associated with the U(1) gauge symmetry and q the charge of the Higgs field under this U(1). Under a U(1) gauge transformation the Higgs field thus transforms as

$$\Phi(x) \rightarrow e^{i\alpha(x)q}\Phi(x). \qquad (7.71)$$

In the following, we will direct our attention to the role played by the potential $V(\Phi^\dagger\Phi)$. This contains the mass term as well as self-interaction terms. To be specific, let us consider

$$V(\Phi^\dagger\Phi) = m^2|\Phi|^2 + \lambda|\Phi|^4. \qquad (7.72)$$

In quantum field theory, we would derive the mass of the Φ-particle as well as the interaction vertices from this. Here, let us first focus on the classical field theory implications of this potential.

We plot the potential Eq. (7.72) for both cases $m^2 \geq 0$ and $m^2 < 0$ in Fig. 7.2 as a projection onto the Re(Φ)-V plane.

Consider the cases $m^2 \geq 0$ and $m^2 < 0$ in turn.

Positive mass squared: $m^2 \geq 0$: In this case the potential has one minimum at $\Phi_0 = 0$. With a static solution (no derivatives) the field takes the classical value $\langle\Phi\rangle = \Phi_0 = 0$ everywhere in spacetime. The quantity $\langle\Phi\rangle$ is also referred to as the *vacuum expectation value (vev)* of Φ. Quantum field theory is then the study of

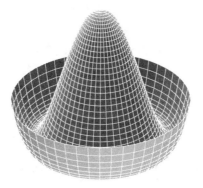

Figure 7.2 The continuous set of minima arranged on a circle with radius $|\Phi| = \sqrt{\frac{-m^2}{2\lambda}}$.

quantum fluctuations (particles) about this background value of $\langle \Phi \rangle = \Phi_0 = 0$. One important observation is that the vev is invariant under a U(1) gauge transformation:

$$\langle \Phi \rangle \rightarrow \langle \Phi \rangle' = e^{i\alpha(x)q}\langle \Phi \rangle = 0, \tag{7.73}$$

that is, $\langle \Phi \rangle = \langle \Phi \rangle' = 0$.

Strictly negative mass squared: $m^2 < 0$: In this case we have more than one minimum. The minima are given by

$$\frac{\partial V}{\partial \Phi} = (m^2 + 2\lambda|\Phi|^2)\Phi^\dagger|_{\Phi_0} = 0 \quad \Rightarrow \quad |\Phi| = \sqrt{\frac{-m^2}{2\lambda}}. \tag{7.74}$$

Thus we have a continuous set of minima arranged on a circle with radius $|\Phi| = \Phi_0 = \sqrt{\frac{-m^2}{2\lambda}}$. This is shown in Fig. 7.2. In the early Universe, Φ will be excited by the high temperatures. As the Universe cools, Φ will relax to the vacuum and thus 'pick' one of those minima at random (since they are all energetically equivalent). Thus we have a vev $\langle \Phi \rangle = \Phi_0 e^{i\beta}$ for some $\beta \in \mathbb{R}$. In quantum field theory we would then study fluctuations about this field value. However, note that now the Higgs vev is not invariant under a gauge transformation:

$$\langle \Phi \rangle = \Phi_0 e^{i\beta} \rightarrow \langle \Phi \rangle' = \Phi_0 e^{i\beta+i\alpha(x)q} \neq \langle \Phi \rangle. \tag{7.75}$$

We thus find the following situation: the Lagrangian density (Eq. (7.70)) is invariant under the gauge symmetry, but for $m^2 < 0$ the vacuum of the theory is not. This is referred to as *spontaneous symmetry breaking*.

Let us derive the observable consequences of a spontaneously broken gauge symmetry. As a first step, it will be convenient to perform a gauge transformation with $\alpha(x) = -\beta/q$ such that the vev becomes $\langle \Phi \rangle = \Phi_0 = \sqrt{\frac{-m^2}{2\lambda}}$. Let us now include fluctuations about the vev. In particular, it will be convenient to write

$$\Phi(x) = \left(\Phi_0 + \frac{h(x)}{\sqrt{2}}\right) e^{i\frac{\chi(x)}{\Phi_0}}. \tag{7.76}$$

The field $h(x)$ is a radial fluctuation while $\chi(x)$ is a fluctuation tangential to the circle of vacua. Let us now insert this into the Lagrangian density (Eq. (7.70)). We will do this step by step. First, let us insert this ansatz into the covariant derivative of Φ:

$$D_\mu \Phi = \left[\partial_\mu - iqA_\mu\right]\left(\Phi_0 + \frac{h}{\sqrt{2}}\right) e^{i\frac{\chi}{\Phi_0}}$$

$$= e^{i\frac{\chi(x)}{\Phi_0}}\left[\frac{\partial_\mu h}{\sqrt{2}} - iqA_\mu\left(\Phi_0 + \frac{h}{\sqrt{2}}\right)\right] + ie^{i\frac{\chi(x)}{\Phi_0}}\left(\Phi_0 + \frac{h}{\sqrt{2}}\right)\frac{1}{\Phi_0}\partial_\mu \chi. \tag{7.77}$$

Now note the following. We can simplify this by redefining the vector field A_μ as

$$A_\mu \to A_\mu + \frac{1}{q\Phi_0}\partial_\mu \chi. \tag{7.78}$$

As this takes the form of a gauge transformation (Eq. (7.11)), this redefinition leaves the vector field kinetic term $-\frac{1}{4}F_{\mu\nu}F^{\mu\nu}$ unchanged. We find

$$D_\mu \Phi \to e^{i\frac{\chi(x)}{\Phi_0}}\left[\frac{\partial_\mu h}{\sqrt{2}} - iqA_\mu\left(\Phi_0 + \frac{h}{\sqrt{2}}\right)\right], \tag{7.79}$$

that is, all terms with $\partial_\mu \chi$ have disappeared. Next, let us insert our ansatz into the potential V. A short calculation gives

$$V(\Phi^\dagger \Phi) = m^2 \Phi^\dagger \Phi + \lambda(\Phi^\dagger \Phi)^2$$

$$= -\frac{m^4}{4\lambda} - m^2 h^2 + \sqrt{-\frac{\lambda m^2}{2}}h^3 + \frac{\lambda}{4}h^4$$

$$= -\frac{\mu^4}{4\lambda} + \mu^2 h^2 + \sqrt{\frac{\lambda \mu^2}{2}}h^3 + \frac{\lambda}{4}h^4, \tag{7.80}$$

where we introduced the positive quantity $\mu^2 \equiv -m^2$. The first term is a vacuum energy density, the second term is a mass term for the h-particle, the third term is a cubic h-particle vertex and the fourth term is a quartic h-particle vertex.

Putting everything together, the Lagrangian density (Eq. (7.70)) becomes

$$\mathcal{L} = -\frac{1}{4}F_{\mu\nu}F^{\mu\nu} + \frac{1}{2}\partial_\mu h \partial^\mu h + q^2\left(\Phi_0 + \frac{h}{\sqrt{2}}\right)^2 A_\mu A^\mu \tag{7.81}$$

$$+ \frac{\mu^4}{4\lambda} - \mu^2 h^2 - \sqrt{\frac{\lambda \mu^2}{2}}h^3 - \frac{\lambda}{4}h^4.$$

We can make the following observations:

- Note that the above contains a term $q^2 \Phi_0^2 A_\mu A^\mu$ which takes the form of a mass term for the vector bosons, that is, it is of the form $\frac{1}{2} M^2 A_\mu A^\mu$. We can read off the mass as

$$M = \sqrt{2} q \Phi_0. \tag{7.82}$$

As $M \propto \Phi_0$ the generation of a gauge boson mass is in one-to-one correspondence with spontaneous symmetry breaking, that is, vacua with $\Phi_0 \neq 0$. This is in fact a general result: a spontaneously broken gauge symmetry is always associated with the generation of a mass for the corresponding gauge bosons.

- In addition to the massive gauge boson, we are left with one *real* scalar field $h(x)$ with *real positive* mass $m_H = \mu$. The particle associated with the field $h(x)$ is known as the *Higgs boson*. It has interaction vertices with itself and with the now massive gauge boson A_μ.

7.3.4.2 The Higgs Mechanism in the SM

Let us now return to the SM and apply the lessons from the previous section. In the SM the Higgs field Φ is a complex scalar field that transforms non-trivially under $SU(2) \times U(1)_Y$. The fact that the Higgs field transforms under $SU(2)$ implies that it is a doublet, that is, it is a 2-component object:

$$\Phi = \begin{pmatrix} \Phi^1 \\ \Phi^2 \end{pmatrix}. \tag{7.83}$$

Note that Φ_1 and Φ_2 are complex numbers with two real components, so there are in total four real degrees of freedom for Φ. The transformation of Φ under $U(1)_Y$ is specified by its hypercharge. For the SM Higgs this is given by $Y_\Phi = +1$.

As in our toy model we take the Higgs potential to be

$$V(\Phi^\dagger \Phi) = m^2 |\Phi|^2 + \lambda |\Phi|^4. \tag{7.84}$$

The question now is whether in our Universe we have $m^2 \geq 0$ or $m^2 < 0$. Interestingly, all experimental data suggests that $m^2 < 0$. In other words, all experimental results in our Universe are consistent with a vacuum with spontaneously broken symmetry, that is, $\langle |\Phi| \rangle \neq 0$ in our vacuum:

- As the SM Higgs is charged under $SU(2) \times U(1)_Y$ we expect that, in a vacuum with $\langle \Phi \rangle \neq 0$, the gauge bosons associated with the $SU(2) \times U(1)_Y$ gauge symmetry should become massive. In fact, performing the analysis in detail, one finds that the gauge bosons W_μ^+, W_μ^- and Z_μ should become massive while A_μ should remain massless.

- As in our toy model there should be a real scalar field in the theory with its corresponding particle. This is the Higgs boson which was duly discovered at the LHC in 2012.

A massive vector field has three degrees of freedom in comparison to the two polarisations of the massless photon. The extra degrees of freedom for $W^\pm$ and Z come from three of the four degrees of freedom of the Higgs. We tend to say that the gauge bosons 'eat' the Higgs. The remaining one degree of freedom can be thought of as the radial excitation away from v in Fig. 7.2 and is the massive Higgs boson. The measured values of the gauge boson masses are consistent with a Higgs vev $\langle |\Phi| \rangle = \frac{v}{\sqrt{2}}$ with $v = 246$ GeV. The Higgs mass is determined by the second derivative of V around the minimum in the radial direction and is measured to be $m_H = 125$ GeV. Thus our world is indeed consistent with a spontaneous breaking of $SU(2) \times U(1)_Y$ due to a non-vanishing Higgs vev.

7.3.4.3 The Higgs Field and Fermion Masses

Let us now address the open problem of how fermions acquire mass in the Standard Model. To endow fermions with a mass the Lagrangian density has to contain a term of the form $\sim m \bar{\psi} \psi$. However, as we observed before, this is not invariant under an SM gauge transformation, and hence such a term is forbidden. The only way that such a term would be allowed is if the mass m also transformed under a gauge transformation. This can only happen if m is not a parameter but a field.

The Higgs field Φ plays exactly that role in the SM. In particular, the Higgs field transforms in such a way that there exists a set of Yukawa couplings of the form

$$\sim y \Phi \bar{\psi}_L \psi_R + y^* \Phi^\dagger \bar{\psi}_R \psi_L, \tag{7.85}$$

which are consistent with the SM symmetries. We can then make the following observation. Once the Higgs field develops a non-zero vev everywhere in spacetime, that is, $\langle |\Phi| \rangle = \frac{v}{\sqrt{2}} \neq 0$, the Yukawa couplings will give rise to a fermion mass term

$$y \Phi \bar{\psi}_L \psi_R + y \Phi^\dagger \bar{\psi}_L \psi_R = \frac{yv}{\sqrt{2}} (\bar{\psi}_L \psi_R + \bar{\psi}_L \psi_R). \tag{7.86}$$

Without loss of generality we have assumed that both the coupling constant y and the vev v are real.[11] In particular, we identify $m_f = yv/\sqrt{2}$ as the fermion mass, that is, the mass is set by the value of the Yukawa coupling and the Higgs vev. Thus the spontaneous breaking of $SU(2) \times U(1)_Y$ symmetry in the SM also gives rise to fermion masses.

So far, we have only shown schematically how the Higgs couples to SM fermions. For completeness, let us write down the full set of Yukawa couplings. In particular, the following Yukawa couplings are allowed by gauge invariance (recall that $Y_\Phi = +1$):

$$\mathcal{L}_{\text{Yukawa}} = \sum_{m=1}^{3} \sum_{n=1}^{3} \left(y_{mn}^L \Phi \bar{L}_{L,m} L_{R,n} + y_{mn}^D \Phi \bar{Q}_{L,m} D_{R,n} + y_{mn}^U \tilde{\Phi} \bar{Q}_{L,m} U_{R,n} + \text{h.c.} \right), \tag{7.87}$$

with $\tilde{\Phi} = -i\sigma_2 \Phi^*$. The sums over m, n are over the three generations. The quantities y_{mn}^L, y_{mn}^D and y_{mn}^U are coupling constants, and their values are not explained by the SM. Once the Higgs field Φ acquires a vev, the first term in the above generates lepton masses, the second gives rise to masses for down-type quarks (d, s, b) and the third is responsible for masses for up-type quarks (u, c, t).

---

[11] We can always compensate a phase in yv by redefining the spinors ψ by a phase.

144   CHAPTER 7

QUIZ

    i. Which of the following are fundamental particles in the standard model? (Select multiple answers.)
- a. Electron.
- b. Kaon.
- c. Up quark.
- d. Gluon.
- e. Proton.
- f. Higgs boson.
- g. Neutralino.

    ii. A global U(1) symmetry leads to:
- a. Conservation of momentum.
- b. Conservation of charge.
- c. Interactions between fermions and gauge bosons.
- d. A finite Lagrangian density.

    iii. A gauge symmetry is:
- a. An Abelian symmetry.
- b. A non-Abelian symmetry.
- c. A local symmetry.
- d. A global symmetry.

    iv. The U(1) gauge symmetry in the standard model:
- a. Forbids the existence of fermion masses.
- b. Forbids the existence of gluon masses.
- c. Is spontaneously broken by the Abelian Higgs.
- d. Forbids the existence of the photon mass.

    v. A gauge symmetry for matter fields requires the introduction of:
- a. Gauge bosons.
- b. The Higgs field.
- c. Chiral projections.
- d. Spin.

    vi. The $SU(2)_L$ symmetry of the standard model (multiple correct answers):
- a. Is Abelian.
- b. Is non-Abelian.
- c. Is chiral.
- d. Leads to the W and Z bosons.
- e. Only interacts with quarks.
- f. Interacts with all standard model particles.

vii. The Higgs boson (multiple correct answers):
  a. Has not been observed.
  b. Gives mass to (some) fermions.
  c. Gives mass to W and Z gauge bosons.
  d. Does not interact directly with quarks.
  e. Carries colour charge.
  f. Spontaneously breaks symmetries of the standard model.

viii. Left- and right-handed fermions:
  a. Are required for (Dirac) mass terms.
  b. Interact in the same way with the electroweak force.
  c. Are 4-component spinors.
  d. Are each other's anti-particles.

ix. Non-Abelian gauge symmetries (multiple correct answers):
  a. Have interacting gauge bosons.
  b. Interact only with quarks.
  c. Have non-commuting group generators.
  d. Arise from the Clifford algebra.
  e. Are all generated by the Pauli matrices.
  f. Include the groups SU(N).

x. How many degrees of freedom does the Higgs field have?

Answers

ix. a, c, f; x. 4

i. a,c,d,f; ii. b; iii. c; iv. c; v. d; vi. b, c, d; vii. b, c, f; viii. a;

Chapter Eight

Massive Neutrinos as Dark Matter?

We now move on to the particle physics of dark matter, with the goal of understanding one of the most popular candidates in the literature: the **weakly interacting massive particle** (WIMP), where 'weakly interacting' may refer to interactions under the weak force or, more generally, with a small, that is, weak, coupling strength.

Questions about the Dark Matter Particle

The most fundamental questions surrounding the nature of dark matter, when we assume it to be a particle, are:

1. What is the **mass**, m, of the particle?
2. What non-gravitational **interactions** does it have? *i.e.*
3. Do these interactions produce the **correct relic abundance**? *i.e.* is $\Omega_{DM}h^2 = 0.12$, consistent with cosmological measurements?

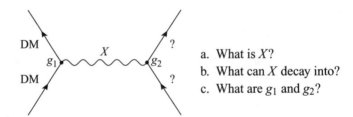

a. What is X?
b. What can X decay into?
c. What are g_1 and g_2?

8.1 MASSIVE NEUTRINOS

8.1.1 Higgs Mechanism for Fermions and the Problem with Neutrino Mass

The SM is a theory of fermions charged under a gauge group. Consider the interactions of the charged leptons:

$$e_R \to e^{iq_{e_R}\alpha(x)} e_R \qquad \text{Right-handed electrons have hypercharge;} \quad (8.1)$$

$$L = \begin{pmatrix} \nu_L \\ e_L \end{pmatrix} \to e^{i\alpha(x)_a \sigma^a} \begin{pmatrix} \nu_L \\ e_L \end{pmatrix} \qquad \begin{array}{l} \text{Weak interactions mix left-handed} \\ \text{electrons and neutrinos.} \end{array} \quad (8.2)$$

These are represented by the Feynman interaction vertices in Fig. 8.1. They do not involve right-handed neutrinos, and so the SM tells us nothing about right-handed neutrinos.

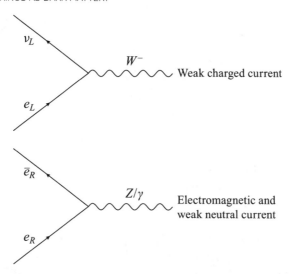

Figure 8.1 Electroweak interactions of electrons and neutrinos.

We now need to discuss how, in the SM, the Higgs gives mass to fermions and does not give mass to neutrinos. We saw in the last section the example of the Higgs giving mass to a fermion for a toy model with a chiral U(1) symmetry. The SM works with the chiral electroweak group, $SU(2)_L \times U(1)_Y$, and it turns out this cannot give mass to neutrinos due to the right-handed neutrinos being neutral under this group. We begin with the Dirac mass term:

$$m_\nu\left(\bar{\nu}_L \nu_R + \bar{\nu}_R \nu_L\right),\tag{8.3}$$

which is not invariant under $SU(2)_L$. We therefore would like to introduce a Higgs field to do this.

The SM gives mass to the charged leptons in the following way. We can use the terms in Eqs. (8.1) and (8.2) along with the Higgs to build a mass term for the electron that is invariant under $SU(2)_L$ and $U(1)_Y$. The term is $\bar{L}\phi e_R$ (plus Hermitian conjugate), and the mechanism works because we can choose charges q_Y for L and ϕ such that $\bar{L}\phi$ rotates with $e^{-iq_{e_R}\alpha}$.

However, ν_R is neutral under the electroweak gauge group, $SU(2)_L \times U(1)_Y$, and thus there is no possible invariant Higgs coupling involving right-handed neutrinos. We cannot get neutrino masses from the Higgs using only the SM particles. For instance, the analogous coupling for neutrinos:

$$y_\nu \bar{L}\phi \nu_R + \text{h.c.}\tag{8.4}$$

is not a singlet under $SU(2)_L \times U(1)_Y$ since ν_R is not charged under $U(1)_Y$. However,

$$y_e \bar{L}\phi e_R\tag{8.5}$$

is in fact invariant under $SU(2)_L \times U(1)_Y$ as e_R has hypercharge, that is, under a $U(1)_Y$ transformation:

$$e_R \to \exp(i\theta)e_R$$
$$\overline{L}\phi \to \exp(-i\theta)\overline{L}\phi \tag{8.6}$$
$$\Rightarrow \overline{L}\phi e_R \to \overline{L}\phi e_R,$$

while on the other hand for the uncharged ν_R:

$$\overline{L}\phi \nu_R \to e^{-i\theta}\overline{L}\phi \nu_R, \tag{8.7}$$

which is not invariant.

There are, however, models that *do* give mass to neutrinos, for example, 'Type I see-saw' or 'Majorana masses', but we won't discuss them further. They require extensions of the SM and introduce *lepton flavour mixing* (see the PDG, Ref. [37], section 14).

8.1.2 Mass and Flavour Eigenstates

An important consideration applied to neutrinos is the difference between the mass and flavour eigenstates. We begin by considering the quark sector, which contains the same basic ideas but is a bit simpler than neutrinos since it only relies on the SM Higgs and not the specific mechanism to generate neutrino mass outside the SM. The Lagrangian term for quark masses and mixing is

$$
\begin{aligned}
\mathcal{L}_{SM} \supset -y_{ij}^d \overline{Q}_{L\,i}^I \phi d_{Rj}^I \quad &\to \quad \text{down-type mass terms,} \\
-y_{ij}^u \overline{Q}_{L\,i}^I \epsilon \phi^* u_{Rj}^I \quad &\to \quad \text{up-type mass terms,}
\end{aligned} \tag{8.8}
$$

where we must remember that Q_L is an $SU(2)$ doublet, as is the Higgs ϕ, and the right-handed fields are singlets (so that $\overline{Q}_L \phi d_R$ overall gives a scalar), ϵ is the antisymmetric tensor acting on $SU(2)$ space, and the indices $i, j = 1, 2, 3$ represent the three generations of the SM. These terms are written in the weak *interaction* basis (hence the 'I' superscript), that is,

$$i\overline{Q}_L^I \slashed{D} Q_L^I \tag{8.9}$$

with $D_\mu = \partial_\mu + igW_\mu$. After $\phi \to \langle \phi \rangle$ the Higgs gives masses to the quarks. However, in this basis, the y_{ij} are not diagonal. Moving to the *mass eigenbasis* y becomes diagonal with mass terms:

$$m_d(\overline{d}_L d_R + \overline{d}_R d_L). \tag{8.10}$$

The change of basis is achieved by a unitary rotation $q_i = R_{ij} q_j^I$. This rotation introduces flavour mixing in the weak interaction:

$$\mathcal{L} = -\frac{g}{\sqrt{2}} (\bar{u}_L, \bar{c}_L, \bar{t}_L) \gamma^\mu W_\mu V_{\text{CKM}} \begin{pmatrix} d_L \\ s_L \\ b_L \end{pmatrix} + \text{h.c.}, \qquad (8.11)$$

where V_{CKM} arises from the rotation matrices between the two bases and is a 3×3 mixing matrix known as the Cabibbo–Kobayashi–Maskawa (CKM) matrix.

The same thing happens to electrons, muons, taus and their corresponding neutrinos, *only if the ν's have mass*.

Mass basis	ν_1, ν_2, ν_3
Weak basis	ν_e, ν_μ, ν_τ.

This happens regardless of how the neutrino mass is generated. In this case, the mixing matrix is known as the Pontecorvo–Maki–Nakagawa–Sakata (PMNS) matrix.

8.1.3 Observed Neutrino Masses

The sum of the neutrino masses,

$$\sum_{i=1}^{3} m_{\nu_i}, \qquad (8.12)$$

is known to be non-zero. How is this known?

Neutrino oscillations: Solar neutrinos: ν's are produced in flavour eigenstates, for example, in the β-decay of nuclei in the Sun. However, they propagate in the mass eigenbasis (since it is the mass that appears in the Dirac equation of motion). The flavour basis eigenstate for the electron neutrino ν_e is

$$\nu_e = c_1 \nu_1 + c_2 \nu_2 + c_3 \nu_3, \qquad (8.13)$$

that is, some linear combination of mass eigenstates, ν_i. In Dirac notation, the state vectors are related by

$$|\nu_\alpha\rangle = \sum_i U_{\alpha i}^* |\nu_i\rangle, \qquad (8.14)$$

where $\alpha = e, \mu, \tau$ are the flavour basis labels, U the mixing matrix (unitary rotation) and $i = 1, 2, 3$ the mass eigenbasis labels.

After some time, t,

$$|\nu_\alpha(t)\rangle = \sum_i U_{\alpha i}^* |\nu_i(t)\rangle$$
$$(8.15)$$
$$\text{where } |\nu_i(t)\rangle = \exp\left(-iE_i(t)\right) |\nu_i\rangle$$

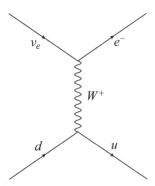

Figure 8.2 Electron neutrinos scatter off d quarks in nuclei, producing an electron via the charged current weak interaction.

and $E_i = \sqrt{p_i^2 + m_i^2}$, since it is the Hamiltonian $H = E$ that generates propagation in time: as this involves m_i it acts on the mass eigenstates. Therefore, the propagation in time mixes up the $|v_\alpha\rangle$ flavour basis from the initial values.

We can directly detect neutrinos in the weak interaction basis. For instance, early v-detectors exploited the interaction in Fig. 8.2 to detect electrons produced when electron neutrinos scatter off nuclei. If you can only detect electrons, then you will only be able to detect electron neutrinos via this interaction.

If $m_v \neq 0$ then Eq. (8.15) mixes up the initial v_e to some combination v_e, v_μ, v_τ in the final state. In the 1960s an experiment at the *Homestake Mine* in the USA (a mine we will meet again later) measured the solar v_e abundance as $1/3$ the predicted value from weak interaction production of electron neutrinos in the Sun. The 'missing' electron neutrinos could be accommodated only if $m_v \neq 0$, since then propagation leads to a final state that is in the mass eigenbasis and an equal mix of v_e, v_μ, v_τ, with only the v_e detected at Homestake. Further experiments, for example, at Sudbury Neutrino Observatory, detected the total $v_e + v_\mu + v_\tau$ abundance consistent with the massive neutrino model. Thus, *the measurement of solar neutrino oscillations directly implies that neutrinos have mass.*

8.1.4 Constraints on Neutrino Mass

What is the neutrino mass? In principle, it could be anything from zero to the Planck scale, but in practice we have experimental constraints. In addition to solar neutrino observatories, neutrino oscillations are also measured by reactor and atmospheric neutrino measurements. These measurements constrain the mass splittings, for example, $\Delta m_{ij}^2 = m_i^2 - m_j^2$. Nuclear decays, for example, ^{3}He in the KATRIN experiment, constrain the 'flavour effective mass',

$$T_{\max} = Q - m_{v_e}^{\text{eff}}, \tag{8.16}$$

where $T_{\max}$ is the temperature, Q is the maximum kinetic energy of the electron (approximately 18.6 keV for ^{3}He) and $m_{v_e}^{\text{eff}}$ is the effective mass of the electron neutrino. The PDG give current bounds on the masses at

$$
\begin{array}{ll}
m_{\nu_e}^{\text{eff}} < 1.1 \text{ eV} & {}^{3}\text{He} \\[2mm]
m_{\nu_\mu}^{\text{eff}} < 190 \text{ keV} & \pi^- \rightarrow \mu^- + \bar{\nu}_\mu \\[2mm]
m_{\nu_\tau}^{\text{eff}} < 18.2 \text{ MeV} & \tau^- \rightarrow \pi^- + \nu_\tau.
\end{array}
$$

In addition, the LEP experiment (the predecessor to the LHC) further constrains new neutrinos beyond the three at $m_\nu \lesssim 45$ GeV. The mixings are also bound,

$$
\Delta m_{21}^2 = 7.5 \times 10^{-5} \text{eV}^2
$$
$$
|\Delta m_{31}|^2 = 2.5 \times 10^{-3} \text{eV}^2. \tag{8.17}
$$

These mass splittings lead to a lower limit on the summed masses,

$$
\sum m_\nu > 0.06 \text{ eV} \qquad \text{(normal ordering)},
$$
$$
\sum m_\nu > 0.1 \text{ eV} \qquad \text{(inverted ordering)}, \tag{8.18}
$$

where

- 'Normal ordering' is $m_1 < m_2 \ll m_3$.
- 'Inverted ordering' is $m_3 \ll m_1 < m_2$.

8.2 NEUTRINO RELIC ABUNDANCE: FREEZE-OUT

The key point to remember here is that, as we stated in section 5.1.5, neutrino decoupling occurs for $T_D \approx 1$ MeV $> m_\nu$ (this is derived in Problem II.5), which implies neutrinos decouple while they are relativistic and thus *neutrinos are hot relics*.

We now consider the all-important *freeze-out* process or, as Kolb and Turner call it, the cosmological 'origin of species'. In thermal equilibrium, recall from chapter 5:

$$
n = \int \frac{d^3p}{(2\pi)^3} f(\mathbf{p}) \qquad \text{number density}
$$
$$
f(\mathbf{p}) = \left[\exp\left(\frac{E-\mu}{kT} \right) \pm 1 \right]^{-1}
$$
$$
E^2 = |\mathbf{p}^2| + m^2
$$
$$
n^{\text{eq}} \sim (mT)^{\frac{3}{2}} e^{-\frac{m}{T}} \qquad T \ll m
$$
$$
n^{\text{eq}} \sim T^3 \qquad T \gg m
$$
$$
s \sim g_{*,S}(T) T^3 \sim a^{-3}.
$$

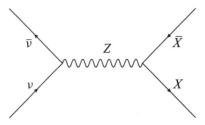

Figure 8.3 Annihilation of neutrinos via the Z boson to final states $X = e, d, \mu, t \cdots$ in the SM.

For the n_{eq} result, if you haven't derived it already, you are once again invited to. If you are feeling more adventurous, evaluate the result numerically and check the analytical limits.

We compute the number density of neutrinos as a function of temperature as they leave equilibrium by solving the collisional Boltzmann equation, Eq. (5.45). For $T < T_D$ the solution departs from the equilibrium value. We define the *abundance*, Y, as

$$Y = \frac{n}{s}, \tag{8.19}$$

where the use of s scales out the expansion of the Universe from the number density, since $s \sim a^{-3}$. We see this from the Boltzmann equation left-hand side, which is

$$a^{-3} \frac{d}{dt} na^3 = (\dot{n} + 3Hn)$$

$$sa^3 = \text{constant} \Rightarrow \dot{n} + 3Hn = s\dot{Y}. \tag{8.20}$$

Thus, $s\dot{Y}$ is equal to the collision term on the right-hand side of the Boltzmann equation, that is:

$$s\dot{Y} = C[f]. \tag{8.21}$$

For the collision term, we consider annihilations $\nu\bar{\nu} \leftrightarrow X\bar{X}$ for all possible final states, X, as shown in Fig. 8.3. Neutrinos annihilate via the Z boson, so the final states are known and we can compute the total cross section in the SM.

We ignore the Pauli blocking and Bose enhancement for low occupation numbers, $f \ll 1$, that is, we approximate $(1 \pm f) \approx 1$. Then $C[f]$ contains a term

$$\left[f_\nu f_{\bar{\nu}} - f_X f_{\bar{X}} \right]. \tag{8.22}$$

We assume that $X, \bar{X}$ are in thermal equilibrium, that is:

$$f_X = f_X^{eq} = e^{-\frac{E_X}{T}} \tag{8.23}$$

$$f_{\bar{X}} = f_{\bar{X}}^{eq} = e^{-\frac{E_{\bar{X}}}{T}}. \tag{8.24}$$

This is justified since, for instance, e^+e^- have EM interactions stronger than νs and so remain in equilibrium with the rest of the thermal bath. Energy conservation, that is, $E_\nu + E_{\bar{\nu}} = E_X + E_{\bar{X}}$, implies

$$
f_X f_{\bar{X}} = \exp\left[-\frac{\left(E_X + E_{\bar{X}}\right)}{T} \right]
$$

$$
= \exp\left[-\frac{\left(E_\nu + E_{\bar{\nu}}\right)}{T} \right] \tag{8.25}
$$

$$
= f_\nu^{eq} f_{\bar{\nu}}^{eq}
$$

$$
\Rightarrow \left[f_\nu f_{\bar{\nu}} - f_X f_{\bar{X}} \right] = f_\nu f_{\bar{\nu}} - f_\nu^{eq} f_{\bar{\nu}}^{eq}.
$$

We recall from section 5.1 the *thermally averaged annihilation cross section*, $\langle \sigma v \rangle$, defined as

$$
\langle \sigma v \rangle = (n_\psi^{eq})^{-2} \prod_{i=1}^{4} \int \frac{d^3 p}{(2\pi)^3 E_i}
$$

$$
\times (2\pi)^4 \delta^{(3)}\left(\mathbf{p}_1 + \mathbf{p}_2 - \mathbf{p}_3 - \mathbf{p}_4 \right) \delta\left(E_1 + E_2 - E_3 - E_4 \right) \tag{8.26}
$$

$$
\times |\mathcal{M}|^2 \exp\left[-\frac{\left(E_\nu - E_{\bar{\nu}}\right)}{T} \right],
$$

where we have factored in the equilibrium distributions. This useful shorthand takes up all of the messiness in the collision term in the Boltzmann equation. It can be difficult to evaluate, and we will simply quote or estimate numbers for it.

Using a change of time variables from t to $x = m/T$ (where m is the neutrino mass) via the Friedmann equation, the collisional Boltzmann equation is now

$$
\frac{dY}{dx} = -\frac{xs(T)}{H(m)} \left(Y^2 - Y_{eq}^2 \right) \langle \sigma v \rangle, \tag{8.27}
$$

where

$$
H^2(m) = \frac{\pi^2}{90} g_{\star,R}(m) \frac{m^4}{M_{Pl}^2}, \tag{8.28}
$$

such that $H(x) = H(m)x^{-2}$ if $g_{\star,R}$ is constant during the temperature range of interest. We have cast the collisional Boltzmann equation as a non-linear ODE for Y, the neutrino abundance. To solve it, we require the temperature dependence of the quantities $s(T)$, $Y_{eq}(T)$ and $\langle \sigma v \rangle$.

Let's get a little intuition. We can rewrite the Boltzmann equation once more:

$$\frac{x}{Y_{\text{eq}}}\frac{dY}{dx} = -\underbrace{\frac{n_{\text{eq}}\langle\sigma v\rangle}{H(x)}}_{\text{effectiveness of annihilations}}\overbrace{\left[\left(\frac{Y}{Y_{\text{eq}}}\right)^2 - 1\right]}^{\text{deviation from equilibrium}}. \tag{8.29}$$

The effective rate, Γ, of interactions is $n_{\text{eq}}\langle\sigma v\rangle$. Notice the units: $[\sigma] = L^2$ (cross-sectional area), $[n] = L^{-3}$ (number density) and $[v] = 1$ ($c = 1$); thus $[n\sigma v] = L^{-1} = t^{-1}$ so it makes sense because a rate $[\Gamma] = t^{-1}$.

At early times, $\Gamma \gg H$ and there are many interactions in a Hubble time. The equation can be solved if $Y \approx Y_{\text{eq}}$ (the assumed initial condition) to balance the interaction rate and maintain RHS of Eq. (8.29) to be *approximately* (but not exactly) zero. That is, if we start with $Y \approx Y_{\text{eq}}$, and $dY/dx \approx 0$, then this can be maintained. We say that the annihilations can maintain equilibrium if the rate is larger than the Hubble rate. Expressed as a comparison of time scales, $\Gamma > H$ implies that the interaction time $\tau = 1/\Gamma$ is shorter than a Hubble time, $\tau_H = 1/H$. Since the Hubble time is approximately the age of the Universe, this tells us that equilibrium is maintained if there are many interactions on average per Hubble time, which seems to make sense.

The rates H and $n\langle\sigma v\rangle$ both fall over time, but Hubble falls more slowly ($\langle\sigma v\rangle$ could have a power law dependence on T to compensate the fall in $n \sim T^{-3}$, but eventually $n \sim e^{-m/T}$ and the term has to drop below H). We notice that when $H \sim \Gamma$ the RHS of Eq. (8.29) can be non-zero and then Y will evolve away from equilibrium. Shortly after this time, Hubble will become much larger than the interaction rate. Then the RHS of Eq. (8.29) falls to zero independently of what Y does, and the abundance *freezes out*, that is, $dY/dx \to 0$. This occurs at some $x_f = m/T_f$, which is approximately when

$$H(T_f) = \Gamma(T_f). \tag{8.30}$$

Some thinking is required to follow this logic: after all, we are approximating solutions to a non-linear differential equation. First, we'll consider the example of neutrinos, and then we will approximately solve a more complicated case. The JUPYTER notebooks give the numerical solution.

Let's find the neutrino abundance, assuming an initial equilibrium, and freeze out at some T_f. For v's, as already stated, this happens at $T_f := T_D \sim 1$ MeV, the decoupling temperature of the weak interactions. Furthermore, we know $m_v \lesssim$ MeV, and so we can consider the limit $T \gg m$. The equilibrium abundance is

$$Y_{\text{eq}} = \text{constant} \times (g_\star)^{-1}, \tag{8.31}$$

and so the final abundance

$$Y(x \to \infty) \approx Y_{\text{eq}}(x_f), \tag{8.32}$$

which is *not* the same as the equilibrium abundance as $x \to \infty$, $T \to 0$. For relativistic fermions, the equilibrium number density is

$$n_{\rm eq} = \frac{3}{4}\zeta(3)\frac{g}{\pi^2}T^3 . \tag{8.33}$$

Recall the entropy density:

$$s = \frac{2\pi^2}{45}g_{\star s}T^3 . \tag{8.34}$$

Thus the equilibrium abundance, $Y_{\rm eq} = n_{\rm eq}/s$, at freeze out, $x = x_f$, is

$$\boxed{Y_{\rm eq}^{\nu}(x_f) = 0.21g/g_{\star s}(x_f).} \tag{8.35}$$

Therefore, the number density today at $T_0 \ll T_f$ is

$$n_{\nu 0} = s_0 Y_{\infty} \tag{8.36}$$

$$\approx 0.092\frac{g_{\star s}(T_0)}{g_{\star s}(T_f)}T_0^3 g , \tag{8.37}$$

and the energy density is

$$\rho_{\nu 0} = m_{\nu}n_{\nu 0}. \tag{8.38}$$

For fermions, $g = 2$ (particle and anti-particle are created in the annihilation), and so the total neutrino density today comes from the sum over all of the mass eigenstates:

$$\Rightarrow \rho_{\nu 0} = \sum_{i=1}^{3} m_{\nu,i} \times 0.018 \times \frac{g_{\star s}(T_0)}{g_{\star s}(T_f)}T_0^3 . \tag{8.39}$$

Using the entropic degrees of freedom at 1 MeV and today:

$$g_{\star s}(T_f) = 10.75, \quad g_{\star s}(T_0) = 3.90. \tag{8.40}$$

Then the relic abundance is

$$\Omega_{\nu}h^2 = \frac{\rho_{\nu 0}}{3M_H^2 M_{\rm Pl}^2} = \frac{0.022\sum m_{\nu}T_0^3}{M_H^2 M_{\rm Pl}^2} . \tag{8.41}$$

We plug in the value for the temperature today:

$$T_0 = 2.725\,{\rm K} = 2.3 \times 10^{-4}\,{\rm eV}, \tag{8.42}$$

and finally we arrive at the classic and very useful approximation for the neutrino relic abundance:

$$\boxed{\Omega_{\nu}h^2 \approx 0.01 \sum m_{\nu} = \frac{\sum m_{\nu}}{100\,{\rm eV}},} \tag{8.43}$$

Neutrinos could have the observed DM relic density, Eq. (5.39), if $\sum m_{\nu} \approx 12$ eV. Neutrinos with $\sum m_{\nu} > 12$ eV would result in too much dark matter. The so-called

'Tremaine-Gunn' bound for fermionic DM, such as neutrinos, requires

$$m_f \gtrsim 100 \text{ eV} \tag{8.44}$$

in order to successfully form galaxies. This bound comes from 'Pauli blocking' and galaxy phase space, and we will discuss it further and give a rough derivation in section 13.1. However, if neutrinos could be heavy enough to satisfy this bound, then they would give too much DM from Eq. (8.43).

Neutrinos freeze out when $T_f > m_\nu$, which implies that neutrinos are *hot dark matter* (HDM), produced with relativistic velocities. Given that we need $\sum m_\nu = 12$ eV if we want the correct relic abundance, such neutrinos would only become non-relativistic shortly before matter-radiation equality at $T \sim 1$ eV. The gravitational instability model of chapter 6 is inconsistent with HDM, and current CMB constraints on HDM imply

$$\sum m_\nu < 0.24 \text{ eV} \tag{8.45}$$

$$\Rightarrow \Omega_\nu h^2 < 0.002. \tag{8.46}$$

We therefore conclude that

> Standard Model neutrinos are excluded from being the dominant component of DM.

The argument we've just outlined applies so long as the neutrino freeze out when they are relativistic. If we make the neutrinos much heavier, then they will freeze out when cold, and one instead arrives at a lower bound on the neutrino mass of around 2 GeV known as the *Lee-Weinberg* bound [46]. At GeV masses and cold freeze out, we enter the realm of the weakly interacting massive particle, which we discuss in the next chapter.

QUIZ

i. Right-handed neutrinos are:
   a. Charged only under SU(2) in the Standard Model.
   b. A 4-component spinor.
   c. Neutral under all standard model interactions.
   d. Interact only with the Higgs.

ii. The Higgs cannot give Dirac masses to neutrinos because (select all that apply):
   a. The relevant Yukawa interaction is not a Standard Model singlet.
   b. The Higgs only gives masses to gauge bosons.
   c. The Higgs only gives masses to quarks.
   d. The right-handed neutrino is a Standard Model singlet.

iii. Name the interaction $\lambda\phi\bar{\psi}\psi$ where ψ is a fermion, ϕ a scalar and λ a constant.
   a. Dirac interaction.
   b. Fermi interaction.
   c. Weinberg interaction.
   d. Yukawa interaction.

iv. The weak interaction basis and the mass eigenbasis are:
   a. The same.
   b. Related by an SU(2) rotation.
   c. The same only for quarks.
   d. Related by a unitary rotation.

v. 'Missing' solar electron neutrinos at the Homestake experiment in the 1960s provided evidence for non-zero neutrino mass because (select all that apply):
   a. The experiment could not detect mass.
   b. The experiment could only detect electron neutrinos in the weak basis.
   c. The Sun produces massive neutrinos less efficiently.
   d. Neutrinos propagate in the mass eigenbasis.

vi. Neutrinos are 'hot' relics because:
   a. They decouple after freeze out.
   b. They decouple at temperatures above the electroweak phase transition.
   c. The CMB measures their temperature precisely.
   d. They decouple before they become non-relativistic.

vii. The 'abundance' Y is defined as:
   a. The number density divided by the entropy density.
   b. The number density times the temperature.

c. The relic density in the asymptotic past.

d. The equilibrium number density.

viii. Which statements are true about neutrino DM in the Standard Model? (Select all that apply.)

a. Neutrinos require a total mass 12 eV to explain the relic density.

b. Neutrinos require a mass less than 1 eV to be consistent with the CMB.

c. Neutrinos require a mass greater than 100 eV to explain galaxy formation.

d. Neutrinos with mass 100 eV give too much relic abundance.

e. Neutrinos with mass 1 eV give too little relic abundance.

Answers

i. c; ii. a, d; iii. d; iv. d; v. a; vi. b,d; vii. d; viii. a; viii. a,b,c,d,e;

Chapter Nine

Weakly Interacting Massive Particles

9.1 RELIC DENSITY

9.1.1 Equilibrium

We have seen how neutrinos can be produced via the weak interactions from the hot plasma of the early Universe but that constraints on their mass prevent them from being a successful DM candidate.

To solve the problems faced by the neutrino and produce a consistent model of DM we will use the same idea of a particle produced via freeze out. However, we will consider a new particle (the WIMP), χ, with mass, m_χ, and annihilation cross section, $\langle \sigma v \rangle$.

There are two mechanisms via which WIMPs can interact with SM particles. Elastic scattering, shown in Fig. 9.1(left), is analogous to pool balls bouncing off of each other and does not change the number of WIMPs but only redistributes their momentum. The annihilation of a pair of WIMPs into SM particles, shown in Fig. 9.1(right), does change the number of WIMPs.

For both of these interactions, the backwards process is also possible. For elastic scattering at sufficiently high pressures the rate of the backwards process is equal to the forwards process, and the interaction maintains *kinetic equilibrium*. For annihilation, high temperatures are required to enable the rate of the backwards reaction to balance the forwards reaction. This balance maintains *chemical equilibrium*.

As seen previously, when in chemical equilibrium (defined by $\mu = 0$), the number density of a fermion of mass m is

$$n_{eq} \sim \begin{cases} T^3 & \text{if } T \gg m, \\ T^{3/2}e^{-m/T} & \text{if } T \ll m. \end{cases} \tag{9.1}$$

When kinetic equilibrium holds, the temperature T_χ of the WIMPs is equal to that of the relativistic bath of SM particles, that is,

$$T_\chi = T_{SM} \equiv T. \tag{9.2}$$

Recall that the abundance, $Y = n/s$, of a particle species is given by the Boltzmann equation, which we can write as

$$\frac{x}{Y_{eq}} \frac{dY}{dx} = \frac{n_\chi^{eq}\langle \sigma v \rangle}{H(x)} \left[\left(\frac{Y}{Y_{eq}} \right)^2 - 1 \right], \tag{9.3}$$

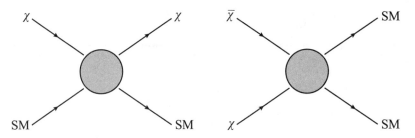

Figure 9.1 Interactions between a WIMP, χ (anti-particle $\bar{\chi}$), and the Standard Model (SM). The 'blob' represents that we do not specify the exact interaction. *Left*: elastic scattering. *Right*: annihilation.

where $x = m/T$. The abundance is equal to the equilibrium value ($Y = Y_{eq}$) when the rate of inelastic (number changing) collisions, $\Gamma_{inelastic}$, is much greater than the Hubble rate of expansion

$$\Gamma_{inelastic}(x) = n_\chi^{eq}(x)\langle\sigma v\rangle \gg H(x). \tag{9.4}$$

Chemical decoupling is the point at which chemical equilibrium ends. This 'freeze-out' occurs at x_f which is defined as the point where these two rates are equal:

$$\Gamma_{inelastic}(x_f) \equiv H(x_f). \tag{9.5}$$

When we studied the freeze out of neutrinos in the last section, we studied chemical decoupling; the phrases are equivalent.

Similarly, *kinetic decoupling* marks the end of thermal equilibrium. This happens at x_k, which is defined by the point at which the rate of elastic interactions is equal to the Hubble rate

$$\Gamma_{elastic}(x_k) \equiv n_{SM}\langle\sigma v\rangle = H(x_k). \tag{9.6}$$

We will not study kinetic decoupling. In order to do so one needs to solve for the evolution of the temperature, T_χ, not just the number densities.

Here we have written things as if $\langle\sigma v\rangle$ is the same for each process, elastic or inelastic, by 'rotating the Feynman diagram' in our heads, but this need not necessarily be the case. While σ is the same, what differs is which number density, χ or SM, multiplies the cross section to give the rate in the thermal average over the reactants.

For cold dark matter, by definition, chemical decoupling occurs while the DM is already non-relativistic, $T_f < m_\chi$. This means it has a number density which is exponentially suppressed as $n_\chi \sim e^{-m/T}$ for some period while the number density of the relativistic bath of SM particles continues to fall only as $n_{SM} \sim T^3$. Therefore the number density of DM particles falls much more quickly than that of SM particles implying that the rate of inelastic collisions falls faster than the rate of elastic collisions. Thus, *for CDM, kinetic decoupling happens after chemical decoupling*, allowing us to safely assume that the CDM temperature is still equal to that of the SM bath during the (chemical) freeze-out process, that is, $T_\chi = T$.

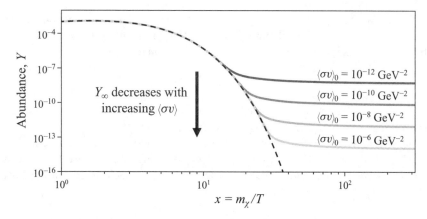

Figure 9.2 **Freeze out of a massive particle**. Calculated for a WIMP with a temperature-independent annihilation cross section, for four different values of the cross section. The result is found by numerically solving the Boltzmann equation as given in Eq. (9.3) with equilibrium initial conditions. We use the full temperature dependence of number density, Hubble rate, entropy and relativistic degrees of freedom $g_{\star,R}(T)$ and $g_{\star,S}(T)$. The black dashed line shows the equilibrium value for the abundance. See colour insert.

9.1.2 Freeze out of a Massive Particle

We will start by looking at the result and then derive it afterwards. As shown in Fig. 9.2, once the temperature has dropped below the mass, defined by $x = 1$, the equilibrium number density n_χ^{eq} falls exponentially. This fall is caused by annihilations which keep the $\chi\bar{\chi} \leftrightarrow$ SM SM process in equilibrium. However, eventually these interactions are unable to keep up with the expansion of the Universe, causing χ to chemically decouple. The number density no longer follows $n_\chi \sim e^{-m/T}$. Instead, when $H \gg \Gamma_{\mathrm{inelastic}}$, we have $dY/dx = 0$. The number density therefore loses its exponential suppression given in Eq. (9.1) and simply redshifts as T^3 (since $Y = n/s$ and $s \sim T^3$).

More efficient interactions, meaning larger $\langle \sigma v \rangle$, are able to keep up with the expansion rate of the Universe for longer, meaning $n\langle \sigma v \rangle = \Gamma > H$, and therefore delay the onset of freeze out. This leads to the relic abundance being reduced as the number density is exponentially suppressed for a longer period of time. *A larger annihilation cross section means lower relic abundance.*

A full numerical solution of the Boltzmann equation requires us to know the temperature dependence of the equilibrium abundance $Y_{\mathrm{eq}}(T)$, the entropy $s(T)$, the Hubble rate $H(T)$ and the interaction cross section $\langle \sigma v \rangle$. While the equilibrium number density can be computed from a single integral, the entropy and Hubble rate are dependent on the effective number of degrees of freedom for the energy density $g_{\star,R}$ and for the entropy density $g_{\star,S}$. Numerical fits to $g_\star$ can be found easily online, for example, Refs. [47, 19], and typically one uses a table. The cross section has to be calculated using QFT and thermally averaged numerically. To simplify the process we will look for approximate solutions.

Firstly, we know that for early times $x \ll x_f$ the inelastic interaction rate is much greater than the Hubble rate ($\Gamma(x) \gg H(x)$). In this limit the solution to the Boltzmann equation is given by the equilibrium abundance

$$Y(x \ll x_f) = Y_{eq}(x). \tag{9.7}$$

This is because we have assumed boundary conditions which require chemical equilibrium, that is, $Y = Y_{eq}$, at high temperature $x \to 0$.

Secondly, we know that for $x \gg x_f$, the inelastic interaction rate is much lower than the Hubble rate ($\Gamma(x) \ll H(x)$). We see from Fig. 9.2 that we expect that the derivative $\frac{dY}{dx}$ tends towards zero. Therefore, we are able to match the final abundance to value of Y_{eq} near x_f to find

$$Y(x > x_f) \approx Y_{eq}(x_f). \tag{9.8}$$

For now we will assume that the interaction cross section has a power law dependence on x:

$$\langle \sigma v \rangle = \langle \sigma v \rangle_0 x^{-n}, \tag{9.9}$$

which we will derive shortly. Additionally, we will assume that the effective number of relativistic degrees of freedom does not change during freeze out, thus allowing us to write down the entropy and Hubble rate as

$$s = s_0 x^{-3}, \tag{9.10}$$

and

$$H(x) = H(m)x^{-2}, \tag{9.11}$$

respectively, where $s_0 = s(x = 1)$ and $H(m) = H(m, x = 1)$. If we approximate $Y(x > x_f) \gg Y_{eq}(x > x_f)$ we can write down an approximate form of the Boltzmann equation

$$\frac{d\Delta}{dx} \approx \frac{\lambda}{x^{n+2}} \Delta^2, \tag{9.12}$$

where $\Delta = Y - Y_{eq}$ and

$$\lambda = \frac{\langle \sigma v \rangle_0 s_0}{H(m)}. \tag{9.13}$$

Next, we can rearrange and integrate this equation from $x = x_f$ to $x = \infty$:

$$\int \frac{d\Delta}{\Delta^2} = -\lambda \int \frac{dx}{x^{n+2}}. \tag{9.14}$$

To calculate this integral we need to know the values of Δ at $x = \infty$ and $x = x_f$ which are

$$\Delta(x = \infty) = Y(x = \infty) - Y_{eq}(x = \infty)$$
$$\approx Y(x = \infty), \tag{9.15}$$

and

$$\Delta(x = x_f) = Y(x = x_f) - Y_{eq}(x = x_f)$$
$$\approx 0,$$
(9.16)

respectively from our assumption $Y(x > x_f) \gg Y_{eq}(x > x_f)$ and our initial condition. This allows us to perform the integral in the left-hand side of Eq. (9.14) and find that

$$Y_\infty \approx \frac{(n+1)x_f^{n+1}}{\lambda}.$$
(9.17)

Finally, we need to estimate x_f as defined by Eq. (9.5). Recall that the full number density for a non-relativistic fermion in chemical equilibrium is given by

$$n_{eq}(x) = \frac{g_\chi}{(2\pi)^{3/2}} m_\chi^3 x^{-3/2} e^{-x},$$
(9.18)

where g_χ is the number of χ degrees of freedom (typically $g_\chi = 2$ since we have χ and $\bar{\chi}$ with χ its own anti-particle). From the Friedmann equation we see that

$$H(x) = \frac{\pi}{3\sqrt{10}} \frac{m_\chi^2}{M_{pl}} g_\star(x)^{1/2} x^{-2}.$$
(9.19)

Substituting Eqs. (9.18) and (9.19) into Eq. (9.5) we find that[1]

$$\frac{x_f^{1/2-n} e^{-x_f}}{g_\star(x_f)^{1/2}} = \frac{2^{3/2}}{3\sqrt{10}} \pi^{5/2} \frac{1}{g_\chi m_\chi M_{pl} \langle \sigma v \rangle_0}.$$
(9.20)

Eq. (9.20) is an implicit equation for x_f; once we find x_f, we can find the relic abundance from Eq. (9.17). Consider the case where $g_\star$ is constant. The left-hand side of Eq. (9.20) decreases monotonically with x. We also see that the right-hand side decreases with an increase in $\langle \sigma v \rangle_0$, therefore corresponding to an increase in x_f and hence a decrease in Y_∞, as expected.

What value of Y_∞ do we require to get the correct relic abundance of dark matter? Since the abundance remains constant for $x \gg x_f$, we can calculate the number density today by using the entropy density and the DM number density:

$$n_{\chi 0} = s_0 Y_\infty$$
$$= \frac{2\pi^2}{45} g_{\star,s}(T_0) T_0^3 Y_\infty,$$
(9.21)

where $T_0 = 2.73$ K is the temperature of the Universe today (most of the entropy is contained in photons of the CMB). The number density is linked to the matter

---

[1] We can find an alternative definition of freeze out directly from the $d\Delta/dx$ equation, which we take as $\Delta(x_f) = cY_{eq}(x_f)$, where c is a constant of order unity we can fit to the numerical solution.

density by $\rho_\chi = m_\chi n_\chi$, while by definition of $\Omega_\chi h^2$ it is also given by

$$\rho_\chi = 3M_H^2 M_{pl}^2 \Omega_\chi h^2, \tag{9.22}$$

and therefore

$$Y_\infty = \frac{45}{2\pi^2} \frac{3M_H^2 M_{pl}^2}{T_0^3} \frac{1}{g_{\star,S}(T_0)} \frac{\Omega_\chi h^2}{m_\chi}, \tag{9.23}$$

where we recall the definition of the Hubble mass scale, M_H, given in section 5.1.4. As discussed previously, we know from measurements of the CMB that $\Omega_{DM} h^2 = 0.12$, and therefore, inputting the relevant values, we find that we require

$$\boxed{Y_\infty \approx 0.4 \left(\frac{\text{eV}}{m_\chi}\right).} \tag{9.24}$$

Next, we can ask which combinations of WIMP mass m_χ and interaction cross section $\langle \sigma v \rangle_0$ produce this required abundance. We will consider an example case in which the WIMP mass is $m_\chi = 100$ GeV and guess that the temperature at freeze out is in the range 100 MeV $< T_f <$ 100 GeV. For these temperatures $g_\star \approx 90$, and we approximate this to be constant (we are above the QCD phase transition). We will additionally assume that the cross-section power law $n = 0$ meaning that the cross section is also constant (we will return to this later). Lastly, we will take χ to be a Dirac fermion that is its own anti-particle, meaning that $g_\chi = 2$.

Substituting the value of Y_∞ necessary for the correct abundance from Eq. (9.24) into Eq. (9.17), we find that we require

$$x_f \approx 4 \times 10^{-12} \lambda. \tag{9.25}$$

Additionally, from Eq. (9.20), we find that

$$x_f^{1/2} e^{-x_f} \approx \frac{300}{\lambda}. \tag{9.26}$$

Solving these two equations for x_f and λ (e.g. with a simple plotting and inspection), we find that $x_f \approx 25$ corresponding to a freeze-out temperature of $T_f \approx 4$ GeV, thus validating our previous guess for the temperature range. Next, solving for λ and using the fact that $\lambda \approx 2.53 M_{pl} m_\chi \langle \sigma v \rangle_0$, we find the corresponding cross section to be

$$\boxed{\langle \sigma v \rangle_0 \approx 10^{-26} \text{ cm}^3 \text{ s}^{-1}.} \tag{9.27}$$

This is the benchmark WIMP total annihilation cross section required to get the correct DM relic abundance. Remember this number! Also, check that you can reproduce it!

This combination of m_χ and $\langle \sigma v \rangle_0$ is known as the *WIMP miracle*. This is because the value of the interaction cross section is approximately equal to what we

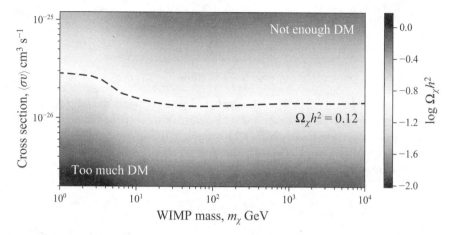

Figure 9.3 WIMP relic density—calculated for a WIMP with temperature-independent anni-hilation cross section. The result is found by numerically solving the Boltzmann equation as given in Eq. (9.3) with equilibrium initial conditions. We use the full temperature dependence of number density, Hubble rate, entropy and relativistic degrees of freedom $g_{*,R}(T)$ and $g_{*,S}(T)$. The black dashed line shows the $(m_\chi, \langle \sigma v \rangle)$ combinations which give the observed relic density, $\Omega_\chi h^2 = 0.12$. See colour insert.

would expect if we set all of the constants at the weak scale (as we show shortly), and furthermore because our chosen mass of 100 GeV is approximately equal to the electroweak scale defined by the Higgs vacuum expectation value.

To calculate the WIMP relic density properly as a function of m_χ and $\langle \sigma v \rangle$, we need to solve Eq. (9.3) numerically with the full temperature dependence of all of the relevant quantities. This is done in the accompanying JUPYTER notebooks. The result is shown in Fig. 9.3.

We see that the numerical result in Fig. 9.3 is in fairly close agreement with our approximate calculation outlined above, where we estimated that the correct relic abundance occurs for $\langle \sigma v \rangle_0 \approx 10^{-26}$ cm^3 s^{-1}. One particular feature of interest is that, while the relic density is mostly insensitive to the WIMP mass, for masses below around 10 GeV the relic density increases slightly for fixed $\langle \sigma v \rangle$. This is because as mass decreases, freeze out occurs later. Once the mass is sufficiently low, freeze out occurs after the QCD phase transition, leading to a large drop in $g_{*,R}(T)$ and $g_{*,S}(T)$. This leads to an increase in the equilibrium abundance and hence an increase in the final abundance and therefore the relic density.

There is a fundamental limit on the annihilation cross section for a massive particle imposed by *unitarity*, which fixes $\langle \sigma v \rangle \leq \mathcal{O}(1)/m^2$ [48]. Taking $x_f = 10$ and $g_* = 100$ in Eq. (9.24), we can estimate a lower limit on the DM particle mass known as the unitarity bound:

$$m \gtrsim 50 \text{ TeV}. \qquad (9.28)$$

For a heavier particle, only an increase $\langle \sigma v \rangle$ beyond the unitarity limit can reduce the freeze-out relic abundance, and thus any heavier particle would give too much

DM if it was ever in thermal equilibrium. This argument strongly suggests that stable, heavy particles in thermal equilibrium are excluded.

9.1.3 Velocity Dependence of the Cross Section

The mechanisms outlined in Fig. 9.1 were introduced to explain the relic density of WIMPs from freeze out. However, these also provide us with the interactions with the SM that we can use to search for DM both directly and indirectly.

Before we derive an estimate for the WIMP cross section, we first need to ascertain the nature of 'v' in $\langle \sigma v \rangle$. By definition, the total cross section is

$$(2\pi)^4 \int \frac{d^3 p_3}{(2\pi)^3 2E_3} \frac{d^3 p_4}{(2\pi)^3 E_4} |\mathcal{M}|^2 \delta \left(\sum_i E_i \right) \delta^{(3)} \left(\sum_i p_i \right)$$

$$= 4g^2 \sigma \sqrt{(p_1 p_2)^2 - (m_1 m_2)^2}, \tag{9.29}$$

where p_3 and p_4 are the momenta of the outgoing particles, and p_1 and p_2 are the momenta of the ingoing particles. Substituting this definition into the collision term in the Boltzmann equation we find two terms:

$$- \int \sigma v_{\text{Møl}}^{12} (dn_1 \, dn_2) - (dn_3 \, dn_4) \sigma v_{\text{Møl}}^{34}, \tag{9.30}$$

where $dn_1 = \frac{d^3 pf}{(2\pi)^3}$, and $v_{\text{Møl}}$ is the Møller velocity given by

$$v_{\text{Møl}}^{ij} = \frac{\sqrt{(p_i p_j)^2 - (m_i m_j)^2}}{2 E_i E_j}. \tag{9.31}$$

When a $1 + 2 \longleftrightarrow 3 + 4$ process is in equilibrium (we take 3 and 4 to be SM particles),

$$\langle \sigma v \rangle_{12} n_1^{\text{eq}} n_2^{\text{eq}} = \langle \sigma v \rangle_{34} n_3^{\text{eq}} n_4^{\text{eq}}, \tag{9.32}$$

which implies:

$$\langle \sigma v \rangle = \frac{\int \sigma v \exp[-(E_1 + E_2)/T] d^3 p_1 d^3 p_2}{\int d^3 p_1 d^3 p_2 \exp[-(E_1 + E_2)/T]}. \tag{9.33}$$

Using the variable $s = 2m^2 + 2E_1 E_2 - 2\vec{p_1} \cdot \vec{p_2}$ and expanding in the non-relativistic limit ($x = m/T > 1$), it can be shown that (see Eq. 2.8 in Ref. [21])

$$\langle \sigma v \rangle = b_0 + \frac{3}{2} b_1 x^{-1} + \dots, \tag{9.34}$$

where b_i are constants to be computed. We will consider only the first two constants b_0 and b_1. The b_0 term is independent of x and is therefore referred to as an 's-wave'.

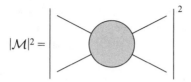

Figure 9.4 A matrix element is the square of a Feynman diagram.

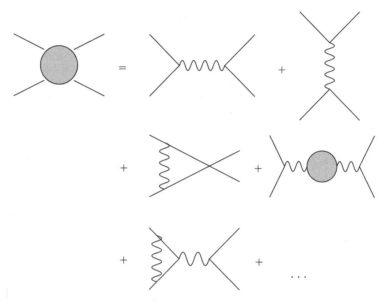

Figure 9.5 An amplitude is given by a sum of all possible Feynman diagrams that contribute to the given process.

On the other hand, the b_1 term is inversely proportional to x and is hence referred to as a 'p-wave' (since we assume thermal velocities). We see that this justifies our previous approximation that $\langle \sigma v \rangle = \langle \sigma v \rangle_0 x^{-n}$. In practice, normally just one of the terms, b_0 or b_1, dominates the cross section.

Now that we have defined v and $\langle \sigma v \rangle$, we need to estimate σ.

9.1.4 Estimating a Cross Section

The starting point for calculating the cross section is the square matrix element $|\mathcal{M}|^2$. The square matrix element is a probability and therefore a real number. It can be expressed diagrammatically as shown in Fig. 9.4. The amplitude inside the absolute value is a complex number, and so a sum of many such amplitudes can have interference in the phases. In order to account for interference, we need to consider all possible processes which contribute to the square matrix element. An example is shown in Fig. 9.5.

This process is derived from a perturbation series for the path integral in QFT. Here, the first, second and third terms are called the s-channel, t-channel and u-channel respectively. Additional terms beyond this are referred to as 'radiative corrections'.

We want to compute the cross section for the process $1 + 2 \longrightarrow 3 + 4$. However, to do that we need to calculate $1 + 2 \longrightarrow$ anything, where 'anything' includes all of the processes that are allowed by conservation of energy. That is,

$$\vec{p_1} + \vec{p_2} = \sum_i p_i^{\text{final}}, \tag{9.35}$$

$$E_1 + E_2 = \sum_i E_i^{\text{final}}, \tag{9.36}$$

$$\sigma_{\text{total}} = \sum_i \sigma_i, \tag{9.37}$$

where, in Eq. (9.37), we sum over all allowed final states.

We compute the diagrams for each of the possible processes using the Feynman rules. In QFT, these rules are derived from the Lagrangian $\mathcal{L}$. We now show just some basic examples, so you know what this looks like if you haven't seen it before, although we won't actually do any computations. For example, an external fermion has the rule:

initial $u^s(p)$

final $\bar{u}^s(p)$

and an external anti-fermion,

initial $\bar{v}^s(p)$

final $v^s(p)$,

in which the arrows on the legs indicate the direction of negative charge flows (since they originate for electrons, where particles are negatively charged and anti-particles positive), the unattached arrows show the direction of momentum flow, and u and v are particle and anti-particle spinors.

The rules for Abelian and non-Abelian vertices are

$$= iQe\gamma^\mu \quad \text{Abelian } U(1)$$

and

$$= ig\gamma^\mu T^a \quad \text{non-Abelian } SU(N),$$

respectively in which μ and a are the Lorentz index of the gauge boson and the group generator index respectively. Internal lines are called 'propagators' and have the rules:

$$= -\frac{i\eta_{\mu\nu}\delta_{ab}}{p^2 + m^2 + i\varepsilon}$$

$$= \frac{i(\gamma^\mu p_\mu + m)}{p^2 - m^2 + i\varepsilon},$$

where δ_{ab} sums over group indices, and ϵ is a gauge parameter, which appears since we are in 'Feynman gauge'. In the case of no divergence of the diagram, we can take $\epsilon \to 0$.

In a Feynman diagram, we read along the lines following the charge, writing down the relevant terms from the rules and inserting propagators. Group theory factors, spin and polarization make computing even a simple process, such as that shown in the diagram on the top of p. 170, rather complicated even in the Abelian case (see e.g. Ref. [20] chapters 1 and 5). This is because we need to trace over the spinors and γ matrices. Ultimately all of this leads to factors of the momenta in the process and constants.

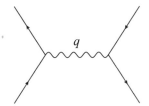

For a cross section, we typically sum over spins and integrate over all of the allowed final states. While this process can be very complicated and tedious, most of the details 'come out in the wash' and we can make use of dimensional analysis to provide good estimates without performing the full calculation.

We know that (with $\hbar = c = 1$)

$$[\sigma] = [A] = [L^2] = [M^{-2}], \tag{9.38}$$

where A is area, L is length and M is mass. Therefore, since the square matrix element does not have units (since it is a probability), we can infer that

$$\sigma \sim \frac{|\mathcal{M}|^2}{\Lambda^2}, \tag{9.39}$$

where Λ is some energy scale.

Hence, if we take the QED process,

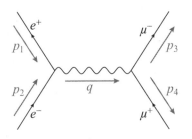

The centre of momentum frame is defined by

$$\vec{p_1} = \vec{p_2} = \vec{p}, \tag{9.40}$$

$$E_1 = E_2 = \sqrt{s}/2, \tag{9.41}$$

and

$$s = 2\left(|\vec{p}|^2 + m_e^2\right), \tag{9.42}$$

that is, s is the square of the total energy in the centre of momentum frame $\sqrt{s} = E_{\text{c.o.m}}^2$. In this frame we find that

$$\mathcal{M} \sim \frac{e^2}{|q|^2}, \tag{9.43}$$

where the factors of e arise from the vertices, and q arises from the propagator and in the COM frame is given by $\vec{q} = (\sqrt{s}, \vec{0})$ and $q^2 = s^2$.[2] This then gives us

$$\sigma \sim |\mathcal{M}|^2 \sim \frac{e^4}{s^2}. \tag{9.44}$$

However, we see that currently this estimate has inconsistent dimensions on the left- and right-hand sides of the equation. The only quantity we have available to fix this is the squared centre of mass energy, s. Recalling the fine structure constant $\alpha = e^2/4\pi$, we then write down our dimensionally consistent estimate of

$$\sigma \sim \frac{\alpha^2}{s}. \tag{9.45}$$

In theory, we could have fixed the dimensions with m_e or m_μ. However, in the limit $E_{\text{c.o.m.}} \gg m_e, m_\mu$ the centre of momentum energy dominates and is the only quantity we could use. This is therefore the limit within which our estimate should work.

A full calculation shows that in this limit the cross section is given by

$$\sigma = \frac{4\pi}{3} \frac{\alpha^2}{s}. \tag{9.46}$$

This is the high energy limit for fermions scattering via a massless gauge boson. Seeing the result of the full calculation shows the strength of our estimate based on dimensional analysis.

---

[2] This is why the photon in question must be 'virtual': it does not satisfy the requirement $q^2 = 0$ for photon squared four momenta on the lightcone, and so it cannot be a real propagating particle. Instead, it is a quantum fluctuation, which is allowed to violate the $q^2 = 0$ condition due to the uncertainty principle.

QUIZ

i. For a cold relic, which happens first?
   a. Chemical decoupling.
   b. Kinetic decoupling.

ii. The term 'freeze out' refers to what process?
   a. The epoch when a particle becomes effectively cold.
   b. The epoch when interactions become efficient, and the distribution approaches equilibrium.
   c. The epoch when a symmetry is spontaneously broken.
   d. The epoch when interactions become inefficient, and the distribution departs from equilibrium.

iii. The number density of a massive particle is (select all that apply):
   a. Always proportional to T^3.
   b. In equilibrium, falls exponentially when $T \ll m$.
   c. Proportional to T^3 after freeze out.
   d. Falls exponentially once out of equilibrium.

iv. Increasing the WIMP annihilation cross section does what? (Select all that apply.)
   a. Makes freeze out happen earlier.
   b. Makes freeze out happen later.
   c. Increases the relic abundance.
   d. Decreases the relic abundance.

v. The 'miracle' WIMP cross section $\langle \sigma v \rangle \approx 10^{-26}$ cm^{3}s$^{-1} \approx 8.5 \times 10^{-10}$ GeV^{-2}. Take the cross section to be $\langle \sigma v \rangle \sim \alpha^2 / \Lambda^2$. If α is of order the fine structure constant, $1/137$, what is Λ?
   a. 6 GeV.
   b. 250 GeV.
   c. 4 TeV.
   d. 70 MeV.

Answers

i. a; ii. d; iii. b,c; iv. b,d; v. b

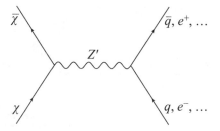

Figure 9.6 Annihilation of a WIMP, χ, via a hypothetical Z'-boson to Standard Model final states.

$$\underset{Z'}{\sim\!\sim\!\sim} = \frac{1}{q^2 - M_{Z'} + i\Gamma_{Z'}M_Z}$$

Figure 9.7 Massive vector boson, for example, Z', propagator.

9.2 MODELS OF WIMPs

9.2.1 'The WIMP Miracle'

We can build a very simple model of DM using $(\chi, \overline{\chi})$ WIMPs and an additional boson, Z'. The annihilation of two WIMPs via the Z' is shown in Fig. 9.6. The propagator for the Z' is shown in Fig. 9.7.

Since Z' is heavy, it is unstable. When $q^2 = m_{Z'}^2$, it is 'on-shell' and decays with a width of $\Gamma_{Z'}$. The width 'regulates' the process for $q^2 \simeq m_{Z'}^2$, and the propagator does not diverge. We can thus estimate the cross section in the centre-of-mass frame as

$$\sigma_{cm} \sim \frac{1}{(4\pi)^2} \frac{g_\chi^2 g_{SM}^2 \, s}{\left(s - M_{Z'}^2\right)^2 + M_{Z'}^2 \Gamma_{Z'}^2}, \tag{9.47}$$

where we assumed that all of the SM decay products are relativistic, so the only scale we had to get the units right was s. We are interested in the cross section for a cold relic, that is, $T \ll m_\chi$, so that all momenta are small compared to m_χ:

$$\Rightarrow s = E_{cm}^2 \sim m_\chi^2. \tag{9.48}$$

Defining a new 'Fermi' constant,[3] $G_{F'}$, we can write the cross section as $\sigma \sim G_{F'}^2 m_\chi^2$, and thus

$$\langle \sigma v \rangle \sim G_{F'}^2 m_\chi^2 \sum_i \left(1 - z_i^2\right)^{\frac{1}{2}}, \tag{9.49}$$

[3] Recall that in the weak interactions of the Standard Model, a four-fermion interaction arises for energy scales below the W mass, and the Fermi constant is $G_F = g^2/M_W^2$, where g is the weak force gauge coupling constant.

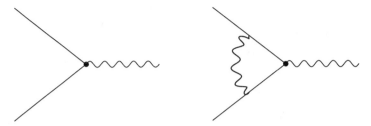

Figure 9.8 The charge is defined by the coupling, g, in the vertex of the left-hand diagram. However, one must also consider loops contributing to the same process and therefore $g \to g(Q)$, where Q is the loop momentum in the right-hand figure.

where $z_i = m_i/m_\chi$ for decay products of the Z' labelled by i and arises from the Møller velocity (see Ref. [49] Eq. 5.52 for the full result). For m_χ in our range of interest, $z \ll 1$ for all SM particles except the top quark, that is, decay products are relativistic $\Rightarrow v_{\text{Møller}} \sim v_{\text{relativistic}} \sim c = 1$.

Estimating $G_{F'} \sim \frac{\alpha}{v_{\text{EW}}^2}$ with $v_{\text{EW}} \sim 200 \, \text{GeV}$ and taking $m_\chi \sim v_{\text{EW}}$,

$$\Rightarrow \langle \sigma v \rangle \sim 1.3 \times 10^{-9} \, \text{GeV}^{-2}$$

$$\sim 5 \times 10^{-37} \, \text{cm}^2 \qquad (9.50)$$

$$\sim 1.5 \times 10^{-26} \, \text{cm}^3 \text{s}^{-1}.$$

This is almost exactly what we needed for the relic density. This coincidence is known as the *WIMP miracle*.

We might ask if this works without some exotic Z' but just with the simple Z-boson of the SM. Unfortunately, as we will see in chapter 12, the Z, with couplings exactly as in the SM, is excluded by experiment as the particle by which the WIMP annihilates. We need the freedom to change the couplings while keeping the total annihilation cross section the same. What is this new particle χ? There are many proposed candidates, but we will spend most of our time with the most famous model.

9.2.2 Supersymmetry

To motivate supersymmetry, we will take a little detour. In QFT, what we think of as constants of nature actually depend on the energy we observe them at due to a phenomenon called *renormalisation*. We can understand it by asking what we really mean by 'charge' and 'mass'.

What is charge? We can think of charge as defined by the vertex shown in Fig. 9.8, that is, the thing that appears in Feynman diagrams and gives us probabilities for an interaction. However, in QFT, we have 'loop corrections': these are additional processes that appear at higher order in perturbation theory and involve virtual particles. The overall probability depends on the energy scales, Q, of the particles in the

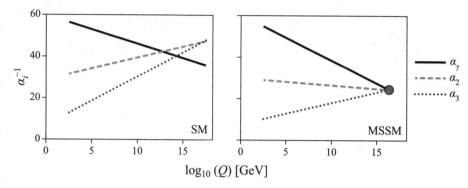

Figure 9.9 The running couplings of the SM gauge group. *Left*: the particle content of the SM. *Right*: the particle content of the MSSM. We notice gauge coupling unification occurs at $\sim 10^{16}$ GeV in the MSSM. In this figure, the SUSY scale is fixed at 1 TeV, where the plots begin. If the scale were moved higher, then we would have SM running below the SUSY scale and MSSM running above it.

loops, which we can think of as the scale at which we are making a measurement (we already saw how the propagator momentum is $\sim E_{cm}$). We can think of this dependence on Q as changing the effective charge, that is, $g = g(Q)$, where Q is the *loop momentum* and thus the 'charge' changes with energy scale. You can picture this by imagining a 'cloud' of virtual quantum particles surrounding our real particle. Another way of thinking of this is that we probe the particle at some energy and try to measure its charge. Due to quantum uncertainty, there is a cloud of virtual particles surrounding the 'bare' charge that we need to get through. Thus we can think of the vacuum itself as a polarised medium that screens (or anti-screens in the case of non-Abelian symmetries) the charge of a particle.

The name for this phenomenon in QFT is renormalisation. The charges are 'renormalised', and we say that the couplings 'run', that is, they change depending on the energy at which we measure them. The running of the Standard Model couplings is depicted in Fig. 9.9, left panel. We notice that the couplings approach a common value at high energies, but they don't exactly meet.

The precise running of the coupling, that is, how fast it changes with energy, depends on the types of particles in the loops, as sketched in Fig. 9.10. It turns out that adding *new scalar particles* at an energy scale $Q \sim$ TeV results in the *unification of couplings* at $Q \sim 10^{16}$ GeV. The energy scale 10^{16} GeV is known as the 'Grand Unified Theory', or GUT, scale.

SUSY predicts that the fermions of the Standard Model have scalar 'superpartners'. If SUSY was an exact symmetry, then the scalar and fermion partners would weigh the same. However, we haven't yet seen these superpartners in experiments, so we know they must be heavier than any energy we have made measurements at. The mass splitting is given by the energy scale, m_{SUSY}, below which SUSY is broken. If the SUSY scale is $m_{SUSY} \approx 1$ TeV, then the couplings of the SM unify at the GUT scale. This unification is aesthetically pleasing; however it is not necessary for a working model of SUSY, and it may turn out that m_{SUSY} is much higher than 1 TeV.

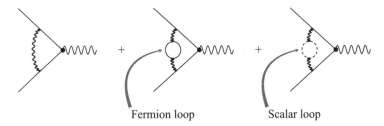

Figure 9.10 The running couplings are dependent on the particle content of the considered model.

What is 'mass'? As we can think of charge as defined by a vertex in a Feynman diagram that gives a probability for a process, so we can also think of the mass m as defined by the Feynman diagram propagator which gives the energy dependence of a process. The propagator is shown for fermions and scalars in Fig. 9.11. The mass is the location of the pole (divergence) in the propagator, which occurs when $p^2 = m^2$: for a virtual particle this is when it goes 'on-shell' and becomes a real particle rather than just a fluctuation allowed by the uncertainty principle.

When we add loops into the propagator, the mass thus defined is also a running quantity with Q. For instance we have measured a scalar mass, the Higgs boson, at $m_H = 125\,\text{GeV}$. Renormalisation then takes the measured value and includes loop corrections,

$$m_H^2 = \overline{m}_H^2 + \delta m_H^2, \tag{9.51}$$

where, very schematically:

$$\delta m_H^2 \sim \int_0^\Lambda \left(\text{loops}\right) dQ. \tag{9.52}$$

In the SM, δm_H^2 is finite with $\Lambda \to \infty$. We say the model is *UV complete*. However we know there are more particles than just the SM, for example, DM. New particles, therefore, will appear at some scale Λ, and so

$$\boxed{\delta m_H^2 \sim \left(\frac{\alpha}{2\pi}\right)\Lambda^2,} \tag{9.53}$$

where we got the α from a one-loop process (two couplings), the 2π is conventional, and Λ^2 comes from dimensionality.

If we have, for example, *quantum gravity* we expect spacetime is quantised at the Planck scale. In this case $\Lambda \sim M_{Pl} \sim 10^{18}\,\text{GeV}$ and so

$$\delta m_H^2 \gg \overline{m}_H^2. \tag{9.54}$$

This is known as the *hierarchy problem*, the question of why $\overline{m}_H^2 \ll \delta m_H^2$: why is the measured Higgs mass much smaller than the scale of quantum corrections to the Higgs mass?

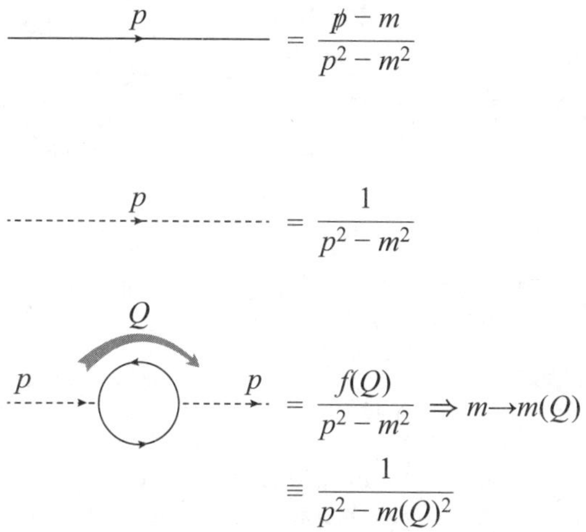

$$\frac{p}{} = \frac{\not{p} - m}{p^2 - m^2}$$

$$\frac{p}{} = \frac{1}{p^2 - m^2}$$

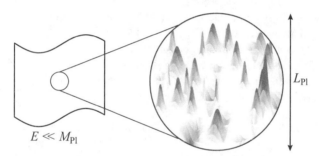

$$= \frac{f(Q)}{p^2 - m^2} \Rightarrow m \to m(Q)$$

$$\equiv \frac{1}{p^2 - m(Q)^2}$$

Figure 9.11 Fermion (top) and scalar (middle) propagators. The mass is defined as the location of the pole in the propagator. If we include virtual particles in a loop with energy Q (bottom), then the propagator is corrected by some function of $Q, f(Q)$, which may shift the location of the pole and can be absorbed in a 'renormalisation' of the particle mass.

Figure 9.12 At large scales, spacetime is homogenous; however, at the Planck scale, L_{Pl}, quantum effects like potential barriers and wells become visible. This is often called the *spacetime foam*. See colour insert.

By introducing *bosons* with equivalent SM charges as the fermions, the problem is altered, as shown in Fig. 9.13. These corrections can be written mathematically as

$$\Rightarrow \delta m_H^2 \sim \left(\frac{\alpha}{2\pi}\right)\left(\Lambda^2 + m_{boson}^2\right) - \left(\frac{\alpha}{2\pi}\right)\left(\Lambda^2 + m_{fermion}^2\right) \qquad (9.55)$$

$$= \left(\frac{\alpha}{2\pi}\right)\left(m_{boson}^2 - m_{fermion}^2\right). \qquad (9.56)$$

In the SM, there is no 'scalar electron' (selectron) with $m_s = m_e$. But if we take

$$m_{boson}^2 = m_{fermion}^2 + m_{SUSY}^2 \qquad (9.57)$$

then we have $\delta m_H^2 \approx m_{SUSY}^2$. Now if we set the SUSY scale $m_{SUSY}^2 = \overline{m}_H^2$, then there is no 'fine-tuning' required to keep the Higgs mass small. We have to do the same

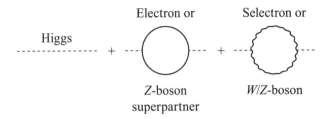

Figure 9.13 The Higgs propagator, with boson and fermion loop contributions. Bosons and fermions have different contributions to the running Higgs mass.

with fermion partners of the W-boson, Z-boson and so on. Thus, we need new *fermions related to the Z-boson with* $m_{\text{fermion}} \sim m_{\text{H}} \sim 100$ GeV.

Introducing bosonic partners for the SM fermions and fermionic partners for the SM gauge bosons solves the Higgs 'hierarchy problem'. The masses need to be

$$m_{\text{new}} \sim m_{\text{SM}} + m_{\text{H}} \sim 100 \text{ GeV}. \tag{9.58}$$

The fermionic partners of the W/Z-bosons are called the 'wino' and 'zino' respectively, those of the hypercharge gauge field and photon are called the 'bino' and 'photino', and the partner of the Higgs is the Higgsino.

There is one neutral, linear combination called the *neutralino*. The neutralino has $m_\chi \sim 100$ GeV and couples to the Z with strength $\mathcal{O}(G_F)$. Therefore: *supersymmetry can solve the hierarchy problem and give a 'natural' candidate for WIMP DM.* In addition, SUSY also helps unify the gauge couplings.

Brief details on SUSY: these details are here to demystify just a bit what we said qualitatively earlier. SUSY extends the Poincaré group with new generators $\hat{\mathcal{Q}}$, that is,

$$\hat{\mathcal{Q}}|\text{fermion}\rangle = |\text{boson}\rangle$$
$$\hat{\mathcal{Q}}|\text{boson}\rangle = |\text{fermion}\rangle, \tag{9.59}$$

which implies that $\hat{\mathcal{Q}}$ is spin-$\frac{1}{2}$. As a generator, $\hat{\mathcal{Q}}$ has spin index a and is conventionally a 'Majorana' fermion, that is,

$$\left\{ \mathcal{Q}_a, \underbrace{\overline{\mathcal{Q}}_b}_{(\mathcal{Q}^\dagger \gamma_0)} \right\} = \overbrace{\gamma^\mu_{ab}}^{\text{Dirac matrix}} \underbrace{P_\mu}_{\text{boost generator}}, \tag{9.60}$$

$$\left\{ \mathcal{Q}_a, P_\mu \right\} = 0, \tag{9.61}$$

$$\left\{ \mathcal{Q}_a, \underbrace{M^{\mu\nu}}_{\text{element of Lorentz group}} \right\} = \sigma^{\mu\nu}_{ab} \mathcal{Q}_b, \tag{9.62}$$

$$\underbrace{\sigma^{\mu\nu}_{ab} = \frac{i}{4}[\gamma^{\mu}, \gamma^{\nu}]_{ab}.}_{\text{tensor spin operator}} \qquad (9.63)$$

Lorentz transformations act on spacetime coordinates as $x^{\nu} = \Lambda^{\nu}{}_{\mu}x^{\mu}$. SUSY acts on a 'fermionic space' coordinate, θ, which is a Weyl two-component spinor. Thus, 'superspace' has coordinates $\{\mathbf{x}, \theta, \bar{\theta}\}$. The coordinates θ anticommute, that is,

$$\{\theta, \bar{\theta}\} = 0. \qquad (9.64)$$

A superfield Φ has coordinates in superspace and has bosonic and fermionic particles in it, for example, electrons and selectrons–the 'chiral superfield'–gluon and gluino–the 'vector superfield'.

The *minimal supersymmetric SM* (MSSM) gives one superpartner to each SM field. It is required to have two Higgs fields (an up-type and a down-type) such that electroweak symmetry breaking preserves supersymmetry.

The stability of the proton requires something called a *discrete R-symmetry*:

$$R = (-1)^{3B+L+2s}, \qquad (9.65)$$

where B is the baryon number, L is the lepton number, and s is the spin. All SM particles have $R = 1$. Their superpartners (sparticles) have $R = -1$. We make the assertion that R-symmetry is conserved, which means that no process exists that changes the value of R. As a consequence, the lightest supersymmetric particle (LSP) is stable and turns out to play a key role as the dark matter candidate for the MSSM. If it is to be the DM, then the LSP should be neutral and thus, in the MSSM, we have two candidates: the *sneutrino* and the *neutralino*.[4] The neutralino is a linear combination of the particles we mentioned above, the bino, the wino or one of the two Higgsinos.

The fields are $\tilde{B}, \tilde{W}_3, \tilde{H}_1^0, \tilde{H}_2^0$ with mass matrix

$$M_N = \begin{pmatrix} m_1 & 0 & -m_Z c_\beta s_W & m_Z s_\beta s_W \\ 0 & m_2 & m_Z c_\beta c_W & -m_Z s_\beta c_W \\ -m_z c_\beta s_W & m_Z c_\beta c_W & 0 & -\mu \\ m_Z s_\beta s_W & -m_Z s_\beta c_W & -\mu & 0 \end{pmatrix}, \qquad (9.66)$$

where $c_\beta = \cos\beta, s_\beta = \sin\beta$, and β is defined by the vevs of H_1^0, H_2^0 as $\tan\beta = \frac{v_1}{v_2}$, $c_W = \cos\theta_W, s_W = \sin\theta_W$ with θ_W the Weinberg angle, m_1, m_2 are the bino/wino masses respectively, and μ is a parameter in the Higgs potential.

The mass eigenstates are eigenvalues/vectors of M_N, and so the *composition* and *interactions* of the LSP are determined by

$$\chi = N_{11}\tilde{B} + N_{12}\tilde{W}_3 + N_{13}\tilde{h}_1^0 + N_{14}\tilde{h}_2^0, \qquad (9.67)$$

where N is the matrix diagonalising M_N and χ is the LSP.

For this model, all cross sections have been computed and the relic density has been found by scanning over parameters. See, for example, Refs. [51, 52, 53, 54, 55].

---

[4] The sneutrino is disfavoured by experiments, see Ref. [50].

The MSSM has $\mathcal{O}(100)$ free parameters. This can be reduced by applying certain theoretical constraints, for example, by demanding unification in *minimal supergravity* (mSUGRA).

9.2.3 Kaluza–Klein WIMPs

If there are *extra dimensions* (for instance, as suggested by string theory) of space then particles obtain a spectrum of *Kaluza–Klein* (KK) *partners* with masses,

$$m_{KK}^n = m_0 + c\left(\frac{n}{R}\right), \tag{9.68}$$

where c is a constant, $n \in \mathbb{Z}^+$, and R is the size of the extra dimension. Why does this occur?

To illustrate this, we take the wave equation of a scalar field ϕ, that is, the Klein–Gordon equation,

$$\Box\phi = \frac{\partial V}{\partial \phi}, \tag{9.69}$$

where $\Box$ is the D'Alembert operator, $\Box = -\partial_t^2 + \nabla^2$. For a massive scalar,

$$\frac{\partial V}{\partial \phi} = m^2\phi + \text{interactions}$$

$$\Rightarrow E^2\phi = (p^2 + m^2)\phi \tag{9.70}$$

$$\because \partial_t = i\hat{E}, \nabla = i\hat{p}.$$

Now, however, $\nabla = \nabla_x + \nabla_y$, where y is the extra dimensional coordinate, for example, if we take the additional dimension to be a line, the coordinate y simply represents the distance along the total length R. We can write any solution for ϕ using the eigenfunctions of the line, which are just cosine modes labelled by n:

$$\phi^{(n)}(x^\mu, y) = \phi(x^\mu)\cos\left(\frac{n\pi y}{R}\right) \tag{9.71}$$

$$\Rightarrow E_n^2 = p_x^2 + \left[m^2 + \frac{n^2\pi^2}{R^2}\right], \tag{9.72}$$

which looks like an effective higher mass. Each 'mode', n, of the field has an increasing 'mass', and we say that ϕ now has a 'tower of states'. The example we just gave was for a single scalar field, but the analysis generalises to all of the particles in the SM.[5]

Since the $n > 0$ modes in the tower have mass larger than the $n = 0$ mode but are necessarily coupled to the lightest mode, then the higher KK modes are all unstable

---

[5] This is called *universal extra dimensions* with all particles 'living' in the extra dimensions. Alternative models, called 'brane worlds', restrict some types of particles to live in particular places in the extra dimensions. Such 'branes' occur in string theory.

and can decay into the lightest mode. Making chiral fermions–that is, left-/right-handed as in the SM–requires a new parity in the additional dimension. This is known as 'orbifolding' the dimension, for example, for a circle $S^1 \rightarrow S^1/\mathbb{Z}_2$ where $\mathbb{Z}_2$ is just the reflection group. Like R-parity in SUSY, this $\mathbb{Z}_2$ symmetry stabilises the lightest KK mode partner. The lightest KK partner (LKP) is $B^{(1)}$, the partner of the B_μ weak hypercharge vector boson. As this comes directly from the SM, we can compute all of the interaction cross sections,

$$\langle \sigma v_{B^{(1)}} \rangle = 1.8 \times 10^{-26} \mathrm{cm}^3 \mathrm{s}^{-1} \left(\frac{100 \text{ GeV}}{m_{\mathrm{KK}}} \right)^2, \tag{9.73}$$

and, therefore, setting the correct relic density gives[6]

$$m_{\mathrm{KK}} \approx 400 \text{ GeV} \tag{9.74}$$

$$\Rightarrow R \approx 5 \times 10^{-19} \text{ m}. \tag{9.75}$$

---

[6] This result and the preceding discussion are from Ref. [50].

QUIZ

i. The coupling strengths of the Standard Model:
  a. Are constants of nature.
  b. Vary depending on the energy scale of the measurement.
  c. All go to zero at high energy.
  d. Are observed to unify at high energy.

ii. Supersymmetry solves the 'hierarchy problem' of the Higgs mass when:
  a. Boson and fermion superpartners have the same masses.
  b. Boson and fermion superpartners are split by approximately the Higgs mass.
  c. Quantum gravity becomes important.
  d. Boson and fermion superpartners are split by the Planck scale.

iii. What are the four components of the lightest neutralino?
  a. (Higgsino, wino, sneutrino, gluino).
  b. (Higgsino 1, Higgsino 2, gravitino, gluino).
  c. (Higgsino, wino, bino, sneutrino).
  d. (Higgsino 1, Higgsino 2, bino, wino).

iv. What is the spin of the SUSY generator, Q?
  a. 0.
  b. 1/2.
  c. 1.
  d. 3/2.
  e. 2.

v. Kaluza–Klein partners have masses of order what?
  a. The inverse length scale of the extra dimensions.
  b. The length scale of the extra dimensions.
  c. The Planck scale.
  d. The supersymmetry scale.

Answers

i. b; ii. b; iii. d; iv. b; v. a;

Chapter Ten

Axions: The Prototype of Wavelike DM

10.1 VACUUM REALIGNMENT PRODUCTION

We now discuss a different type of DM production: *from a scalar field condensate.* A condensate is a state of *high occupation number* that obeys the classical equations of motion, for example, analogous to the EM field and Maxwell's equations.

10.1.1 Scalar Field Lagrangian

We begin with a single scalar field with action,[1]

$$S = \int d^4x \sqrt{-g} \underbrace{\left[-\frac{1}{2}(\partial_\mu \phi \partial^\mu \phi) - \frac{1}{2}m^2\phi^2 \right]}_{\mathcal{L}}, \tag{10.1}$$

where $g = \det[g_{\mu\nu}]$ is required in general relativity such that the integral measure $d^4x\sqrt{-g}$ is invariant under coordinate transformations. This is the action for a scalar field with harmonic potential energy $V(\phi) = m^2\phi^2/2$. The potential is minimised at the vacuum where $\phi = 0$.

We note that ϕ is a *real* scalar field, that is, there can be no local or global U(1) symmetry: there is no ϕ^* in the Lagrangian, and real fields alone thus cannot carry charge. We also take ϕ to have very weak interactions with the SM, and so it is *not* in thermal equilibrium (gravity is too weak to establish thermal equilibrium). We then proceed using the Euler–Lagrange equations,

$$\partial_\mu \frac{\partial \sqrt{-g}\mathcal{L}}{\partial(\partial_\mu \phi)} = \frac{\partial \sqrt{-g}\mathcal{L}}{\partial \phi}$$

$$\Rightarrow \partial_\mu\left(\sqrt{-g}\partial^\mu \phi\right) = \sqrt{-g}m^2\phi. \tag{10.2}$$

We define the D'Alembertian operator, $\Box$, on a curved background as

$$\Box = \frac{1}{\sqrt{-g}}\partial_\mu\left(\sqrt{-g}g^{\mu\nu}\partial_\nu\right), \tag{10.3}$$

---

[1] Note, we switch to the 'cosmology convention' for the metric with signature $(-,+,+,+)$; e.g., the Minkowski metric would be $\eta_{\mu\nu} = \text{diag}(-1,1,1,1)$.

where we have used the convention that indices are raised with the inverse metric $g^{\mu\nu}$, that is, $\partial^\mu \phi = g^{\mu\nu} \partial_\nu \phi$. This gives the Klein–Gordon equation,

$$\boxed{\Box \phi - m^2 \phi = 0,} \tag{10.4}$$

which is the *classical* wave equation for a scalar field on any background metric. You should think of this classical equation as akin to Maxwell's equations: we can find classical solutions for what the axion field is doing, which apply on large scales in some average way.

Above we have used a generic background metric $g_{\mu\nu}$; however we can also specify, for example, the FRW metric

$$g = \begin{pmatrix} -1 & & & \\ & a^2 & & \\ & & a^2 & \\ & & & a^2 \end{pmatrix}. \tag{10.5}$$

For the *relic density* we are interested in the homogeneous and isotropic spatial average and so consider $\phi = \phi(t)$ only, that is, no spatial perturbations. Taking the initial conditions,

$$\begin{aligned} \phi(t_i) &\equiv \phi_i \\ \dot{\phi}(t_i) &\to 0 \end{aligned} \tag{10.6}$$

and

$$\Box = -\partial_t^2 - 3H\partial_t, \tag{10.7}$$

we derive

$$\boxed{\ddot{\phi} + 3H\dot{\phi} + m^2 \phi = 0.} \tag{10.8}$$

You should be able to derive this using the metric Eq. (10.5) and the definition of $\Box$, Eq. (10.3). This is just the equation for a *damped harmonic oscillator* where $H = \dot{a}/a$ gives the damping and is referred to as the *Hubble friction*.

Recalling that H depends on time and decreases as the Universe expands, we see that the motion of ϕ will be

$H \gg m$ field overdamped: 'frozen'.
$H \ll m$ damped oscillations.

We have seen this type of behaviour in chapter 5, where the expansion of the Universe, H, acts like a friction term in cosmological perturbation theory. For simplicity let's start at late times when $g_\star = $ constant and use t as the time coordinate (rather than temperature as we did earlier),

$$3H^2 M_{\text{Pl}}^2 = \frac{\rho_r}{a^4} + \frac{\rho_m}{a^3} + \rho_\Lambda. \tag{10.9}$$

Taking t early enough such that $a \ll a_{eq}$ and only radiation where $\rho \propto a^{-4}$, the Friedmann equation is solved by

$$\Rightarrow a = \left(\frac{t}{t_i}\right)^{\frac{1}{2}}$$
$$H = \frac{1}{2t}.$$

(10.10)

Taking $t_i \ll \frac{1}{m}$ then our initial condition $\dot{\phi}(t_i) \to 0$ is consistent since $H(t_i) \gg m$. Thus, Eq. (10.8) becomes

$$\boxed{\ddot{\phi} + \frac{3}{2t}\dot{\phi} + m^2\phi = 0.}$$

(10.11)

A change of variables brings this into the form of *Bessel's equation*, which has the general solution

$$\phi = a^{-\frac{3}{2}} \left(\frac{t}{t_i}\right)^{\frac{1}{2}} \phi_i \left[c_1 J_n(mt) + c_2 Y_n(mt) \right],$$

(10.12)

with $n = \frac{1}{4}$. Imposing the boundary conditions as $t \to 0$ gives the simple solution

$$c_2 = 0$$
$$\phi = \phi_i \left(\frac{2}{mt}\right)^{\frac{1}{4}} \Gamma\left(\frac{5}{4}\right) J_{\frac{1}{4}}(mt),$$

(10.13)

where Γ is the Euler gamma function.

To find the relic density we also need an asymptotic expression for $\phi(t)$ at late time. For this, we can use the Wentzel–Kramers–Brillouin (WKB) approximation. We write the solution as $\phi = A(t) \cos \omega t$. WKB assumes that the time variation of A is much shorter than the time scale ω, that is, $\dot{A} \ll \omega A$. WKB works order by order, first assuming $\dot{A} = 0$ (zeroth order) to find ω, then solving for $A(t)$ at first order taking ω as fixed.

Taking $mt \gg 1$, the equations of motion at zeroth order become

$$\ddot{\phi} + m^2\phi = 0$$

(10.14)

$$\Rightarrow \omega = m.$$

(10.15)

Substituting $\phi = A(t) \cos(mt)$, we can solve for the envelope function $A(t)$. We use our Bessel function solution to match the WKB envelope when $mt \sim 1$. Defining

$$3H(t_{osc}) = m$$
$$\phi(t_{osc}) \approx \phi_i,$$

(10.16)

we find

$$\boxed{\phi \simeq \phi_i \left(\frac{a}{a_{osc}}\right)^{-\frac{3}{2}} \cos(mt).}$$

(10.17)

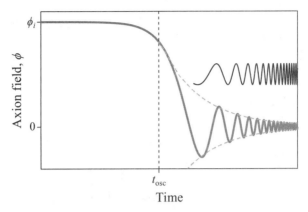

Figure 10.1 Evolution of the scalar field ϕ according to Eq. (10.11). The purple line shows the exact solution, Eq. (10.13), the black line shows a cosine with frequency m, and the dashed grey line shows the envelope $a^{-3/2}$. Oscillations begin at t_{osc}, taken here as $tm = 1$ and thus $H \sim m$. Note that the x axis is on a log scale.

The solutions are plotted in Fig. 10.1. We see that ϕ oscillates with frequency m and decays like $a^{-3/2}$. Oscillations begin at t_{osc} when $tm = 1$, and the subsequent evolution is well fit by the WKB approximation.

The limit $mt \gg 1$ leaves us independent of the specific time dependence of $a(t)$, and so we can extrapolate all the way to arbitrarily late times. The final ingredient required for the relic density is the expression for the density itself. The field is a harmonic oscillator, so we use

$$\rho_\phi = \frac{1}{2}\dot{\phi}^2 + \frac{1}{2}m^2\phi^2. \tag{10.18}$$

Note that this expression for ρ can be derived in general relativity from the general expression for the energy-momentum tensor, which we will turn to briefly later. Substituting in the WKB approximation of Eq. (10.17) leads to

$$\Rightarrow \rho_\phi \approx \frac{1}{2}m^2\phi_i^2\left(\frac{a}{a_{osc}}\right)^{-3}, \tag{10.19}$$

where we used the $\approx$ sign because we have dropped an extra term that oscillates and decays more rapidly. Given that a_{osc} is fixed by m, there are only two free parameters for the relic density: the mass of the field, m, and the initial field value, ϕ_i.

The solution for the energy density using our Bessel function of Eq. (10.13) is shown in Fig. 10.2. We notice that it is constant at early times when the field is frozen and scales like a^{-3} at late times. If we consider that radiation energy density goes like a^{-4}, then the scalar field energy density will eventually equal the radiation at matter-radiation equality, a_{eq}, just like we need DM to do. This method of production of DM from a scalar field with some initial value ϕ_i that relaxes to zero over time is known as *vacuum realignment*.

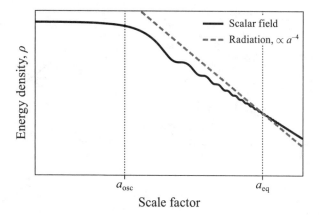

Figure 10.2 Evolution of the scalar field ϕ energy density, according to Eq. (10.11). The purple line shows the exact solution, Eq. (10.13), substituted into the expression Eq. (10.18). Oscillations begin at a_{osc}, after which $\rho \propto a^{-3}$. Radiation, which scales like a^{-4}, is sketched in orange, and the approximate epoch of equality is indicated. Note that the x and y axes are on a log scale.

We could now write

$$\rho_\phi = mn_\phi \tag{10.20}$$

with

$$n_\phi = \frac{1}{2}m\phi_i^2\left(\frac{a}{a_{osc}}\right)^{-3}, \tag{10.21}$$

and notice that this represents a *conserved particle number density* (i.e., it goes down only as the spatial volume increases while the Universe expands), as shown in Fig. 10.3, along with other quantities relating to vacuum realignment. This is somewhat surprising: *an oscillating scalar field behaves like decoupled non-relativistic matter!* However, we never mentioned particles: this is just an oscillating uniform *classical* field. In classical perturbation theory and QFT, 'particles', that is, occupation number of $\phi_k \neq 0$, are produced when ϕ starts oscillating.

10.1.2 Axion Relic Density

From here on, let's call our field ϕ an *axion* (or axion-like particle, ALP). We will talk more about axion theory in the next subsection. Using Eq. (10.19), the relic abundance is given by

$$\Omega_a h^2 = \frac{\rho_\phi}{3M_{Pl}^2 M_H^2} = \frac{1}{6}\left(\frac{\phi_i}{M_{Pl}}\right)^2\left(\frac{m}{M_h}\right)^2\left(\frac{a_0}{a_{osc}}\right)^{-3}. \tag{10.22}$$

Remember

$$\left(\frac{a_0}{a_{osc}}\right)^{-3} = \left(\frac{g_{\star,S}(T_0)T_0}{g_{\star,S}(T_{osc})T_{osc}}\right)^3 \tag{10.23}$$

by the conservation of entropy.

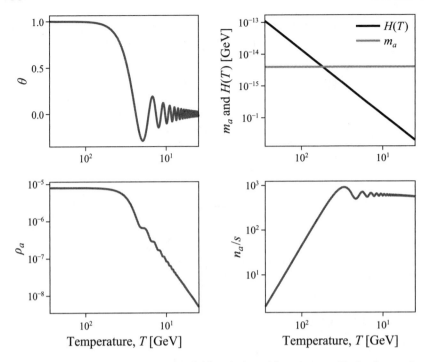

Figure 10.3 Quantities relevant to axion field evolution with constant m. Notice the number density, which remains constant after field oscillations begin.

So we need to find $T_{\rm osc}$: this is the axion equivalent of finding the freeze-out temperature. It is conventionally defined as

$$3H(T_{\rm osc}) = m, \tag{10.24}$$

which comes from equating the damping and mass terms in the Klein–Gordon equation, Eq. (10.8), taking $d/dt \approx m$. This provides an estimate for the time when the mass term dominates over damping and when oscillations of the axion field begin. Thus the 3 in Eq. (10.24) is not exact and should really be thought of as a fitting parameter. We can solve this equation for $T_{\rm osc}$ by using the Friedmann equation for H:

$$3H^2 M_{\rm Pl}^2 = \frac{\pi^2}{30} g_\star T^4. \tag{10.25}$$

As an example, let's consider a mass $m = 10^{-15}$ eV (axions are extremely light!) and assume $g_\star \approx 3.4 = g_\star(T_0)$. It follows that

$$T_{\rm osc} = \left(\frac{10}{g_\star}\right)^{\frac{1}{4}} \frac{(mM_{\rm Pl})^{\frac{1}{2}}}{\sqrt{\pi}} \tag{10.26}$$

and so

$$\Rightarrow T_{\rm osc} = 1 \text{ MeV} \left(\frac{m}{10^{-15} \text{ eV}}\right)^{\frac{1}{2}}. \tag{10.27}$$

The redshift factor is

$$\left(\frac{a_0}{a_{\text{osc}}}\right)^{-3} = \left(\frac{T_0}{T_{\text{osc}}}\right)^3. \tag{10.28}$$

Substituting into the expression above we find that

$$\Omega_a h^2 = 0.12 \left(\frac{\phi_i}{1.5 \times 10^{15} \text{ GeV}}\right)^2 \left(\frac{m}{10^{-15} \text{ eV}}\right)^{\frac{1}{2}}, \tag{10.29}$$

which applies for masses below about 10^{-15} eV such that $g_\star \approx 3.4$ is a constant.[2] We notice that the right relic density is obtained for very small masses and very large initial field values.

There doesn't seem to be an obvious 'axion miracle' here that fixes the vastly different values of m and ϕ_i, which appear arbitrary. However, looking ahead slightly, we find that in string theory there are axions with masses:

$$m \approx \frac{\Lambda M_{\text{Pl}}}{\phi_i} \underbrace{e^{-c/\alpha_{\text{GUT}}}}_{\text{exponentially suppressed masses}} \tag{10.30}$$

(the exponential factor is typical and is a consequence of symmetries in the extra dimensions). How are the quantities in Eq. (10.30) related to more fundamental ones? We came across the idea of a SUSY-GUT earlier:

SUSY scale: $E_{\text{SUSY}} \sim 10$ TeV.
Unification scale: $E_{\text{GUT}} \sim 10^{15}$ GeV.
Coupling at unification: $\alpha_{\text{GUT}} \sim \frac{1}{25}$.

Using these quantities in Eq. (10.30),

$$\begin{aligned} \Lambda &= E_{\text{SUSY}} \\ \phi_i &= E_{\text{GUT}}, \end{aligned} \tag{10.31}$$

then with $c = 3$ (which is somewhat 'natural', that is, order unity) we get the reference values in Eq. (10.29) and thus the correct relic density. Clearly, however, many other choices can also give the right value because we have two unknowns and only one quantity to fix them.

10.1.3 Initial Conditions: Spontaneous Symmetry Breaking

There are still several unresolved questions we can now ask:

- What is the purpose of this particle in the SM?
- What sets up the ϕ condensate with initial value ϕ_i?
- What sets the particle mass, m?

---

[2] It also applies assuming that the mass is constant with respect to temperature. This may seem a natural thing to assume, but it does not hold for the canonical 'QCD axion' explored shortly.

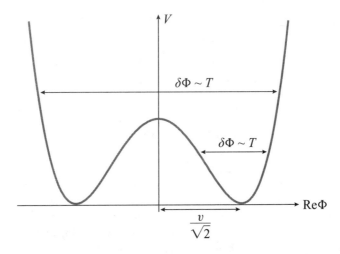

Figure 10.4 Thermal fluctuations of the Peccei–Quinn field, $\delta\Phi \sim T$. If fluctuations are smaller than the vev, then the field can 'see' the minimum. If fluctuations are large, the field can't see the minimum and will be minimized at zero instead, with an effective thermal mass $m_{\text{therm}} \propto T^2$.

To begin to answer these, we build on the toy model outlined above.

Consider a complex scalar field with a *global* U(1) *symmetry*,

$$\mathcal{L} = -\partial_\mu \Phi^* \partial^\mu \Phi - \lambda \left(|\Phi|^2 - \frac{v^2}{2} \right)^2. \tag{10.32}$$

The symmetry is spontaneously broken in the minimum of the potential, where $|\Phi| = v/\sqrt{2}$. A cross section through the potential along the real Φ axis is shown in Fig. 10.4. If this field is coupled to a thermal bath, then it fluctuates with

$$\Phi \rightarrow \Phi + \delta\Phi \tag{10.33}$$

$$\text{with} \qquad \delta\Phi \sim T.$$

As we see in Fig. 10.4, if $T \gg \frac{v}{\sqrt{2}}$ the field does not 'see' the minimum at $|\Phi| = \frac{v}{\sqrt{2}}$. We can write the thermal term as an effective potential, that is,

$$V(\Phi) = \lambda \left(|\Phi|^2 - \frac{v^2}{2} \right)^2 + \frac{\lambda}{6} |\Phi|^2 T^2. \tag{10.34}$$

There are now two regimes for our potential:

- $T \gg v$: here the potential is effectively quadratic

$$\Rightarrow \langle |\Phi| \rangle = 0, \qquad \text{that is, no SSB.} \tag{10.35}$$

- $T \ll v$: here

$$\Rightarrow \langle |\Phi| \rangle = \frac{v}{\sqrt{2}} e^{i\theta}, \qquad \text{that is, SSB.} \qquad (10.36)$$

Thus, in the early Universe at high T, the symmetry is unbroken, but when the temperature drops below approximately v, the symmetry is broken. After SSB, we can write the field as $\Phi = v e^{i\theta}/\sqrt{2}$ and the Lagrangian is

$$\mathcal{L} = -\frac{1}{2} v^2 \partial_\mu \theta \partial^\mu \theta. \qquad (10.37)$$

θ, the angular d.o.f. of Φ, is a *real scalar Goldstone boson*. θ appears in an exponential, so it has no units: it is just a number, a canonical scalar field with mass dimension 0. We can define a 'canonically normalised field' $\phi = v\theta$ with mass dimension 1, and it follows that

$$\mathcal{L} = -\frac{1}{2} \partial_\mu \phi \partial^\mu \phi. \qquad (10.38)$$

Since

$$\theta \in [-\pi, \pi] \qquad (10.39)$$

$$\Rightarrow \phi \in [-\pi v, \pi v]. \qquad (10.40)$$

The canonically normalised angular degree of freedom ϕ is called the axion: *the Goldstone boson of the spontaneously broken global* U(1) *symmetry*.

When SSB occurs, θ takes a random value. We can think of spontaneous symmetry breaking as like a pen falling off its point: while the pen is balanced, the situation is rotationally symmetric, but when it falls, it has to fall in one direction in particular, at random, which spontaneously breaks the rotational symmetry. For the axion θ we model this as a uniform random variable:

$$P(\theta) = U[-\pi, \pi]. \qquad (10.41)$$

So far, ϕ has no mass or potential and $\mathcal{L}$ is invariant under

$$\phi \to \phi + \text{constant}, \qquad (10.42)$$

since it only contains derivatives, which don't care about the constant piece. We say that it has a *shift symmetry*. This shift symmetry in θ is a consequence of the global U(1) symmetry, since a U(1) rotation of Φ that is independent of x corresponds to a shift in θ by a constant. If there is an exact shift symmetry, the actual value of θ is meaningless: what's significant is the *distribution*.

Two possibilities for SSB:

When SSB happens, θ takes a random value in regions which are causally disconnected.

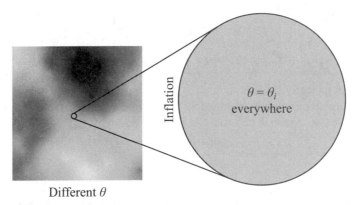

Different θ

Figure 10.5 If SSB occurs before inflation, θ takes a random value in different patches. One of these patches, that is, a fixed value of θ, is then inflated and $\theta = \theta_i$ everywhere. See colour insert.

1. $T < v$: **before or during inflation.**

   Inflation blows up single patches with fixed $\phi_i = v\theta$ to cover the whole observable Universe (see Fig. 10.5). To find the conditions under which this occurs, we need to ask 'what is the temperature, T, during inflation?' An effective temperature arises due to quantum fluctuations of the inflaton field, the so-called Gibbons–Hawking temperature, which is $T_{\mathrm{GH}} = H_{\mathrm{inf}}/2\pi$ where H_{inf} is the (approximately constant) Hubble scale during inflation.

   Inflation also induces tensor-type vacuum fluctuations of the metric (so-called 'primordial gravitational waves'), which in turn induce curl-type polarisation in the CMB (known as 'B-modes'). This type of polarisation is not (yet) observed, which allows us to set a limit on H_{inf}:

   $$H_{\mathrm{inf}} \lesssim 10^{14} \text{ GeV}. \tag{10.43}$$

   So our toy model with

   $$\phi_i = \theta_i v/\sqrt{2} \approx 10^{15} \text{ GeV} \tag{10.44}$$

   requires $v \approx 10^{15}$ GeV, and since we know that $H_{\mathrm{inf}} < v$ then the Gibbons–Hawking temperature is small enough that the symmetry is broken, that is, $\delta\Phi \sim T \ll v$. Inflation then blows up our causal patch of ϕ, making it smooth across the observable Universe (removing field gradients), such that $\phi = \phi(t)$ only, giving us the initial condition that we need. We pay the price, aesthetically, that there is a random number, θ_i, appearing in our prediction of the relic density, which makes this scenario unpredictive, although typically one might assume $\theta_i \sim 1$.

2. $T < v$: **after inflation.**

   If $H_{\mathrm{inf}} > v$ *and* T_{max} after inflation (the maximum thermalisation temperature) is also larger than v then SSB occurs *after* inflation. Now the Universe is a mess of causally disconnected patches of θ (each patch being the size of the horizon at symmetry breaking). This is depicted in Fig. 10.5 as the image on the left with no inflation occurring to smooth out the lumps and bumps. We must average $\Omega_a h^2$ such that

$$\langle \Omega_a h^2 \rangle = \int_{-\pi}^{\pi} d\theta \ \overbrace{U(\theta)}^{} \overbrace{\Omega_a h^2(\theta)}^{\text{single patch abundance}}. \tag{10.45}$$

$$\underbrace{}_{\text{uniform distribution}}$$

The relic density is quadratic in the initial field value. Thus, performing the averaging over the uniform distribution:

$$\phi_i^2 = v^2 \theta_i^2 \quad ; \quad \int_{-\pi}^{\pi} d\theta \ \theta^2 = \frac{\pi^2}{3} \tag{10.46}$$

$$\Rightarrow \langle \phi_i^2 \rangle = v^2 \frac{\pi^2}{3}. \tag{10.47}$$

In this case, we can estimate the relic density by getting rid of the random free parameter θ_i and replacing it with a fundamental parameter, v. This scenario is attractive from a predictivity point of view. In practice, this scenario is much messier due to the highly inhomogeneous field distribution: the relic density requires an intensive numerical solution and is subject to large uncertainties. This scenario involves the production of objects called 'cosmic strings' and 'miniclusters', which we unfortunately have no space to discuss.

To summarise, in both cases (1) and (2), the effective value of ϕ_i is set by the SSB scale, v. In (1) we have a uniform field and the KG equation is easy to solve as before with a single additional free parameter, θ_i. In (2), however, we get a very messy field distribution which needs to be solved for ϕ with *topological defects* and sets of *PDEs*. However, the estimate using Eq. (10.47) is a decent rule of thumb.

10.1.4 The Axion Mass

For a scalar field, mass is defined as the coefficient of the quadratic term of the canonically normalised field $\phi = v\theta$ in a Taylor expansion of the potential around any given point on the potential (normally taken to be the vacuum), divided by two (for a multifield potential, this is the mass matrix, or Hessian, and the masses are given by the eigenvalues). The potential should still be written in the complex plane of Φ. If we give the field θ a mass, then the potential depends explicitly on θ and it clearly violates the global U(1) symmetry and, therefore, the θ shift symmetry. This symmetry breaking should be 'small' and appear at late times: we say that U(1) is an approximate symmetry that is softly broken.

A potential for θ selects a preferred value of θ at the minimum; consider, for example,

$$V(\Phi) = \lambda \left(|\Phi|^2 - \frac{v^2}{2} \right)^2 - \mu^4(T) \frac{\text{Re}[\Phi]}{v/\sqrt{2}}. \tag{10.48}$$

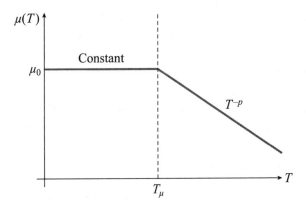

Figure 10.6 Expected temperature dependence of symmetry breaking parameter μ with temperature. The symmetry breaking switches on with some power law from zero at high T and saturates at some fixed value μ_0 when the phase transition is over.

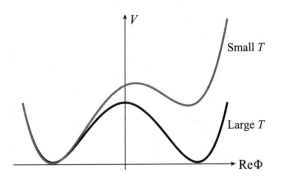

Figure 10.7 'Switching on' the mass term for the axion at some temperature, T, tilts the potential.

Choosing the explicit symmetry breaking potential along the real axis minimizes θ at 0. For now this is clearly arbitrary, and we could always redefine θ such that this holds true (the location of the real axis in the complex plane is just a choice of coordinates in field space). The scale μ, which has dimensions of mass, is the scale of explicit breaking of the U(1). For this to be 'small' we need $\mu \ll v$.

In general, we might expect that the symmetry breaking 'switches on' at some T, as sketched in Fig. 10.6, and we see that this 'tilts' the potential, as shown in Fig. 10.7. This is a phase transition. Phase transitions can break continuous global symmetries as in the case of a liquid freezing to a solid, which breaks continuous translational symmetry down to the discrete symmetry group of the solid crystal.

Fixing the Φ modulus to its vev at $v/\sqrt{2}$, the resulting potential for θ is

$$V(\theta) = -\mu^4(T) \cos \theta \qquad (10.49)$$

$$\Rightarrow V(\phi) = -\mu^4(T) \cos\left(\frac{\phi}{v}\right) \qquad (10.50)$$

and Taylor expanding

$$
V = \frac{1}{2}\mu^4\theta^2 \Rightarrow m = \frac{\mu^2}{v}. \tag{10.51}
$$

As v appears in the denominator, we can get a very small mass if $v \gg \mu$ without making μ itself quite so small; for example, $m = 10^{-15}$ eV with $v \approx 10^{15}$ GeV implies $\mu \approx 10$ keV.

In the next section, we look at understanding:

- What provides the explicit symmetry breaking?
- How does $m(T)$ change the relic density calculation?
- How do the values (m, v) determine interactions with the SM?

QUIZ

i. Vacuum realignment requires thermal equilibrium as an initial condition.
   a. True.
   b. False.

ii. In the Klein–Gordon equation, what role does the Hubble expansion play?
   a. The expansion of the Universe does not affect the Klein–Gordon equation.
   b. The Hubble expansion increases the amplitude of the field.
   c. The Hubble expansion provides a potential for the field.
   d. The Hubble expansion acts as a friction term, slowing down the field.

iii. At early times, what is the approximate solution to the Klein–Gordon equation, when $H \gg m$?
   a. The field remains nearly constant.
   b. The field undergoes coherent oscillations.
   c. Interactions imply thermal equilibrium.
   d. The number density falls exponentially.

iv. Increasing axion mass does what (select all that apply)?
   a. Increases relic abundance.
   b. Decreases relic abundance.
   c. Causes field oscillations to happen later.
   d. Causes field oscillations to happen earlier.
   e. Increases the frequency of field oscillations.
   f. Keeps the axion in thermal equilibrium for longer.

v. Select which statements apply to which type of U(1) symmetry breaking, spontaneous (s), explicit (e), neither (n).
   a. Fixes the radial mode of the field.
   b. Fixes the angular mode.
   c. Occurs when the temperature drops below the vacuum expectation value.
   d. Is forbidden by gauge invariance.
   e. Gives a potential to the angular mode.
   f. Leads to a uniform random distribution for the angular mode.

Answers

i. b; ii. d; iii. a; iv. a,d,e; v. s,e,s,n,e,s

10.2 MODELS OF AXIONS

In the previous section we computed the vacuum realignment relic density of a scalar condensate, dependent on the initial value, ϕ_i, and mass, m, which were required to be large and small respectively, that is,

$$\phi_i \sim 10^x \text{ GeV}, \quad m \sim 10^{-y} \text{ eV}, \quad x, y \gg 1. \tag{10.52}$$

In this model DM is a *coherently oscillating classical field*:

$$\boxed{\phi \sim \phi_0(x, t) \cos(mt).} \tag{10.53}$$

We want to know what sets $m(T)$ and v and how this field couples to the SM.

10.2.1 QCD Axion

This first archetypical model is to axions as SUSY is to WIMPs: a theory from the 1970s/1980s invented to solve another problem with a 'free' DM candidate. As with WIMPs and SUSY, we need to start with symmetry in the SM and QFT.

10.2.1.1 *The QCD Vacuum and CP-Symmetry*

The QCD vacuum corresponds to minimum energy field configurations of the gluons. The gluons are matrix valued fields, $A_\mu = A_\mu^a T^a$, where T^a are the SU(3) generators for whatever representation the quarks are in (in the SM, we only use the fundamental three-dimensional representation, so T^a are the Gell-Mann matrices). We can think of a Lie group like SU(3) as a manifold, and it is possible to map this manifold onto spacetime, that is, for the gluons it is possible to associate every point in space to a point on SU(3). This *topological* mapping leads to non-trivial vacuum configurations of the gluons labelled by an integer, n_w, called the *winding number*, which tells us how many times we orbit the SU(3) manifold making a single rotation in real space. It can be expressed as

$$n_w = \frac{ig^3}{24\pi^2} \int d^3x \text{Tr}\varepsilon_{ijk} A^i A^j A^k, \tag{10.54}$$

where we used the temporal gauge, that is, $A^0 = 0$, and we recall that for a non-Abelian symmetry like SU(3) the field A_μ is matrix valued and thus the trace is taken over this matrix.

Since it is a topological property of the field configuration, any field configuration with a given n_w cannot smoothly change to another. These configurations are said to be *topologically distinct*.

The QCD vacuum we observe is found by a superposition of all possible n_w values given by

$$|\Theta\rangle = \sum_n \exp(-in_w\Theta)|n_w\rangle \tag{10.55}$$

and often known as the 'Θ-vacuum'. Θ is an arbitrary parameter which cannot change and fixes the energy of the vacuum state. The parameter Θ itself is not measurable, but the parameter $\overline{\Theta}$ given by

$$\overline{\Theta} = \Theta + \arg \det \mathcal{M}_Q \tag{10.56}$$

is measurable, where $\mathcal{M}_Q$ is the quark mass matrix.[3] It shows up in the Lagrangian as

$$\mathcal{L}_\Theta = \bar{\Theta} \frac{g_3^2}{32\pi^2} \text{Tr}\left[G_{\mu\nu} \tilde{G}^{\mu\nu} \right], \tag{10.57}$$

where $G^a_{\mu\nu} = \partial_\mu A^a_\nu - \partial_\nu A^a_\mu + i f_{abc} A^b_\mu A^c_\nu$ is the gluon field strength tensor, and $\tilde{G}^{\mu\nu} = \epsilon^{\mu\nu\alpha\beta} G_{\alpha\beta}/2$ is called the 'dual field strength'.

In classical field theory, and in perturbation theory in QFT, the $\bar{\Theta}$ term has no effect because

$$G^a_{\mu\nu} \tilde{G}^{a\,\mu\nu} = \partial_\mu K^\mu, \tag{10.58}$$

$$\text{where} \qquad K^\mu = \epsilon^{\mu\alpha\beta\gamma}\left[A^a_\alpha G^a_{\beta\gamma} - \frac{g_3}{3} f^{abc} A^a_\alpha A^b_\beta A^c_\gamma \right], \tag{10.59}$$

that is, a total derivative, such that

$$\int d^4x \tilde{G}^a_{\mu\nu} G^{a\,\mu\nu} = \int d^4 x \partial_\mu K^\mu \tag{10.60}$$

$$= \underbrace{\int_{S_3} d\sigma_\mu K^\mu}. \tag{10.61}$$

Integration by parts/Stokes' theorem, over boundary

Thus, this has no effect if it vanishes on the boundary. For the configuration with index n_w we simply shift the action by the winding number constant; therefore there is *no effect on the classical equations of motion* ($\delta S = 0$ for a constant shift to the action).

At the quantum level, $\bar{\Theta}$ has an effect due to *CP-violation*. We can understand why by rewriting the term in the Lagrangian as

$$G_{\mu\nu} \tilde{G}^{\mu\nu} \propto \vec{E}_{\text{colour}} \cdot \vec{B}_{\text{colour}}, \tag{10.62}$$

where $\vec{E}_{\text{colour}}$ and $\vec{B}_{\text{colour}}$ are the colour electric and colour magnetic fields defined from $G_{\mu\nu}$ in analogy to electromagnetism. This term violates *CP*, the combination of charge conjugation and parity inversion. We can show easily that this holds for the electromagnetic field strength tensor, $F_{\mu\nu}$ and $\tilde{F}^{\mu\nu}$, and the same is true for the gluons by direct analogy. The charge conjugation and parity operators C and P act

---

[3] CP-violation in the CKM matrix in the SM means quark masses, in general, are complex. $\bar{\Theta}$ depends on the complex phase of the quark mass.

on the electromagnetic field as

$$C\vec{E} = -\vec{E}$$
$$C\vec{B} = -\vec{B}$$
$$P\vec{E} = \vec{E}$$
$$P\vec{B} = -\vec{B} \quad \text{left-handed} \to \text{right-handed},$$

(10.63)

that is, $\vec{E}$ is a pseudovector (charges don't change under parity) and $\vec{B}$ is a vector (currents do change under parity). They both change under C since the sign of the charge, and thus the effective direction of current, does change. We thus find that

$$\Rightarrow CP(\vec{E} \cdot \vec{B}) = -\vec{E} \cdot \vec{B},$$

(10.64)

where the term changes sign under CP, that is, it violates CP-symmetry. This can also be derived directly from the field strength; from $F_{\mu\nu} = \partial_\mu A_\nu - \partial_\nu A_\mu$, we see that

$$CF_{\mu\nu} = -F_{\mu\nu}$$
$$\because A_\mu = (\nabla\phi, \vec{A}) \quad \text{that is, flip signs of potentials under } C$$
$$PF_{\mu\nu} = F_{\mu\nu},$$

(10.65)

but $P\epsilon_{\mu\nu\alpha\beta} = -\epsilon_{\mu\nu\alpha\beta}$.

In quantum theory, if a term in the Lagrangian violates a given symmetry, then it will generate all other possible terms that violate that symmetry for particles interacting with it under quantum corrections. Thus the presence of the CP-violating Θ term, Eq. (10.57), induces CP-violation for all particles interacting under the strong force. Electric dipole moments (EDMs) violate CP (this follows from considering the transformations in Eq. (10.63) and the Wigner-Eckart theorem in QM that states that all vectors in the rest frame point parallel or anti-parallel to the spin), and one consequence of the Θ term is therefore that it results in a *neutron electric dipole moment*,

$$d_n = 3.6 \times 10^{-16} \cdot \underbrace{\Theta}_{} \; \underbrace{e}_{\text{charge of an electron}} \; \overbrace{\text{cm}}^{\text{length of dipole}}.$$

The calculation of this quantity is too complex for us to derive here, but a simple order of magnitude estimate can be arrived at by estimating the EDM as the elementary charge e times the size of the neutron, $r_n \approx 10^{-15}$ m, and visualising Θ as one angle between the three quarks in the neutron arranged in a triangle.

Experiments such as the 'nEDM' experiment, which tries to measure the neutron EDM by Ramsey interferometry of ultracold neutrons [56], observe a neutron EDM consistent with zero, with very small errors:

$$d_n \lesssim 10^{-26} e\,\text{cm} \tag{10.66}$$

$$\Rightarrow \overline{\Theta} \lesssim 10^{-10}. \tag{10.67}$$

A very small $\overline{\Theta}$ implies a fine-tuning between the unrelated QCD Θ-vacuum and the quark mass matrix (which comes from the Higgs and the electroweak sector). This fine-tuning is theoretically displeasing and is known as the *'strong-CP problem'*.

Could we just do away with this confusing Θ-vacuum? Unfortunately not: without the Θ-vacuum, QCD predicts the mass of the eta prime meson $m_{\eta'} \leq \sqrt{3} m_{\pi^0}$ while experimentally [37]:

$$
\begin{aligned}
m_{\eta'} &= 957 \text{ MeV} \\
m_{\pi^0} &= 134 \text{ MeV}.
\end{aligned}
\tag{10.68}
$$

This puzzle of the eta prime mass was historically called the '$U(1)_A$' problem, and its solution requires the contribution of the Θ-vacuum, which in turn induces the strong-*CP* problem. Therefore, we need some new physics to explain why the observed $\overline{\Theta}$ is small.

10.2.1.2 Instantons and Vacuum Energy

A clue to the solution comes from the fact that in the quantum theory of QCD, the vacuum energy depends on Θ (due to a quantum object called an instanton, which we don't have the need to explain further):

$$\boxed{E_{\text{vac}} \propto - \cos \Theta.} \tag{10.69}$$

Due to its topological nature, Θ cannot change (due to e.g. renormalisation for the strong coupling constant g_3), and in flat space, where constant energies are not observable, we can absorb E_{vac} with an arbitrary constant (in curved space we can't do this, and the QCD vacuum is one part of the cosmological constant problem). However, if we could make Θ dynamical, then E_{vac} would give an effective potential for Θ, which, when minimized, would set $\langle \Theta \rangle = 0$ by minimizing the energy. Instantons, and thus E_{vac}, are what is called a 'non-perturbative' property of QCD: something that doesn't show up by computing Feynman diagrams and is certainly non-classical. In general, computing such effects requires lattice QFT and intensive numerical calculations, although analytic approaches can work in some limits (approximations known as the 'instanton gas' or 'chiral perturbation theory'). Details of the QCD vacuum and topology can be found in the excellent book *'Aspects of Symmetry'* by Coleman, Ref. [57].

10.2.1.3 Peccei–Quinn and Dynamical Θ

Peccei and Quinn introduced a model that solves the strong-*CP* problem by making Θ dynamical. We will discuss the Kim–Shifman–Vainshtein–Zakharov (KSVZ) version of this model here, which remains valid at present. The model extends the SM with two new fields:

Φ: an SM singlet, complex scalar field, that is, no SM gauge interactions.
$Q = Q_L + Q_R$: Dirac 'quark' that interacts with SU(3) and nothing else, that is, SU(2)$_L$ singlet with hypercharge 0.

$$\mathcal{L} = |\partial_\mu \Phi|^2 + \overline{Q} \underbrace{\slashed{D}}_{\text{SU(3) gluon covariant derivative}} Q - \left(y_Q \overline{Q}_L Q_R \Phi + \text{h.c.} \right) - \underbrace{\lambda_\Phi \left(|\Phi|^2 - \frac{v_a^2}{2} \right)^2}_{V(\Phi)}.$$

Note that in the above we again switched the metric convention to the particle physics one, $(+, -, -, -)$, which flips the signs of the kinetic terms of the fields.

This Lagrangian has a global, *chiral* U(1) symmetry, that is, a U(1) symmetry that must act with opposite signs on the left- and right-handed components of Dirac fermions. The Peccei–Quinn (PQ) symmetry transformations are

$$\Phi \rightarrow e^{i\alpha} \Phi$$

$$Q \rightarrow e^{\frac{i\alpha \gamma_5}{2}} Q \tag{10.70}$$

$$\Rightarrow Q_L \rightarrow e^{\frac{i\alpha}{2}} Q_L \quad ; \quad Q_R \rightarrow e^{-\frac{i\alpha}{2}} Q_R.$$

The symmetry is spontaneously broken by the potential $V(\Phi)$

$$\Rightarrow \Phi = \frac{1}{\sqrt{2}} (v_a + \rho_a) e^{\frac{i\phi}{v_a}}, \tag{10.71}$$

where

ρ_a is a radial excitation of Φ, a 'Higgs-like field' with mass $m_{\rho_a} = \sqrt{2\lambda_\Phi} v_a$.
ϕ is the phase of Φ and has no potential[4] since the Lagrangian for Φ is only dependent on the *magnitude* of Φ.

Thus, the Lagrangian is *shift symmetric* for the field ϕ.
The Yukawa term is

$$\mathcal{L}_{Q\phi} = m_Q \overline{Q}_L Q_R e^{\frac{i\phi}{v_a}} + \text{h.c.}, \tag{10.72}$$

with $m_Q = y_Q v_a / \sqrt{2}$. Using the small angle approximation, we can write

$$e^{\frac{i\phi}{v_a}} = 1 + \frac{i\phi}{v_a} + \dots \tag{10.73}$$

---

[4] This is the Goldstone boson of the broken global U(1).

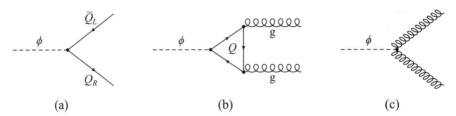

Figure 10.8 Feynman diagrams for the KSVZ axion model. (a) The Yukawa interaction between the axion and the quark Q. (b) Loop-mediated interaction between axions and gluons. (c) The anomaly-induced axion-gluon coupling.

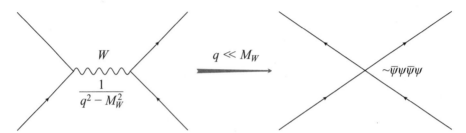

Figure 10.9 The Fermi interaction. Fermions interact via the W-boson, but in the limit of momentum transfer that is small compared to the W mass (i.e. long distances) this appears as a point-like four-fermion interaction.

Taking $\lambda_\Phi \sim y_Q \sim 0.1$ as the values for the couplings (i.e., small coupling regime), we notice

$$\Rightarrow m_{\rho_a} \text{ is large } \mathcal{O}(v_a) \tag{10.74}$$

$$\text{and } m_Q \text{ is large } \mathcal{O}(v_a). \tag{10.75}$$

For the interaction term $\phi \overline{Q}_L Q_R + \text{h.c.}$ we have the Feynman diagram shown in Fig. 10.8(a). Since the quark is charged under SU(3)$_C$, we can connect this via a loop to an interaction with gluons as shown in Fig. 10.8(b). Now recall the Fermi interaction shown in Fig. 10.9, which shows how the W-boson mediated interaction between fermions can, in the limit of energies smaller than the W mass, be thought of as an effective four-fermion interaction. We can similarly imagine the limit $E_\phi \ll m_Q$ and see that the quark loop interaction of the axion gives a tree-level axion-gluon interaction, shown in Fig. 10.8(c). The U(1)$_{\text{PQ}}$ symmetry introduced an interaction between the phase of the field Φ and the gluons, mediated by the quarks. The fact that this interaction is non-zero relies on the *chiral nature* of the quarks under this symmetry: it is known as the *chiral anomaly* (the same triangle diagram vanishes if the interaction treats Q_L and Q_R the same way (a vector interaction)).[5]

---

[5] In QFT there is strictly a difference between a loop-induced effect and an anomaly, and the axion acquires a coupling to gluons irrespective of the limit $E_\phi \ll m_Q$, but the intuitive picture from the Feynman loop and the analogy to the weak interaction is enough for the qualitative picture we need here.

For details on the chiral anomaly, see the books by Zee [58] and Srednicki [59].

The 'anomaly' version of this picture arises by considering a change of variables on $\mathcal{L}_{Q\phi}$ chosen as

$$Q \to \exp\left[i\gamma_5 \frac{\phi}{2v_a}\right]Q$$

$$\Rightarrow Q_L \to \exp\left[\frac{i\phi}{2v_a}\right]Q_L \quad ; \quad Q_R \to \exp\left[\frac{-i\phi}{2v_a}\right]Q_R \tag{10.76}$$

$$\Rightarrow \mathcal{L}_{Q\phi} \to \left(\frac{\partial_\mu \phi}{2v_a}\right)\bar{Q}\gamma^\mu \gamma_5 Q + \mathcal{L}_{\phi G},$$

where

$$\mathcal{L}_{\phi G} = \frac{\phi}{v_a}\frac{g_3^2}{32\pi^2}\mathrm{Tr}[G_{\mu\nu}\tilde{G}^{\mu\nu}]. \tag{10.77}$$

The shift in $\mathcal{L}_{\phi G}$ arises due to a change of measure in the path integral (see Ref. [59]). Although the classical theory is invariant under $U(1)_{PQ}$ rotations, the quantum theory is not: hence the name 'anomaly'. The bottom line is: the quark loops/chiral anomaly introduce an interaction between ϕ and the gluon that looks just like the Θ-term, that is, Eq. (10.77) resembles Eq. (10.57).

We need to go just a little deeper into this story for later use. The model for the QCD axion we just outlined, with a single new heavy quark charged under SU(3) and $U(1)_{PQ}$, is based on the historical 'KSVZ' axion model. It is possible to have more particles charged under the combination of these symmetries, either with more exotic quarks or by charging the Standard Model fermions directly under $U(1)_{PQ}$ as in the historical 'DFSZ' model. When we have more particles running in the anomaly loop the shift to the Lagrangian caused by the chiral rotations is

$$\mathcal{L}_{\phi G} = C\frac{\phi}{v_a}\frac{g_3^2}{32\pi^2}\mathrm{Tr}[G_{\mu\nu}\tilde{G}^{\mu\nu}], \tag{10.78}$$

where C is the *colour anomaly* which counts the degrees of freedom running in the loop and is given by group theory factors relating to the representation of the particles under SU(3) and $U(1)_{PQ}$. It is conventional to define the axion *decay constant*, $f_a = v_a/C$, so that regardless of C we can write the shift to the Lagrangian as

$$\mathcal{L}_{\phi G} = \frac{\phi}{f_a}\frac{g_3^2}{32\pi^2}\mathrm{Tr}[G_{\mu\nu}\tilde{G}^{\mu\nu}]. \tag{10.79}$$

10.2.1.4 Solving the Strong-CP Problem

A couple of changes, in fact, are all we need. First, we have the Lagrangian,

$$\mathcal{L} = \frac{1}{2}(\partial\phi)^2 + \frac{\partial_\mu \phi}{2v_a}\bar{Q}\gamma^\mu \gamma_5 Q + \frac{\phi}{f_a}\frac{g_3^2}{32\pi^2}G\tilde{G} + \Theta\frac{g_3^2}{32\pi^2}G\tilde{G}. \tag{10.80}$$

Upon changing variables, $\phi \to \phi - f_a \Theta$, the derivative terms in ϕ do not change, but the $G\tilde{G}$ terms do:

$$\mathcal{L} = \frac{1}{2}(\partial \phi)^2 + \frac{\partial_\mu \phi}{2v_a}\overline{Q}\gamma^\mu \gamma_5 Q + \frac{\phi}{f_a}\frac{g_3^2}{32\pi^2}G\tilde{G}. \qquad (10.81)$$

The problematic Θ-term is gone. Furthermore, the vacuum energy caused by the instantons of the $G\tilde{G}$ term gives ϕ an effective potential,

$$E_{\mathrm{vac}} \equiv V(\phi) = -\chi(T)\cos\frac{\phi}{f_a}. \qquad (10.82)$$

Lastly, we notice that the coupling of ϕ to the quarks appears with a $1/v_a$: this is typical of axion models, which have very small couplings if the SSB scale, v_a, is large.

The potential energy is fixed by $\chi(T)$, known as the *QCD topological susceptibility*, which is a non-perturbative property as alluded to above. However, we can make some simple estimate by noting that QCD becomes strongly coupled at the quark-hadron phase transition (more properly called the QCD crossover), that is, when

$$T \sim \Lambda_{\mathrm{QCD}} \sim 200\ \mathrm{MeV} \sim m_\pi, \qquad (10.83)$$

which sets the scale of the problem, that is, the scale at which we expect χ to switch over from a small to a constant value. Dimensionally, $[\chi] = M^4$, so naively we expect $\chi = \Lambda_{\mathrm{QCD}}^4$. We can get a better estimate by noting that if the quarks were massless then there would be no chiral anomaly (this is a technical point we don't prove), so we might guess

$$\chi \sim m_q \Lambda_{\mathrm{QCD}}^3, \qquad (10.84)$$

where $m_Q \approx \mathrm{MeV}$ is the scale of the lightest quark mass. Eq. (10.84) has a form that crops up again and again in *effective field theory* [60]: that of 'spontaneous times explicit symmetry breaking'. The symmetry is chiral symmetry (symmetry that applies when all fields are massless). The spontaneous breaking is from Λ_{QCD}; strong coupling gives pions mass. The explicit breaking is from the bare quark masses. The relation is sometimes called the Gell-Mann–Oakes–Renner relation. The explicit calculation in chiral perturbation theory specifies the value of m_q more exactly.

Taylor expanding $V(\phi)$ as $T \to 0$:

$$V(\phi) = \frac{1}{2}m^2(T)\phi^2 + \dots, \qquad (10.85)$$

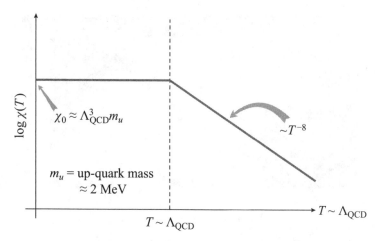

Figure 10.10 Sketch of the temperature dependence of the QCD topological susceptibility. The dependence is a power law at high temperatures far above the QCD crossover at Λ_{QCD}. The power law can be derived using so-called instanton methods and is T^{-8} at leading order. At low temperature, χ approaches a constant, which can be calculated in chiral perturbation theory. The full dependence requires numerical lattice QCD.

where detailed calculations give

$$m(0) = \frac{\chi(0)^2}{f_a} = 5 \times 10^{-6} \text{ eV} \left(\frac{10^{12} \text{ GeV}}{f_a} \right)$$

$$m(T \gg \Lambda_{QCD}) = m(0) \left(\frac{T}{\Lambda_{QCD}} \right)^{-4}$$

(10.86)

The temperature dependence of $\chi(T)$ is sketched in Fig. 10.10.

The field ϕ is called the *QCD axion*, and the field quanta are axion particles. The Klein–Gordon equation means that we can start with any value of ϕ and it will evolve to near $\phi = 0$ with small oscillations. Relaxation to zero implies that ϕ *solves the strong-CP problem and $d_n \sim 0$.*

The oscillations of ϕ carry energy density. The relic density is slightly trickier to solve with $m = m(T)$. In general it is a function of θ_i and f_a and depends on the exact temperature evolution of m. The relic density is given by

$$\rho \approx \frac{1}{2} \langle \theta_i^2 \rangle f_a^2 m_a(T_{\text{osc}}) m_a(0) \left(\frac{a}{a_{\text{osc}}} \right)^{-3},$$

(10.87)

where the factors of $m(T)$ arise from the temperature dependence of m and the conserved number density. Classical oscillations of the QCD axion field can provide the dark matter. This relic density is approximated in Fig. 10.11 for the QCD axion. For $\theta \approx 1$ we obtain the correct relic density for $f_a \approx 10^{12}$ GeV. Note that with this reference value of $f_a \approx 10^{12}$ GeV the QCD axion mass is very small, around 10^{-6} eV,

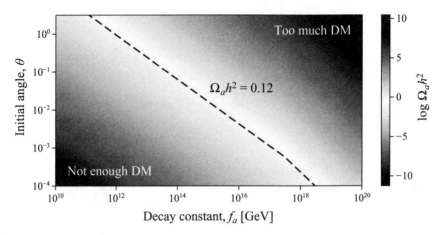

Figure 10.11 Axion relic density for $m_a(T)$ of the QCD axion. The solution here contains no anharmonic corrections. See colour insert.

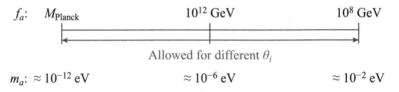

Figure 10.12 The axion has a tiny mass. The decay constant, f_a, is cut off at the Planck scale (*left*), reproduces the correct relic density if $\theta_i = 1$ (*centre*) and is constrained by supernova cooling (*right*).

much smaller than that of a typical WIMP. We will find out in later chapters what the allowed range of mass for the QCD axion is, but the rough picture is previewed in Fig. 10.12. We will also see later that *we can experimentally search for relic ϕ oscillations*.

10.2.2 Axion-Like Particles

There is a whole zoo of related particles with similar cosmology but unrelated to solving the strong-CP problem. An 'axion-like particle' (ALP) satisfies:

1. An (approximate) global symmetry.
2. SSB at some high scale, f_a.
3. Mass from non-perturbative effects at a scale Λ *not* Λ_{QCD},

$$\Rightarrow m \sim \frac{\Lambda^2}{f_a} \quad \text{small.}$$

4. A pseudoscalar (CP-odd),

$$CP\phi \rightarrow -\phi,$$

which specifies that the couplings to the SM are also via *CP*-violating operators, the canonical example being

$$\frac{\phi}{f_a}\frac{\alpha}{8\pi}F_{\mu\nu}F_{\alpha\beta}\epsilon^{\alpha\beta\mu\nu} = \frac{\phi}{f_a}\frac{\alpha}{2\pi}\vec{E}\cdot\vec{B}. \tag{10.88}$$

Since the mass of an ALP does not come from QCD, an ALP has more free parameters; namely the mass and couplings are unrelated. One class of interesting models (see chapter 15 for further information) has

$m \sim 10^{-22}\text{eV} \Rightarrow \lambda_{dB} \sim \text{kpc}$, that is, $\sim$ galactic scales.
$f_a \sim 10^{16}\text{GeV} \Rightarrow$ implies a connection to GUTs?

10.2.3 ALPs from String Theory

A very important situation in which ALPs arise in a unified theory is from *string theory*. String theory is the only theory known that provides a unified description of the particles and forces of the Standard Model and gravity. It is also a consistent quantum theory of gravity. In order to describe fermions as well as bosons in a unified way, string theory must be supersymmetric. It turns out (due to something called 'anomaly cancellation') that a supersymmetric string theory is only self-consistent when the number of spacetime dimensions is ten.[6] Thus, these 10D theories must be 'compactified' on a six-dimensional manifold of small volume in order to reproduce the physics we see around us in 4D.

To try to appreciate what compactification means for physics in 4D, consider the simple example with two extra dimensions folded up in a torus, shown in Fig. 10.13. The metric on the torus is a function of different 'modulus' fields, σ_i, which describe the size and shape of the torus, that is, the relative sizes of the cycles indicated in Fig. 10.13.

10.2.3.1 An Aside on the Torus

The topology of the torus, T_2, is that of a product of two circles, $T_2 = S_1 \times S_1$. We note that the familiar doughnut representation of the torus in Fig. 10.13 is a *curved* metric on the torus topological space, which we find by computing the induced metric or 'pullback' of the constraints defining the doughnut's surface against the flat metric in 3D Euclidean space. If the coordinates on the doughnut are u and v around the circles shown, each with a size σ_1 (for the circle that changes height in

---

[6] There are five different self-consistent 10D string theories, related to each other in different limits. In the strong coupling limit, the non-spacetime degrees of freedom in one of these theories group together to behave like an extra, eleventh dimension of space. When we say 'string theory' we tend to mean this web of interconnected theories including the 11-dimensional 'M-theory'.

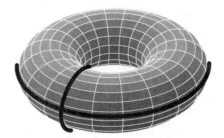

Figure 10.13 A torus and two 'one cycles' on it.

the z-direction) and σ_2 in the (x, y) plane, which defines the eccentricity $e = \sigma_1/\sigma_2$, then the equation of the torus is

$$x = \sigma_2(1 + e \cos u) \cos v, \quad y = \sigma_2(1 + e \cos u) \sin v, \quad z = \sigma_1 \sin u, \quad (10.89)$$

and we find the metric by simply substituting into the flat metric in the Euclidean 'embedding space':

$$ds^2 = dx^2 + dy^2 + dz^2, \quad \text{(Flat metric in } \mathbb{R}^3), \quad (10.90)$$

$$dx = \frac{\partial x}{\partial u} du + \frac{\partial x}{\partial v} dv, dy = \frac{\partial y}{\partial u} du + \frac{\partial y}{\partial v} dv, dz = \frac{\partial z}{\partial u} du + \frac{\partial z}{\partial v} dv, \quad (10.91)$$

which gives

$$ds^2 = \sigma_1^2 du^2 + \sigma_2^2 (1 + e \cos u)^2 dv^2, \quad (10.92)$$

factorising

$$ds^2 = \sigma_1^2 \left[du^2 + \left(\frac{1}{e} + \cos u \right)^2 dv^2 \right]. \quad (10.93)$$

Using the rules of GR, we could then compute, for example, the Ricci curvature, R, and we would find that it is non-zero: the torus thus defined is curved. We also see that the moduli can be organised with one for overall size, σ_1, and one for shape, $e = \sigma_1/\sigma_2$.

However, there are other possible metrics with the same topology, and in particular there is a flat metric which we can find by simply identifying the sides of a square to make a torus, without 'bending it' into a third dimension. The flat torus is a two-dimensional, 'Calabi–Yau manifold'. Six-dimensional Calabi–Yaus play an important role in string theory. The flat torus does not have a simple embedding in our familiar three dimensions, so it is hard to visualise. The embedding in three dimensions turns out to be fractal (search "HEVEA torus" on YouTube for a fly-over of the fractal torus). In four dimensions, the flat torus is embedded trivially as the so-called Clifford torus. The issue of embedding a flat object turns out to place an important roadblock in the way of finding the metrics of six-dimensional Calabi–Yaus, but it does not prevent us knowing about their topology.

It is easy to write down a flat metric on the torus by simply wrapping up the Euclidean plane $\mathbb{R}^2$:

$$ds^2 = \sigma_1^2 du^2 + \sigma_2^2 dv^2, \tag{10.94}$$

where $u, v \in [0, 2\pi]$ are periodic coordinates on each identified side of a square (on each S_1 sub-manifold), and σ_i are the lengths of the sides. We can still think of u, v as the coordinates around the circles in Fig. 10.13 and σ as the radii of the circles, but the metric Eq. (10.94) is *not* the metric on the surface drawn: there isn't a coordinate transformation that relates the metrics Eq. (10.93) and Eq. (10.94). As long as the σ_i in Eq. (10.94) are independent of the u, v it is clear that this metric is flat (all the metric components are coordinate independent), and so the Ricci curvature $R = 0$.

We still haven't been as general as possible with the freedom of size and shape for our torus: we can allow off-diagonal components for the metric, which corresponds to turning our square into a parallelepiped with some shear angle in addition. Such a metric on the flat torus takes the form (this example is taken from Ref. [61]):

$$ds^2 = \frac{\mathcal{V}}{\tau_2} \left[(du + \tau_1 dv)^2 + \tau_2 dv^2 \right], \tag{10.95}$$

where $\mathcal{V}$ is the area of the parallelepiped, and $\tau = \tau_1 + i\tau_2$ defines its shape. In string theory jargon, this is the difference between the Kähler moduli (the size, $\mathcal{V}$) and the complex structure moduli (the shape, τ).

10.2.3.2 Dimensional Reduction

To work out what this means for 4D physics in a higher dimensional world, we 'dimensionally reduce' the Einstein–Hilbert action in d dimensions:

$$S = \int d^d x \sqrt{-g_d} R_d, \tag{10.96}$$

which just means that we integrate over the compact directions given the form of the metric and assuming that it factorises between the compact space and our four large dimensions. In the example with the torus, we would start with writing the Ricci curvature in six dimensions with the metric factorised into $g = M_4 \times T^2$, and then we can explicitly perform the integration in the action over the coordinates u, v on the torus.

In general, we allow that the modulus fields, σ (in our torus example, these were $\mathcal{V}$ and τ), may depend on the coordinates of the four large dimensions (the compact space can change size and shape from place to place), but we won't allow for any 'warping' the other way: the curvature of our dimensions won't depend on where we are in the compact space (this assumption is relaxed in so-called 'brane world' models). The Ricci scalar is a two-derivative function of the metric, and so it has to depend on second derivatives of the moduli fields. Dimensional reduction then looks like (schematically)

$$S = \int d^{10}x\sqrt{-g_{10}}\ \overbrace{R_{10}}^{\text{10D Ricci scalar}}$$

$$= \int d^4x\sqrt{-g_4}\Big\{ R_4 + \sum_{\text{moduli}} f(\sigma)(\partial\sigma)^2 \Big\}, \tag{10.97}$$

where R_4 is the 4D Ricci curvature of the large dimensions. The function $f(\sigma)$ is a schematic representation for the metric on field space for the moduli, which they inherit from the metric on the compact space. In string theory jargon it is called the Kähler metric. In the general case beyond the torus, σ describes the sizes of different parts of the compact space, with the number of such moduli being fixed by the topology of the compact 6D space. We get moduli describing each closed sub-manifold in the space: in the analogy of the torus, imagine adding extra 'handles' and holes.

ALPs arise once we add supersymmetry into the mix. For the Kähler ('size') moduli, supersymmetry demands that the real modulus field σ is paired with its superpartner, which we suggestively call a, into a complex field z:

$$z = a + i\sigma. \tag{10.98}$$

The a field arises in the 4D action not from dimensional reduction of the Ricci scalar R, but from 'p-form' fields in the 10D 'supergravity' action. We won't go into this in any more detail, because the mathematics of p-forms is beyond the scope of this book, but the qualitative story is very much the same as for the Ricci scalar. These extra fields in the gravitational sector are a consequence of supersymmetry. A p-form field is analogous to the field strength tensor, $F_{\mu\nu}$, of electromagnetism, which is a two-form. A p-form is defined as being totally antisymmetric with p indices. Just like we dimensionally reduced the Ricci scalar, we must dimensionally reduce the kinetic term of the p-form (the F^2 term in the action). When we do this, the p-form field gives rise to fields, a, in 4D when it can 'wrap' its indices on a closed p-dimensional sub-manifold (called a p-cycle) of the 6D space, and we get an axion for each distinct sub-manifold. The number of closed p-dimensional sub-manifolds is a topological invariant known as the pth Betti number of the compact space.

For Type-IIB string theory, the relevant p-form is a 4-form called C_4, and the number of axions is given by the fourth 'Betti number' of the compact space, which counts how many different four-dimensional holes the space has (these types of topological invariants are related to the possibly more familiar 'Euler characteristic'). Due to the complexity of the topology of possible 6D backgrounds, it is probable that there are *many axion-like particles*.[7] For us to think about these axions as DM candidates, we need to know, at a minimum, their masses and their decay constants. The decay constants arise from the classical dimensional reduction (the

---

[7] Under the constructions of Calabi–Yaus considered in Ref. [62], it is thought that the overwhelming majority of Calabi–Yaus have 491 such closed string axions.

same Kähler metric that was mentioned above in the modulus kinetic term). The masses arise as for the QCD axion from 'non-perturbative effects', stringy equivalents of the QCD instantons we met already, called things like, for example, D-brane instantons, gravitational instantons and gaugino condensation. The decay constants and masses are approximately given by

$$
f_a \sim \frac{M_{\mathrm{Pl}}}{S}; \; m_a \sim \frac{M_{\mathrm{Pl}}}{f_a} e^{-S},
\tag{10.99}
$$

where S is the action of the non-perturbative effect and is proportional to the Kähler modulus σ that in Type-IIB theory is the volume of the four-cycle on which C_4 was compactified. The important point is that we notice that the axion masses depend exponentially on S (this is always true for the energy scales of non-perturbative effects), while the decay constants depend on it only as a power law. String theory therefore gives us axions with hugely varying masses, but with decay constants not too many orders of magnitude separated from the Planck scale. In explicit constructions, the masses can span from essentially M_{Pl} to many orders of magnitude below H_0, while f_a goes from M_{Pl} down to around 10^8 GeV. The non-gravitational couplings of these axions to Standard Model fields are subject to much more model building uncertainty, but we might parametrically expect them to be controlled by $1/f_a$ in the same way as the QCD axion and other ALPs.

10.3 STRUCTURE FORMATION WITH SCALAR FIELDS

So far we have treated axion dark matter quite differently to particle models, by solving the classical equations of motion of a scalar field. We will now very briefly outline how scalar fields fit into cosmological perturbation theory and how they behave inside galaxies and under their own gravity.

A scalar field, like all other fields, is coupled to gravity by the metric. For our minimally coupled, canonically normalised axion, the metric appears in the D'Alembertian operator as in Eq. (10.3). To treat the scalar field under cosmological perturbations, we simply compute this operator using the perturbed metric given by Eqs. (5.83, 5.85).

In order to do cosmological perturbation theory, it is also necessary to make a perturbation ansatz for the scalar field itself, introducing small spatial perturbations, that is, $\phi(t, \vec{x}) = \bar{\phi}(t) + \delta\phi(t, \vec{x})$. Going to Fourier space, the Klein–Gordon equation, Eq. (10.4), gives rise to second-order equations of motion for the mode functions $\delta\phi_k(t)$, with the metric potentials appearing as a source term. To complete the picture, one must finally compute the perturbed scalar field energy-momentum tensor. The canonical energy-momentum tensor in general relativity is defined by variation of the action as

$$
T_{\mu\nu} = \frac{\delta S}{\delta g^{\mu\nu}}.
\tag{10.100}
$$

You should check that this gives the harmonic oscillator expression for ρ given in Eq. (10.18). Computing $T_{\mu\nu}$ for the perturbations leads to the appearance of $\delta\phi_k(t)$ and its derivatives as a source term for metric potentials on the right-hand side of Einstein's equations.

We could now go through the same exercise as in section 5.2 and find how $\delta\phi_k$ behaves in different limits when coupled to radiation perturbations. Perturbations behave very differently in the two regimes $m \ll H(t)$ and $m \gg H(t)$, which complicates the analytical and numerical treatment. The details, however, are not important here and are covered elsewhere [63, 64]. The most important features can be derived in the non-relativistic limit, to which we now turn.

10.3.1 The Non-Relativistic Limit

We learned in section 5.2.3 that in the absence of relativistic matter the metric potentials in Eqs. (5.83 and 5.85) satisfy $\Psi = -\Phi$. What else enters into our definition of 'non-relativistic'? As we learned in section 10.1, axions, and scalar field DM in general, only behave 'like matter' with $\rho \propto a^{-3}$ in the limit $H \ll m$, so let's set $H = 0$ to begin with (we will restore the expansion of the Universe for sub-horizon modes at the end by a choice of coordinates).

The metric is now

$$g = \mathrm{diag}[-(1+2\Psi), 1-2\Psi, 1-2\Psi, 1-2\Psi]. \tag{10.101}$$

Working to first order in Ψ, the metric inverse and metric determinant are

$$g^{-1} = [g^{\mu\nu}] = \mathrm{diag}[-(1-2\Psi), 1+2\Psi, 1+2\Psi, 1+2\Psi],$$

$$\sqrt{-g} = 1-2\Psi \,.$$

Thus, the D'Alembertian to first order is given by

$$\Box = 4\dot{\Psi}\partial_t - (1-2\Psi)\partial_t^2 + (1+2\Psi)\nabla^2, \tag{10.102}$$

and the Klein–Gordon equation reads

$$-(1-2\Psi)\ddot{\phi} + 4\dot{\Psi}\dot{\phi} + (1+2\Psi)\nabla^2\phi - m^2\phi = 0. \tag{10.103}$$

The first term in this equation looks a little unfamiliar. To make it look more like an ordinary wave equation, we can divide this term off and Taylor expand to first order in Ψ:

$$\ddot{\phi} - 4\dot{\Psi}\dot{\phi} - (1+4\Psi)\nabla^2\phi + (1+2\Psi)m^2\phi = 0. \tag{10.104}$$

The next trick we use is to write the real axion field, ϕ, in terms of a complex field, ψ, and its complex conjugate. In the limit $H \ll m$ the homogeneous axion field simply oscillates with frequency m, so we take the ansatz:

$$\phi = \frac{1}{\sqrt{2m}}\left(\psi e^{imt} + \psi^* e^{-imt}\right). \tag{10.105}$$

To complete the trick, we need to take a similar ansatz for the metric potential, Ψ. The non-relativistic limit in this context means considering only the evolution of the *envelope function* for the scalar field, that is, ψ, *and also for the metric potential,*[8] in the limit that $\dot\psi \ll m\psi$ and $\ddot\psi \ll m\dot\psi$. This limit is related to the first-order WKB approximation and, when the sub-horizon limit is included, is also related to the gradient expansion.

It should come as no surprise that the correctly defined non-relativistic limit of the Klein–Gordon equation gives rise to a Schrödinger-like equation for ψ and that the Newtonian potential Ψ behaves as a potential:[9]

$$i\dot\psi - \frac{1}{2m}\nabla^2\psi + m\Psi\psi = 0. \qquad (10.106)$$

Einstein's equations in the non-relativistic limit just reproduce Newton's, and so the metric potential Ψ is determined by the Poisson equation:

$$\nabla^2\Psi = 4\pi G_N(\rho - \bar\rho). \qquad (10.107)$$

Subtracting the mean density in Eq. (10.107) is easily justified in cosmological perturbative GR; in the context of classical Newtonian physics it is sometimes called the 'Jeans swindle' [5]. If we consider axion DM to dominate the gravitational potential then we have

$$\rho = \rho_a = |\psi|^2, \qquad (10.108)$$

leading to a coupled set of non-linear equations known as the Schrödinger–Poisson equations. The Schrödinger–Poisson equations describe the self-gravity of axion and scalar field DM in the sub-horizon limit.

Lastly, we can reinstate the expansion of the Universe by moving to comoving coordinates, $x \to ax$. Since $\rho_a = |\psi|^2$ the number density is $n_a = |\psi^2|/m$, which tells us that in comoving coordinates ψ must rescale as $\psi \to a^{-3/2}\psi$ if it is going to represent a number density behaving as non-relativistic matter (this also follows if we recall the enveloped solution for $\phi \sim a^{-3/2}$ when $H \neq 0$ and the WKB ansatz Eq. (10.105)). The Schrödinger–Poisson equations in comoving coordinates are

$$i\dot\psi - \frac{3iH}{2}\psi - \frac{1}{2ma^2}\nabla^2\psi + m\Psi\psi = 0. \qquad (10.109)$$

$$\frac{1}{a^2}\nabla^2\Psi = 4\pi G_N(\rho - \bar\rho). \qquad (10.110)$$

As in all cosmological applications, we notice that ∇ appears with $1/a$, and thus in Fourier space we see that the momentum redshifts. The Hubble constant also appears as a sink term for ψ (move the H term to the right-hand side, and multiply by i: the H term contributes a negative term to $\dot\psi$).

A numerical solution of the Schrödinger–Poisson equations is required to fully account for the physics of axion DM on all scales. Fortunately, as we will show, on large scales (to be defined) this is not necessary: axions behave as CDM and N-body

---

[8] We have not emphasised this point before, but we will return to it again shortly below.

[9] You are encouraged to work through this derivation, from Klein–Gordon–Einstein to Schrödinger–Poisson, at least once in your life, for the good of your soul. Niemeyer [65] takes a slightly different route than we have outlined here, via the quadratic action.

techniques, and all their successes discussed earlier in this book apply to axions. On small scales, the Schrödinger–Poisson equations cause axions to differ from CDM, and the ability to explore this with full-scale simulations has developed massively in the last ten years, beginning with the pioneering work of Schive, Chiueh and Broadhurst [66].

10.3.2 Some Schrödinger–Poisson Physics

There is not space or necessity for us to delve into all of the very interesting physics of the Schrödinger–Poisson equations, but we quickly tour a few key aspects.

10.3.2.1 Cosmological Perturbations and the Axion Jeans Scale

We can relate the Schrödinger–Poisson equations to the cosmological perturbation theory of effective fluids using the *Madelung transformation*. We write the 'wavefunction' ψ in polar coordinates:

$$\psi = \sqrt{\rho_a} e^{iS}. \tag{10.111}$$

We define the axion fluid velocity:

$$\vec{v}_a = \frac{1}{m} \nabla S. \tag{10.112}$$

The axion fluid velocity is the gradient of the phase of the 'wavefunction', which has interesting consequences relating to fluid vorticity and turbulence inside non-linear structures such as galaxies, but we will discuss them no further here.

The interpretation of the modulus and phase of the wavefunction in terms of fluid density and velocity can be found by plugging the Madelung transformation into the Schrödinger equation and separating real and imaginary parts, which yields

$$\dot{\rho}_a + 3H\rho_a + \frac{\nabla}{a}(\rho_a \vec{v}_a) = 0, \tag{10.113}$$

$$\dot{\vec{v}}_a + H\vec{v} + \left(\vec{v}_a \cdot \frac{\nabla}{a}\right)\vec{v}_a = -\frac{\nabla}{a}\left(\Psi - \frac{1}{2ma^2}\frac{\nabla^2\sqrt{\rho_a}}{\sqrt{\rho_a}}\right). \tag{10.114}$$

The left-hand sides of the above equations should start to look familiar as the continuity and Euler equations of a fluid, with a few additional pieces due to the expansion of the Universe. For comparison with our CDM Boltzmann equations in chapter 5 we need a few more steps.

Next we split background and perturbations in the density by writing $\rho_a = \bar{\rho}_a (1 + \delta_a)$. For the background, this gives

$$\dot{\bar{\rho}}_a + 3H\bar{\rho}_a = 0. \tag{10.115}$$

The solution to Eq. (10.115) gives $\bar{\rho}_a \propto a^{-3}$, as required. The fluctuations δ_a and v_a obey

$$\dot{\delta}_a + \frac{1}{a}\vec{v}_a \cdot \nabla \delta_a = -\frac{1}{a}(1 + \delta_a)\nabla \cdot \vec{v}_a, \tag{10.116}$$

$$\dot{\vec{v}}_a + \frac{1}{a}(\vec{v}_a \cdot \nabla)\,\vec{v}_a = -\frac{1}{a}\nabla(\Psi + Q) - H\vec{v}_a, \tag{10.117}$$

$$Q \equiv -\frac{1}{2m^2 a^2}\frac{\nabla^2 \sqrt{1+\delta_a}}{\sqrt{1+\delta_a}}, \tag{10.118}$$

where I have defined the 'quantum potential' Q that accounts for the field gradient energy. Finally, we expand the above equations to first order in δ_a and v_a and move to Fourier space:

$$\dot{\delta}_a + \frac{ik}{a}v_a = 0. \tag{10.119a}$$

$$\dot{v}_a + Hv_a + \frac{ik}{a}\Psi = -\frac{ik}{a}\left(\frac{k^2}{4m^2 a^2}\delta_a\right). \tag{10.119b}$$

Eqs. (10.119) are the analogous equations of motion for axions to Eq. (5.128). There are two differences between these equations for axions and the equations for CDM we derived in chapter 5. Firstly, the axion continuity equation in the upper equation is missing the factor $3\dot{\Phi}$: this factor is subleading in the sub-horizon limit and Eqs. (10.119) *are only valid for sub-horizon modes*. The second difference is the appearance of a new term on the right-hand side of the Euler equation for v_a.

We can see what this term does by combining the two first-order equations into a single second-order equation for δ_a. We assume an axion DM-dominated Universe and use the Poisson equation to substitute for δ_a:

$$\frac{1}{a^2}\nabla^2\Psi = 4\pi G_N \bar{\rho}_a \delta_a \Rightarrow \Psi = 4\pi G \bar{\rho}_a \frac{k^2}{a^2}\delta_a. \tag{10.120}$$

The resulting second-order equation for δ_a is

$$\ddot{\delta}_a + 2H\dot{\delta}_a + (k^2 c_{s,\text{eff}}^2 - 4\pi G_N \bar{\rho}_a)\delta_a = 0, \tag{10.121}$$

where we have defined

$$c_{s,\text{eff}}^2 = \frac{k^2}{4m^2 a^2}, \tag{10.122}$$

the axion effective sound speed. The effective sound speed arose from the additional term on the right-hand side of the velocity equation in Eqs. (10.119) and ultimately comes from the gradient term in the classical Klein–Gordon equation for ϕ. The sound speed term competes with $4\pi G \bar{\rho}_a$ in Eq. (10.121). If $k^2 c_{s,\text{eff}}^2 < 4\pi G \bar{\rho}_a$, the solutions of Eq. (10.121) grow in exact analogy to the 'growing mode' solution for CDM found in section 6.2.2. On the other hand, if $k^2 c_{s,\text{eff}}^2 > 4\pi G \bar{\rho}_a$ the solutions of Eq. (10.121) oscillate and do not grow. The condition $k_J^2 c_{s,\text{eff}}^2 = 4\pi G \bar{\rho}_a$ defines the *axion Jeans scale*. Modes with $k < k_J$ behave just like CDM, while modes with $k > k_J$ have suppressed growth of perturbations. This fact will be used to set a lower limit on the axion mass (if it is to compose a significant fraction of the DM) in chapter 15.

10.3.2.2 Axion Stars

An axion star (AS) is a localised distribution of axions, supported by a balance between gravity and scalar field gradients. In the non-relativistic limit, we find such a solution as an energy eigenstate of Eqs. (10.106) and (10.107). A solution can be found assuming spherical symmetry, so that $\psi(r, t) = e^{iEt} \chi(r)$. With an appropriate choice of units, we can fix the central value to be $\chi(0) = 1$, and finite total energy demands $\chi(r \to \infty) = 0$, with general solutions found by rescaling: $\{t, x, \Psi, \psi\} \to \{\lambda^{-2}t, \lambda^{-1}x, \lambda^2\Psi, \lambda^2\psi\}$ for some scaling parameter λ. The solution to this boundary value eigenvalue problem cannot be found analytically but is simple to find numerically. The resulting solution of lowest energy is smooth (no oscillations) and almost Gaussian in shape. The density profile thus has $d\rho_{AS}/dr = 0$ at the origin: if an axion star forms at the centre of a DM halo, then the density profile will have a *core* (rather than an NFW profile which has a *cusp* at the origin, $d \ln \rho_{NFW}/d \ln r = -1$).

We can estimate the properties of the solution by a simple comparison of scales. If the axion star has a total mass M and characteristic radius R then $\Psi \sim GM/R$ while $\nabla^2 \sim 1/R^2$. Therefore, an equilibrium solution balancing the second and third terms in Eq. (10.106) has $GMm/R \sim 1/mR^2$, and so the mass and radius of an axion star are related as

$$R_{AS} \sim \frac{1}{Gm^2} \frac{1}{M_{AS}}. \tag{10.123}$$

This tells us that axion stars have a radius related inversely to their mass: heavier stars are more compact. Axion stars have a maximum mass, fixed by demanding that the radius should be larger than the Schwarzschild radius, such that the star does not collapse to a black hole, that is, $GM/R < 1$; this gives

$$M_{max} \sim \frac{1}{Gm} \sim 10 M_\odot \left(\frac{10^{-11} \text{ eV}}{m} \right). \tag{10.124}$$

We have only considered an axion star so far in the non-relativistic limit, where the density, ρ, and gravitational potential, Ψ, are static, and the field Ψ oscillates on a time scale given by the energy eigenvalue $E \sim 1/R \ll m$. Because of the static nature of ρ, axion stars in this limit are referred to as *solitons*. It is also possible to study axion stars in the full relativistic theory. In this case we use the field ϕ, which oscillates on a time scale m, leading to metric potentials and terms in the energy-momentum tensor which also oscillate with this time scale, which is much faster than the gravitational time scale. In the full relativistic theory, because of this oscillation, axion stars are referred to as *oscillatons*.

Axion stars are seen to form in numerical simulations of the axion field. They can form due to a type of *classical wave condensation*, which is related to Bose–Einstein condensation. They can also form due to the *direct collapse* of density perturbations, which leads to the existence of axion stars in the centres of DM halos. The number of axion stars expected in the Universe and their mass distribution are not exactly known across all cosmic time scales and all axion masses, due to the problem of separation of scales between the axion star mass and typical DM halo masses. They can have some interesting phenomenological consequences, but there is no space to discuss them in this book: some can be found from the references in the appendix and chapter 15.

10.3.2.3 Axion DM Halos

Most DM structures, both the ones we see out there in the Universe, and the ones we see inside our computers from numerical simulations, are not supported by a balance of gravity and field gradients, but rather by the balance of gravity and velocity, that is, virialised motion, balancing the non-linear velocity term on the left-hand side of the Euler equation Eq. (10.114) with the gravitational potential, which gives $v^2 \sim GM/R$. How does the inside of a virialised halo composed of axion waves differ from one composed of particles? Intuitively, we expect differences between particles and waves to emerge on the de Broglie wavelength, $\lambda_{dB} = 1/mv$, where v is the virial velocity in the halo. Taking a typical $v \sim 100$ km s^{-1} $\sim 10^{-3}c$ then we have

$$\lambda_{dB} \sim 0.2 \text{ kpc} \left(\frac{10^{-22} \text{ eV}}{m} \right) \left(\frac{100 \text{ km s}^{-1}}{v} \right). \tag{10.125}$$

The de Broglie wavelength is galaxy sized if the mass is very small, $m \sim 10^{-22}$ eV: this defines a regime known as *fuzzy dark matter*, where wave effects are important for galaxy formation.

Niemeyer (Ref. [65]) treats this more rigorously by defining a quantity similar to the Reynolds number in the Navier–Stokes equation, comparing the non-linear velocity term to the Q term on the right-hand side of Eq. (10.117). We find that wave effects will be important if

$$v^2 \lesssim \frac{1}{R^2 m^2} \Rightarrow R \lesssim \frac{1}{mv}, \tag{10.126}$$

which, unsurprisingly, gives us exactly the de Broglie wavelength. We can think of an axion halo as composed of de Broglie-sized 'granules' or quasi-particles, moving at speed v, which leads to the notion of a coherence time, τ, given by the time it takes a granule to move across its own size, $\tau \sim 1/mv^2$. The axion halo is dynamical on this time scale, which for typical halo velocities $v \sim 10^{-3}$ is of order $1/v^2 = 10^6$ times larger than the axion field oscillation period, $1/m$.

In the non-relativistic limit, an axion halo is given by a superposition of complex valued waves, ψ. The superposition naturally leads to the appearance of *wave interference*: places where the local density is exactly zero. Superposition also gives a more precise view on the meaning of the coherence time. Following Hui [67], consider a wavefunction that is a superposition of plane waves: $\psi = \sum A_k e^{iB_k} e^{ikx - i\omega_k t}$, where each mode k has a frequency, amplitude and phase. The density is given by

$$\rho = |\psi^2| = \sum_k A_k^2 + \sum_{k' \neq k} A_k A_{k'} e^{i(B_k - B_{k'})} e^{i(k-k')x - i(\omega_k - \omega_{k'})t}. \tag{10.127}$$

We see that in taking the modulus squared we end up with cross terms from different modes. The cross terms have non-trivial space and time dependence. If the typical difference in momenta between modes is mv then the typical spatial scale comes out as $1/(k - k') = 1/mv$, the de Broglie wavelength. If the typical difference in energies is mv^2 then the typical time scale variation is $1/(\omega_k - \omega_{k'}) = 1/mv^2$, the coherence time τ that we estimated above.

We have now seen how an axion halo is in fact a dynamical, turbulent environment of waves with scales set by the virial velocity and the particle mass via the de Broglie wavelength. This dynamical environment can have effects on astrophysics, one of which we will meet in chapter 15.

QUIZ

i. What physical observable is fixed by the QCD theta parameter?
   a. The mass of the pion.
   b. The lifetime of the proton.
   c. The neutron electric dipole moment.
   d. The neutron magnetic dipole moment.

ii. Which discrete symmetries are violated by the theta term?
   a. Parity (P).
   b. Charge conjugation (C).
   c. Time reversal (T).
   d. CP.
   e. CPT.
   f. CT.

iii. In the QCD axion model we discussed in the chapter (the KSVZ model), what new particles are added to the standard model? (Select all that apply).
   a. A complex scalar singlet field.
   b. A complex scalar SU(2) doublet.
   c. A lepton SU(2) doublet.
   d. A quark.
   e. A right-handed neutrino.
   f. No new fields are added.

iv. Why is the Peccei–Quinn U(1) symmetry called 'chiral'?
   a. It only affects chiralon fields.
   b. It affects the left- and right-handed components of fermions with different charges.
   c. It affects the real and imaginary components of scalars with different charges.
   d. It only affects the non-Abelian gauge fields.

v. Select all true statements about the QCD axion mass:
   a. The mass becomes constant below the QCD confinement scale.
   b. The mass depends on temperature above the QCD confinement scale.
   c. The mass is of order the QCD scale.
   d. The mass is much smaller than the QCD scale.
   e. The mass arises due to the effect of instantons on the vacuum energy.
   f. The mass breaks the U(1) Peccei–Quinn symmetry explicitly.
   g. The mass breaks the U(1) Peccei–Quinn symmetry spontaneously.

vi. The axion DM relic density is:
   a. Due to vacuum realignment.
   b. Due to thermal freeze out.

Answers

i. c; ii. a,c,d; iii. a,d; iv. b; v. a,b,d,e,f; bf vi. a

Chapter Eleven

Primordial Black Holes

As we saw in chapter 6, ordinary matter is insufficient for structure formation. Prior to $z \approx 1100$, baryons are strongly coupled to photons and the acoustic oscillations prevent the formation of gravitational structures like galaxies. Furthermore, the light element abundances restrict the baryon-to-photon ratio at nucleosynthesis $T \simeq 1$ MeV, so we know there cannot be many more baryons than we see. If, however, we could 'lock up' ordinary matter (i.e., baryons and photons) *inside black holes* at $T \gg 1$ MeV then maybe this could work, as baryons inside BHs cannot interact with the plasma outside. Ordinary black holes form due to the death of stars. Stars form in galaxies, and even the earliest of these would form at, say, $z \ll 1100$. Thus, the black holes necessary for dark matter must *form by some exotic mechanism in the very early Universe*. They are known as

$$\boxed{\textbf{Primordial Black Holes (PBHs)}}$$

11.1 COLLAPSE OF CURVATURE PERTURBATIONS

Consider a curvature perturbation, as shown in Fig. 11.1. For a spherical perturbation, the metric is

$$ds^2 = -dt^2 + a^2 e^{2\zeta} \delta_{ij} dx^i dx^j, \tag{11.1}$$

with ζ the (non-linear) comoving curvature. If we convert to spherical polars and write the radius as

$$R = re^{\zeta(r)}, \tag{11.2}$$

we find

$$ds^2 = -dt^2 + a^2 \left[\frac{dR^2}{1 - K(R)R^2} + R^2 d\Omega^2 \right], \tag{11.3}$$

where

$$K(R) = -\frac{\zeta'(r)}{r} \left[\frac{2 + r\zeta'}{e^{2\zeta}} \right]. \tag{11.4}$$

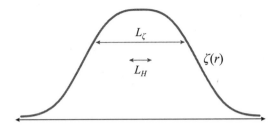

Figure 11.1 A density perturbation.

The metric in Eq. (11.3) resembles a closed Universe with curvature K. If the gradients are small compared to the horizon size, that is,

$$\frac{\partial K}{\partial R} \sim \frac{K}{L_\zeta} \ll \frac{K}{L_H},$$ (11.5)

where L_ζ is the size of the perturbation and L_H the Hubble radius, then we can neglect the gradients at leading order and consider the Friedmann equation in the so-called separate Universe approximation.[1] The Friedmann equation for a closed Universe is

$$H^2 + \frac{K}{a^2} = \frac{8\pi G}{3}\rho.$$ (11.6)

The mean density outside the perturbation, where $K = 0$, is $\overline{\rho}$:

$$\frac{3H^2}{8\pi G} = \overline{\rho},$$ (11.7)

and thus we can identify that the density contrast in ρ sources K:

$$\delta = \frac{\rho - \overline{\rho}}{\overline{\rho}} = \frac{K}{H^2 a^2}.$$ (11.8)

The evolution of the scale factor in a closed Universe is shown in Fig. 11.2. Turnaround has $\dot{a} = 0$ by definition, and so at turnaround we also have

$$\Rightarrow H = 0$$ (11.9)

$$\Rightarrow \frac{8\pi G}{3}\rho = \frac{K}{a^2}.$$ (11.10)

Up to this point, δ grows. When turnaround happens, the separate Universe approximation fails, which happens when gradients in K are important, that is, $\partial K/\partial x \gtrsim$

---

[1] This is very much akin to the spherical collapse which we saw in chapter 2.

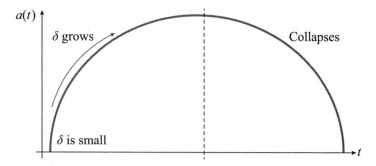

Figure 11.2 Evolution of a closed Universe.

$\mathcal{O}(1/L_H)$. The curvature radius is

$$L_\zeta = \frac{a}{\sqrt{K}}$$

$$(11.11)$$

$$L_H = H^{-1} \qquad L_K = L_H.$$

Thus turnaround happens when

$$\boxed{\Rightarrow \delta \approx 1.}$$

$$(11.12)$$

Collapse happens when $a = 0$, forming a singularity and a black hole. How big does a density perturbation have to be initially for this to occur?

At early times, the Universe is filled with radiation and the density perturbation for sub-horizon modes obeys

$$\boxed{\ddot{\delta}_\gamma + \left(c_s^2 \frac{k^2}{a^2} - H^2\right)\delta_\gamma = 0,}$$

$$(11.13)$$

with $c_s^2 = \frac{1}{3}$ the photon sound-speed and, recalling from section 5.2, $\dot{\delta}_\gamma = 4\Theta_{\gamma,0}$.

The photon perturbations only grow for $k < k_J$, given by

$$c_s^2 \frac{k_J^2}{a^2} = H^2 .$$

$$(11.14)$$

This is the photon 'Jeans scale': on smaller scales, $k > k_J$, radiation pressure is stronger than gravity and solutions for δ_γ oscillate rather than grow (we have met this phenomenon in section 6.2.3, where it led to the baryon acoustic oscillations). If we set $\delta = 1$ at the Jeans scale in order for non-linear collapse to take over, we find

$$\Rightarrow \frac{K}{c_s^2 k^2} = 1 .$$

$$(11.15)$$

Collapse then occurs when

$$\delta > \delta_c = c_s^2 = \frac{1}{3}. \tag{11.16}$$

A density perturbation with $\delta > \delta_c$ initially will collapse to a BH. The mass of the BH formed is the total mass within the Jeans radius, that is,

$$R_J = c_s H^{-1} \approx H^{-1}. \tag{11.17}$$

In other words, all of the mass within the horizon of the perturbation goes into a BH. So, if we can generate large-density perturbations in the radiation fluid, then we can form BHs. How many BHs we form depends on how rare density perturbations with $\delta > \delta_c$ are.

11.2 RELIC ABUNDANCE

11.2.1 PBH Density and Mass

We define the total density of PBHs at the time of collapse as

$$\rho_{\text{PBH}} \approx \beta \rho_r(z_{\text{collapse}}), \tag{11.18}$$

where β is the 'collapse fraction', that is, the fraction of the density field with $\delta > \delta_c$, and z_{collapse} is the redshift of the collapse. After collapse this fraction of the relic density redshifts, not like radiation, but like non-relativistic matter (assuming that this mechanism causes BHs to be 'born cold'). Thus the relic density of PBHs today is simply

$$\Rightarrow \rho_{\text{PBH},0} = \beta \left(\frac{\pi^2}{30} g_{\star,R} T_{\text{collapse}}^4 \right) \underbrace{\frac{g_{\star,S}(T_0) T_0^3}{g_{\star,S}(T_{\text{collapse}}) T_{\text{collapse}}^3}}_{\text{conserved comoving number density}}, \tag{11.19}$$

where $\rho_{\text{PBH},0}$ is the abundance today and

$$\rho_{R,0} = \frac{\pi^2}{30} g_{\star,R}(T_0) T_0^4 = \frac{\rho_{\text{crit}} \Omega_m}{(1 + z_{\text{eq}})}. \tag{11.20}$$

Rearranging, we can write the relic density as

$$\Omega_{\text{PBH}} h^2 = \beta \frac{\Omega_m h^2}{(1 + z_{\text{eq}})} \left(\frac{T_{\text{collapse}}}{T_0} \right) \frac{g_{\star,R}(T_{\text{collapse}})}{g_{\star,R}(T_0)} \frac{g_{\star,S}(T_0)}{g_{\star,S}(T_{\text{collapse}})}, \tag{11.21}$$

where, at high T,

$$g_{\star,R} = g_{\star,S} \tag{11.22}$$

and, at low T,

$$\frac{g_{\star,S}}{g_{\star,R}} = 1.16. \tag{11.23}$$

Setting reference parameters, we then get

$$\Omega_{\text{PBH}}h^2 = 0.12\left(\frac{\Omega_m}{0.3}\right)\left(\frac{h}{0.7}\right)^2\left(\frac{1+z_{\text{eq}}}{3400}\right)^{-1}\left(\frac{T_{\text{collapse}}}{1\,\text{GeV}}\right)\left(\frac{2.725\text{K}}{T_0}\right)\left(\frac{\beta}{6.5\times 10^{-10}}\right).$$

$$(11.24)$$

Getting the correct abundance requires a precise, and very small, value of β, which depends on T_{collapse}. The PBH mass is fixed by T_{collapse} by finding the mass contained within the horizon, M_H:

$$M_{\text{PBH}} \approx M_H \approx \frac{4\pi}{3}\rho(T_{\text{collapse}})H(T_{\text{collapse}})^{-3}$$

$$= \frac{4\pi}{3}\frac{\pi^2}{30}g_{\star,R}T^4\left(\frac{\pi^2}{90}g_{\star,R}\frac{T^4}{M_{\text{Planck}}^2}\right)^{-\frac{3}{2}}$$

$$(11.25)$$

$$\Rightarrow M_{\text{PBH}} = 5\times 10^{-2}M_\odot\left(\frac{g_{\star,R}(T_{\text{collapse}})}{100}\right)^{-\frac{1}{2}}\left(\frac{1\,\text{GeV}}{T_{\text{collapse}}}\right)^2.$$

$$(11.26)$$

This estimate based on the horizon mass is only parametrically correct: in fact 'critical collapse' leads to a dependence on the amplitude of the fluctuation itself, leading to a *mass function* for PBHs (discussed e.g. in Ref. [68]).

Note that if $T_{\text{collapse}} = M_{\text{Planck}}$,

$$\Rightarrow M_{\text{PBH}} \approx M_{\text{Planck}} \approx 10^{-8}\,\text{kg}.$$

$$(11.27)$$

Taking another reference time for PBH formation as nucleosynthesis, that is, $T \approx 1$ MeV, we find another reference mass:

$$\Rightarrow M_{\text{PBH}} \approx 5\times 10^4 M_\odot.$$

$$(11.28)$$

11.3 INFLATIONARY MODEL FOR PBHs

11.3.1 Sketch of the Model

What does a model for producing density perturbations that can give rise to PBHs look like? We can imagine histogramming the density perturbations (i.e. making a probability density function (PDF)), as shown in Fig. 11.3. We first take only those density perturbations of a given radius, by choosing our filter. This is related to a mass scale, M_*. The variance of the density perturbation on this scale is the width of the histogram, σ. Single-field-slow roll inflation gives us almost Gaussian distributions (the quantum field is only weakly interacting), and so we will use a Gaussian distribution to perform illustrative calculations. The collapse fraction is given by the fraction of the PDF in the tail of the distribution, which must be tuned to just the right very small value. Because β must be very small, it is exponentially

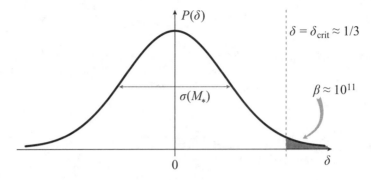

Figure 11.3 Probability distribution of density perturbations. A large density perturbation $\delta > \delta_{\text{crit}}$ should have $\beta \approx 10^{-11}$.

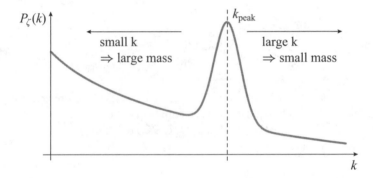

Figure 11.4 Sketch of a power spectrum with a peak at some fixed scale k_{peak}.

sensitive to the value of the variance: if the variance is just slightly too big or slightly too small, the PBHs at this mass scale will give far too much or far too little DM.

We can construct a density field with these properties by starting with a power spectrum of density perturbations, that is, density perturbations on many scales k. To produce PBHs around one fixed mass scale, we introduce a peak in the power spectrum at some scale k_{peak}, with all other values nearby smaller than it. Such a power spectrum is sketched in Fig. 11.4. Furthermore, the value of P near this scale must be much larger than the value at CMB scales, where we know $P_\zeta \sim 10^{-9} \Rightarrow \zeta \sim 10^{-5}$. If for this power spectrum we compute $\sigma^2(R)$ as in Eq. (11.32) we can associate a mass to the scale R using the enclosed mass in the spherical top hat as $M = 4\pi R^3 \bar{\rho}/3$ where $\bar{\rho}$ is the mean cosmic density (everything feels the curvature perturbation). We can thus write the variance as a function of mass scale, $\sigma(M)$, and it would have a peak at a scale corresponding to the peak in $P_\zeta (k \approx 1/R)$. If the perturbations are approximately Gaussian, then they are entirely described by the power spectrum, and we can compute $\beta(M)$ analytically. In general, however, models of inflation that can generate PBHs can deviate strongly from Gaussianity, requiring a numerical solution for $\beta(M)$. If $\sigma(M)$ is strongly peaked, then the DM

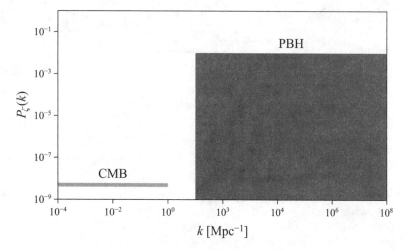

Figure 11.5 Sketch (based on results in Ref. [69]) of the known amplitude of curvature per-turbations. On large scales (small k), the CMB measures the primordial curvature power spectrum at the level of $P_\zeta(k) \approx 10^{-9} \sim \zeta^2$. On smaller scales, the requirement that we do not overproduce PBH DM limits $P_\zeta(k) \lesssim 10^{-2}$.

relic density will be dominated by PBHs at the peak scale. If we precisely tune the value of P_ζ at the peak, then we can ensure that $\beta \approx 10^{-11}$.

Thus, we can get DM from PBHs with mass M by inserting features in the pri-mordial power spectrum at $k(T_{collapse})$. We can then choose the amplitude of the feature to get $\beta \approx 10^{-11}$. The exponential sensitivity of β to the amplitude implies a *severe fine-tuning*, that is, it is very easy to get far too much or far too little DM. On the upside, keeping $\Omega_{PBH}h^2 \leq 0.12$ gives a robust constraint on the power spectrum on small scales. If the power spectrum were too large, we would produce too much PBH DM. This roughly limits $\zeta \lesssim 0.1$ on small scales. The situation is sketched in Fig. 11.5.

As we saw in section 5.3, the theory of inflation explains the primordial curvature power spectrum as being due to quantum fluctuations of a scalar field ϕ rolling in a potential $V(\phi)$. The curvature power spectrum on a scale k at late times is given by the curvature generated from vacuum fluctuations when the scale k left the horizon during inflation, which during slow roll is

$$P_\zeta(k) = \left(\frac{H}{2\pi\dot\phi}\right)^2 \Bigg|_{k=aH}. \qquad (11.29)$$

During inflation, H is almost constant. Due to the shrinking horizon during inflation, large-scale modes (small k) leave the horizon at the start of inflation, while small-scale modes leave the horizon at late times. Thus we can enhance P_ζ on small scales for PBH production, while keeping it at the right value on large scales to match the CMB, if we cause the inflaton field to roll even more slowly for some short period in the late stages of inflation: a smaller $\dot\phi$ at the time when k_{peak} leaves the horizon will enhance $P_\zeta(k_{peak})$. A sketch of a potential that can achieve this is given in Fig. 11.6.

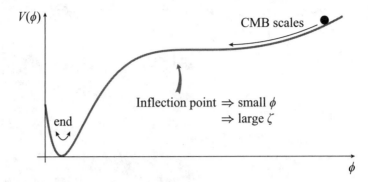

Figure 11.6 Sketch of an inflationary potential for PBH production. The beginning of inflation is somewhere to the right of the potential. Here the inflaton field is slowly rolling and
generating the $\zeta \sim 10^{-5}$ fluctuations seen on large scales in the CMB. At later times during
inflation, the inflaton passes through an inflection point in the potential, where it goes into
ultra-slow roll. The slower roll enhances the curvature power according to Eq. (11.29). If
the potential has just the right shape, then the modes exiting the horizon at this precise time
will have $\zeta \sim 0.1$. Later still, the inflaton reaches the minimum of its potential and oscillates,
reheating the Universe. During the subsequent radiation-dominated epoch, the modes with
large ζ re-enter the horizon, sourcing density perturbations in the radiation fluid that collapse
to form primordial black holes in rare regions where $\delta > \delta_{\text{crit}}$.

A precise reverse engineering of the necessary potential can be done, and you can
read about it in more detail in Ref. [70].

11.3.2 Computing the Collapse Fraction

We can compute β if we know the PDF of the overdensity, δ:

$$\beta = \int_{\delta_{\text{crit}}}^{\infty} \underbrace{P(\delta)\, d\delta}_{\text{PDF of }\delta}, \tag{11.30}$$

with $\delta_{\text{crit}} \approx \frac{1}{3}$.

For Gaussian fluctuations

$$\beta \simeq \int_{\delta_{\text{crit}}}^{\infty} \frac{d\delta}{\sqrt{2\pi}\sigma} \exp\left(-\frac{\delta}{2\sigma^2}\right), \tag{11.31}$$

where σ is the variance of fluctuations evaluated on the mass scale M_{PBH}. This is
defined from the power spectrum as we have seen:

$$\sigma^2(R) = \int d^3k \underbrace{P_\delta(k)}_{\text{power spectrum of }\delta} W^2(kR), \tag{11.32}$$

where $W(kR)$ is the window function Fourier transform. The window function
reflects the shape that we expect the density perturbations to have, and we are asking what the variance of the density field is filtered using this template. Taking a

spherical real space top hat for simplicity (i.e. a uniform sphere), we find

$$\Rightarrow W(kR) = \frac{3}{(kR)^3}\Big[\sin(kR) - kR\cos(kR) \Big]. \tag{11.33}$$

Here, R is the radius of the perturbation, which is the size of the horizon when the PBH forms (see below).

Approximating the integral yields:

$$\beta \approx \frac{1}{\sqrt{2\pi}} \frac{\sigma(M_{\text{PBH}})}{\delta_{\text{crit}}} \exp\Big(-\frac{\delta_{\text{crit}}^2}{2\sigma(M_{\text{PBH}})} \Big). \tag{11.34}$$

So β is *exponentially sensitive to* $\sigma(M_{\text{PBH}})$.

We relate P_δ to the primordial curvature spectrum, that is, the initial conditions (see Ref. [71]).

$$P_\delta(k) = \frac{16}{81}(kR)^4 P_\zeta(k). \tag{11.35}$$

If the perturbation on scale k has sufficient amplitude, the PBH forms when the mode crosses the horizon, that is, when

$$\boxed{a_{\text{form}} H_{\text{form}} = k.} \tag{11.36}$$

Thus, the radius of the perturbation is $R \approx k^{-1}$.

Using conservation of entropy, the scale factor at T_{collapse} is related to the scale factor today as

$$a = \frac{g_{\star,S}(T_0)^{\frac{1}{3}} T_0}{g_{\star,S}(T_{\text{collapse}})^{\frac{1}{3}} T_{\text{collapse}}}. \tag{11.37}$$

Friedmann's equation allows us to also find $H(T_{\text{collapse}})$, and thus we can find a the relationship between the PBH mass resulting from density perturbations with wavenumber given by k.

QUIZ

  i. In the separate Universe approximation, we require the curvature radius
     to be:
     a. Larger than the Hubble radius.
     b. Smaller than the Hubble radius.
     c. Equal to the Hubble radius.

 ii. Critical collapse requires density perturbation Δ:
     a. > 0.
     b. $> 1/3$.
     c. $> 1/2$.
     d. > 1.

iii. The collapse time is given by the time when:
     a. Mode k enters the horizon.
     b. Mode k exits the horizon.
     c. Nucleosynthesis.
     d. $T = 1$ GeV.

 iv. The PBH mass is determined by:
     a. The proton mass.
     b. The mass of the inflaton.
     c. The horizon mass at collapse.
     d. The dark matter density.

  v. The PBH collapse fraction (select multiple):
     a. Is exponentially sensitive to the variance of the density field.
     b. Is insensitive to the variance of the density field.
     c. Determines the PBH relic density.
     d. Is found from the PDF of the density field.
     e. Is found from the curvature radius of the collapsing region.
     f. Can be predicted from the curvature power spectrum.

 vi. During inflation, the inflaton field must do what? (select multiple):
     a. Oscillate in its potential.
     b. Create dark matter from quantum fluctuations.
     c. Drive the expansion of the Universe.
     d. Create primordial black holes.
     e. Have quantum fluctuations, sourcing curvature.
     f. Roll slowly in its potential.

vii. The primordial curvature power spectrum (select multiple):
     a. Is measured on CMB scales.
     b. Is measured on PBH scales.

c. Can be generated successfully by inflation.

d. Must be larger on PBH scales than CMB scales.

e. Must be smaller on PBH scales than on CMB scales.

f. CMB scales are longer than PBH scales.

g. PBH scales are longer than CMB scales.

Answers

i. a; ii. b; iii. a; iv. a; v. c; vi. a,c,d,f; vii. c,e,f; viii. a,c,d,f

Problems on Theories of Dark Matter

Problem II.1. We have seen that the Dirac matrices obey the Clifford algebra given by

$$\{\gamma^\mu, \gamma^\nu\} = \gamma^\mu \gamma^\nu + \gamma^\nu \gamma^\mu = 2\eta^{\mu\nu}, \tag{11.38}$$

where $\eta^{\mu\nu}$ are the components of the Minkowski metric with signature $(+ - - -)$.

a) Show that $\gamma^5 \gamma^\mu = -\gamma^\mu \gamma^5$.
b) Show that $(\gamma^5)^2 = 1$.

Problem II.2. We know that the elements of SU(2) are formed by $M = e^{i\theta_a T^a}$ where the set of matrices T^a are known as the *generators* of the group.

a) For a three-dimensional representation of SU(2), these are given by

$$T^1 = \begin{pmatrix} 0 & 0 & 0 \\ 0 & 0 & -i \\ 0 & i & 0 \end{pmatrix}, \quad T^2 = \begin{pmatrix} 0 & 0 & i \\ 0 & 0 & 0 \\ -i & 0 & 0 \end{pmatrix}, \quad T^3 = \begin{pmatrix} 0 & -i & 0 \\ i & 0 & 0 \\ 0 & 0 & 0 \end{pmatrix}. \tag{11.39}$$

Show that $[T^a, T^b] = i\epsilon^{abc} T^c$ where ϵ^{abc} is the totally antisymmetric Levi-Civita symbol, with $\epsilon^{123} = 1$.

b) The Pauli matrices can be used to write a two-dimensional representation of SU(2). Calculate $(\sigma^3)^2$, $(\sigma^3)^3$, $(\sigma^3)^4$ where σ^3 is the third Pauli matrix.

c) Hence show that

$$e^{i\theta\sigma^3/2} = \begin{pmatrix} \cos(\theta/2) - i\sin(\theta/2) & 0 \\ 0 & \cos(\theta/2) + i\sin(\theta/2) \end{pmatrix} \tag{11.40}$$

$$= 1\cos(\theta/2) - i\sigma^3 \sin(\theta/2). \tag{11.41}$$

This is the SU(2) matrix analogue of $e^{i\theta} = \cos\theta + i\sin\theta$.

Hint: the exponential of a matrix is defined through the Taylor expansion of the exponential function

$$e^M \equiv 1 + M + \frac{1}{2}M^2 + \frac{1}{3!}M^2 + \cdots \tag{11.42}$$

Problem II.3. Show explicitly that $D_\mu D_\nu \psi$ transforms covariantly, that is, $D_\mu D_\nu \psi \to e^{-i\Lambda} D_\mu D_\nu \psi$.
Hint: make the gauge transformations $\psi \to e^{-i\Lambda}$ and $A_\mu \to A_\mu + \partial_\mu \alpha$.

Problem II.4. The complex Klein–Gordon equation, $\partial_\mu \partial^\mu \phi - m^2 \phi = 0$, has a global symmetry $\phi' = e^{-i\Lambda}\phi$. Write down the local, gauged version of the complex Klein–Gordon equation.

Problem II.5. This question is all about neutrino decoupling. We're going to start by making assumptions and quick estimates and then justify them by more explicit calculation.

a) Estimate the neutrino annihilation cross section σ, for the process $\nu\bar{\nu} \longleftrightarrow e^+ e^-$ with coupling constant g, in the Fermi limit, $m_i \ll E_\nu \ll M_Z^2$, where m_i is the electron or neutrino mass, M_Z^2 is the Z-boson mass, and $E_\nu \sim T$ is the incoming neutrino energy. Your result should depend explicitly on the temperature, T.

   Write the cross section in terms of Fermi's constant, $G_F = (\sqrt{2}/8)g^2/M_W^2$. The W-boson mass, M_W, is related to the Z-boson mass by the Weinberg angle $M_Z = M_W/\cos\theta_W$. Finally, write this in terms of the Higgs vev, $v = (\sqrt{2}G_F)^{-1/2}$, otherwise known as the weak scale.

b) Taking the thermally averaged cross section to be equal to $\langle\sigma v\rangle = c_\nu\sigma$ (where c_ν is some constant) find the neutrino freeze-out temperature from $n_{EQ}\langle\sigma v\rangle = H$ (also called neutrino decoupling). You can initially assume $g_\star(T_f) = 10$; then, once you have computed T_f, justify this value by explicit computation of $g_\star$ from the definition and the particle content of the Standard Model.

c) Using the definition of $\langle\sigma v\rangle$ given in chapter 8, Eq. (8.26), compute c_ν from part (a), taking the limit $T \gg m_\nu$.

Problem II.6. a) For complex field Φ and real field ϕ show which of the following potentials have U(1) global symmetry and which do not.

$$V = \frac{1}{2}m^2|\Phi|^2, \tag{11.43}$$

$$V = \frac{1}{2}m^2\phi^2, \tag{11.44}$$

$$V = \frac{1}{2}m^2\Phi^2, \tag{11.45}$$

$$V = \sum_n c_n|\Phi|^n, \tag{11.46}$$

$$V = \sum_n c_n|\Phi^n|, \tag{11.47}$$

$$V = \sum_n c_n\phi^n, \tag{11.48}$$

$$V = \sum_n c_n|\Phi|^n + A\mathrm{Re}[\Phi]^p. \tag{11.49}$$

b) For the potential

$$V = \lambda(|\Phi|^2 - v^2/2)^2, \tag{11.50}$$

first show that there is a U(1) global symmetry. Next, by a change of variables, show that the U(1) symmetry is equivalent to a shift symmetry $\theta \rightarrow \theta + \text{const.}$, where θ is the phase of Φ.

c) Thus show that adding a term $\Delta V = \mu^3 \text{Re}[\Phi]$ to the potential in Eq. (11.50) breaks the U(1) symmetry and the shift symmetry. Substituting $\Phi = v e^{i\theta}/\sqrt{2}$, what is the potential for θ?

d) Repeat for $\Delta V = \mu^3 \text{Im}[\Phi]$. Where is the minimum of the θ potential now? If the only other term in the Lagrangian is $\mathcal{L} = -(\partial_\mu \Phi)(\partial^\mu \Phi^*)$, show by change of variables that the two choices for ΔV give rise to equivalent physics for the phase θ.

Problem II.7. **a)** Assuming a single fluid Universe with constant equation of state w satisfying $\dot{\rho} = -3H(1+w)\rho$, first solve Friedmann's equation in the form:

$$3H^2 M_{\text{Pl}}^2 = \rho \tag{11.51}$$

for $a(t)$ and thus $H(t)$.

b) Using the definition of the D'Alembertian:

$$\Box = \frac{1}{\sqrt{-g}} \partial_\mu \left[\sqrt{-g} g^{\mu\nu} \partial_\nu \right], \tag{11.52}$$

show that $\Box = -\partial_t^2 - 3H\partial_t$ for the Friedmann metric, $g = \text{diag}[-1, a^2, a^2, a^2]$. Thus show that the Klein–Gordon equation $\Box \phi - \partial_\phi V = 0$ for $V = m^2\phi^2/2$ is

$$\ddot{\phi} + 3H\dot{\phi} + m^2\phi = 0. \tag{11.53}$$

c) Substitute your solution for $H(t)$ into the Klein–Gordon equation and derive the solutions for $\phi(t)$ given in chapter 10. First find the change of variables for the Bessel function solution, and express the order, n, in terms of w. Next find the WKB solution for $\phi = A(t) \cos mt$ taking $\dot{A} \ll mA$, and show that in this limit $A(t)$ is independent of w.

d) Given that the energy density and pressure are $\rho = \frac{1}{2}\dot{\phi}^2 + V(\phi)$ and $P = \frac{1}{2}\dot{\phi}^2 - V(\phi)$, derive the behaviour of the equation of state, $w = P/\rho$. What is the asymptotic value of w for $m \ll H$, and $\langle w \rangle$ for $m \gg H$ (brackets denote period average)?

e) Repeat this exercise for $V = \lambda\phi^4$ (you may use the WKB approximation). What do you notice about the result for $\langle w \rangle$?

PART III

Testing Dark Matter

Chapter Twelve

WIMP Direct Detection

As discussed in chapter 9, WIMPs are produced via thermal freeze out in the early Universe. The process $\chi\,\chi \longleftrightarrow \text{SM}\,\text{SM}$ maintains chemical equilibrium (changing number density, inelastic scattering), and the process $\chi\,\text{SM} \longleftrightarrow \chi\,\text{SM}$ maintains kinetic equilibrium (elastic scattering). The WIMP relic abundance depends on the thermally averaged total-annihilation cross section $\langle \sigma v \rangle$, with averaging defined over the distribution function at temperature T.

The same processes also give us the 'holy trinity' of WIMP searches, shown in Fig. 12.1:

- The $\text{SM}\,\text{SM} \to \chi\,\chi$ process is searched for in *collider* experiments such as at the Large Hadron Collider (LHC).
- The $\chi\,\chi \to \text{SM}\,\text{SM}$ process is looked for in *indirect* detection experiments which try to detect high-energy photons produced by the annihilation of WIMPs in areas of high dark matter density.
- The $\chi\,\text{SM} \to \chi\,\text{SM}$ process is looked for in *direct* detection experiments which attempt to detect the recoil of SM nuclei on Earth due to collisions with WIMPs that make up the local dark matter 'wind'.

While the processes are related to those in the early Universe, the averaged cross section can be quite different due to different distributions and energetics for initial and final states.

We begin with direct detection. Indirect detection and other astrophysical limits are discussed in chapter 13. Collider limits are beyond the scope of this textbook since the discovery of the WIMP in this context would not indicate that the WIMP must also be DM.

12.1 WIMP NUCLEAR RECOIL

When a WIMP collides with some SM nucleus N they interact via some 'portal' boson (e.g. Z, W or H from the SM or possibly some yet undetected boson). For elastic scattering, the WIMP and nucleon final states are changed only by their momenta. If the final states are excited, which we show by adding a label χ^*, N^*, shown schematically in Fig. 12.2, then this is known as inelastic scattering. In the non-relativistic limit in the lab frame, the initial momenta of the nucleus and WIMP are given by

$$p_\mu = (m_\text{N}, \mathbf{0}), \tag{12.1}$$

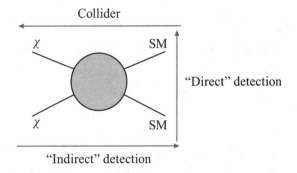

Figure 12.1 WIMP 'holy trinity' searches.

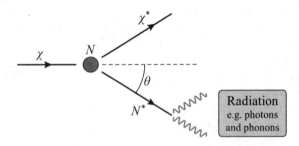

Figure 12.2 WIMP-nucleon recoil event.

and

$$k_\mu = (m_\chi + \frac{1}{2}m_\chi v_i^2, m_\chi v_i, 0, 0), \qquad (12.2)$$

respectively. After the interaction, the pair have momenta given by

$$p'_\mu = (m_N^* + \frac{1}{2}m_N v_{Nf}^2, \mathbf{p'}), \qquad (12.3)$$

and

$$k'_\mu = (m_\chi^* + \frac{1}{2}m_\chi v_{\chi f}^2, \mathbf{k'}). \qquad (12.4)$$

The 4-momentum transfer $q_\mu = p_{\mu'} - p_\mu$ and is given by

$$q_\mu = \left(E_R, \sqrt{2m_N E_R}\cos\theta, \sqrt{2m_N E_R}\sin\theta, 0\right), \qquad (12.5)$$

where E_R is the *nuclear recoil energy*. In the limit of elastic scattering, $m_N^* = m_N$ and $m_\chi^* = m_\chi$, and the recoil energy is given by

$$E_R = \frac{\mu^2 v_i^2}{m_N}, \qquad (12.6)$$

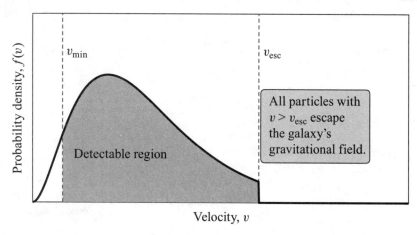

Figure 12.3 DM-speed distribution–Maxwell–Boltzmann distribution truncated at the galactic escape velocity.

where μ is the nucleus-DM reduced mass given by

$$\mu = \frac{m_N m_\chi}{m_N + m_\chi}. \tag{12.7}$$

The final state nucleon eventually loses this recoil energy, for example in the form of photons or phonons that can be detected.

As an example, we will consider the nucleus of xenon with an atomic mass of $A = 131$. Due to the Sun's motion through the galaxy, dark matter is expected to travel through the Earth at a speed of around 0.1% of the speed of light. For the canonical WIMP mass of about 100 GeV we find $E_R \approx 50$ keV. This recoil energy is smaller for lighter dark matter particles. Thus, to detect WIMP nuclear recoil, one needs to be able to resolve recoil energies on the order of 10s of keV.

Additionally, for some minimally detectable recoil energy, there exists a minimum dark matter velocity that we can detect. This is given by

$$v_{\min} = \sqrt{\frac{m_N E_R}{2\mu_N^2}}, \tag{12.8}$$

where again we have assumed elastic scattering. This minimum energy is important as the dark matter passing through the Earth does not have a single velocity but instead has a velocity distribution. Therefore, we expect that some of the dark matter will be moving too slowly to be detectable. The dark matter is expected to have a velocity given by the Maxwell–Boltzmann distribution which is truncated at the galactic escape velocity as shown in Fig. 12.3. For a particular detectable recoil energy and nucleon mass, the minimum detectable velocity will become larger than the escape velocity at some value of m_χ. DM lighter than this value is not detectable via nuclear recoil. This sets the range of masses to which a given detector is sensitive.

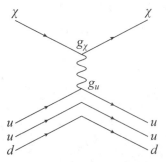

Figure 12.4 WIMP-proton scattering.

To calculate the number of recoil events we would expect in a direct detection experiment, we have to calculate the distribution of events. The recoil rate, R, per unit recoil energy, per unit detector mass, also known as the differential recoil rate, is given by

$$\frac{dR}{dE_R} = \frac{1}{m_N} \left(\frac{\rho_{DM}}{m_\chi} \right) \left\langle v \frac{d\sigma}{dE_R} \right\rangle. \qquad (12.9)$$

On the right-hand side of this equation, the first term normalises the equation to a unit target nucleus, the second term is the dark matter number density, and the last term is an average of the differential cross section over the local dark matter velocity distribution v_{DM}.

The total rate per nucleus in our detector is then

$$R = \frac{\rho_{DM}}{m_\chi} \langle \sigma v \rangle. \qquad (12.10)$$

It is important to note that here $\langle \sigma v \rangle$ is the cross section with a nucleus and not the total annihilation cross section that we met previously, and the angle brackets average the local distribution using the standard halo model and recoil energy spectrum.

12.2 CROSS-SECTION ESTIMATES

To estimate the cross section for this process, we start by considering the interactions of χ with a proton p, shown in Fig. 12.4, and with a neutron n. We assume that the interaction is mediated by some vector boson Z'. The matrix element for the DM–proton interaction then scales like:

$$\mathcal{M} \sim \frac{1}{q^2 - m_{Z'}^2} g_\chi (2g_u + g_d). \qquad (12.11)$$

Similarly, the DM–neutron interaction is found to be

$$\mathcal{M} \sim \frac{1}{q^2 - m_{Z'}^2} g_\chi (2g_d + g_u). \tag{12.12}$$

We're interested in the DM particles scattering off of the entire nucleus, not just individual nucleons. DM can scatter coherently off of the entire nucleus when its de Broglie wavelength is larger than the nuclear radius ($\lambda_{dB} > r_N$). This is encoded into the cross section through the use of a form factor $F^2(q^2)$ through the equation

$$\sigma(q^2) = \sigma_0 F'^2(q^2), \tag{12.13}$$

where σ_0 is the scattering cross section in the limit of vanishing momentum transfer. This form factor ranges from 1 to 0 at very small and very large momenta respectively. Taking the momentum transfer much smaller than the mass of the boson ($q \ll m_{Z'}$) we find the DM proton cross section:

$$\sigma_p \approx \frac{1}{m_{Z'}^4} \left[g_\chi (2g_u + g_d) \right]^2 \mu_p^2 := f_p^2 \mu_p^2, \tag{12.14}$$

where this defines the effective proton coupling f_p, and $\mu_p = m_p m_\chi / (m_p + m_\chi)$ is the proton reduced mass. The spin-independent cross section of the whole nucleus can then be written as

$$\sigma_N = \sigma_p \frac{\mu_N}{\mu_p} \frac{[f_p Z + f_n (A - Z)]^2}{f_p^2}. \tag{12.15}$$

For heavy detector nuclei, the ratio of the WIMP-nucleus and the WIMP-proton reduced masses increases as A^2. Therefore, for these heavy nuclei, the spin-independent nuclear cross section scales as $A^4 \sigma_p$. This is not true for the spin-dependent cross section, thus allowing the spin-independent cross section to dominate. As such, to simplify the calculations, we often assume that the WIMP-nucleon interaction is purely spin-independent.

12.3 HOMESTAKE MINE EXPERIMENT

As a rough example calculation, we will consider the Homestake Mine Experiment. Based in South Dakota, the 1987 Homestake Mine Experiment was one of the first to directly constrain the WIMP DM parameter space. Just like modern WIMP direct detection experiments, the lab was built nearly 1500 m underground to protect the equipment from background radiation sources such as cosmic rays. They attempted to observe ionisation due to WIMP-nucleon interactions in a 135 cm³ sample of germanium ($A = 72$) over a total of 14 days.

For our example calculation take $g_d \sim g_u \sim g_\chi$ and recall our derivation of the WIMP miracle in section 9.2.1. We found that the total annihilation cross section was $\langle \sigma v \rangle_{\text{ann}} \sim G_F^2 m_\chi^2 \approx 10^{-26}$ cm³ s⁻¹ for a process mediated by the Z of the Standard Model. Comparing to Eq. (12.14) we see that $\sigma_p \sim (\mu_p/m_\chi)^2 \langle \sigma v \rangle_{\text{ann}}$. The rate

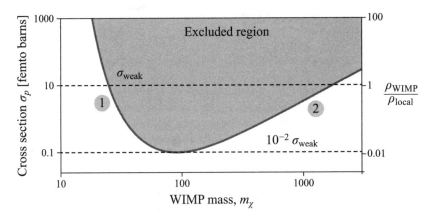

Figure 12.5 WIMP exclusion from direct detection—sketch of the excluded region of WIMP parameter space from the Homestake Mine Experiment. Region 1: reduced sensitivity to low-mass WIMPs from the minimum recoil energy around 10 keV. Region 2: reduced sensitivity to higher-mass WIMPs due to the reduced DM number density.

$R \approx (\rho_{DM}/m_\chi)\sigma_N v_{loc}$ with the local DM velocity $v_{loc} \approx 300$ km s$^{-1} \sim 10^{-3}$. Taking $\rho_{DM} = 0.3$ GeV cm^{-3} and $m_\chi = 100$ GeV we find a recoil rate of $R \sim 10^{-28}$ s^{-1} on every germanium nucleus.

In crystal form, germanium has a density of 5.3 g cm^{-3}. The total number of moles of germanium is therefore $N_{moles} = 135 \times 5.3/72 \approx 10$ mol. Then, multiplying by Avogadro's constant, we find the total number of atoms to be $N_{atoms} \approx 10 N_A \approx 6 \times 10^{24}$. Finally, 14 days of observation is a total of around $\tau = 10^6$ s. We therefore would expect the experiment to measure a total of $N = R\tau N_{atoms} \approx 10^{-29} \times 6 \times 10^{24} \times 10^6 = 600$ events. Instead, the experimenters saw *nothing*.

This doesn't mean that the WIMP is ruled out as a candidate for dark matter, but it does narrow down the range of properties the WIMP can have. Since the expected rate scales with both the WIMP-nucleon cross section σ_p and the local WIMP density ρ_{DM}, either the WIMP-nucleon cross section is around two orders of magnitude less than the electroweak cross section (which could happen if, for example, the mediator is heavier than the Z), or the local WIMP density is around two orders of magnitude less than expected. The excluded region of the WIMP parameter space is sketched in Fig. 12.5.

There are two key features in the mass-cross-section plane which are marked in Fig. 12.5. We can understand these features as follows:

1. There is a minimum recoil energy E_{min} which can be detected. Due to Eq. (12.8), as the WIMP mass decreases, the minimum detectable velocity increases. This then reduces the range of particle velocities that are detectable as shown in Fig. 12.3. Therefore, the detector loses sensitivity for decreasing WIMP masses.
2. Since the local dark matter density is fixed, as the WIMP mass increases, the number density of DM particles incident on our detector decreases, $n_\chi = \rho_{DM}/m_\chi$.

As a result, we expect fewer interactions and are less able to constrain the cross section.

Here we have only considered a heavily simplified calculation for predicting the WIMP-nucleon recoil rate within a detector. To calculate the rate fully, we have to integrate over the detectible range of the velocity dispersion as given by

$$\left\langle v \frac{d\sigma}{dE_R} \right\rangle = \int_{v_{min}}^{v_{max}} d^3 v v \tilde{f}(v) \frac{d\sigma}{dE_R}, \qquad (12.16)$$

where $\tilde{f}(v)$ accounts for the motion of the Earth. The average dark matter velocity we have considered so far comes from the orbit of the Sun around the galaxy. However, in addition to this, the Earth orbits around the Sun, meaning that the dark matter velocity experiences an annual modulation. As we will see later, this modulation may play a critical role in detecting dark matter.

12.4 MODERN DIRECT DETECTION TECHNIQUES

While the Homestake Mine Experiment has ruled out the WIMP cross section being mediated by the Z-boson of the SM, the hunt for WIMPs continues. This is because the interactions with SM fermions could instead be mediated by other particles such as the Higgs boson. The Higgs mediates a similar total annihilation cross section of DM to all SM particles as the Z-boson does. However, the Higgs couples to the SM in proportion to the quark masses and thus couples most strongly to b and t quarks and only weakly to the u and d quarks that dominantly make up nuclei. Thus the 'Higgs portal' can achieve the correct relic abundance for WIMPs, while having a much reduced σ_{SI} spin-independent nucleon cross section.

In a complicated model like SUSY, it is possible to perform 'scans' on the full parameter space and find regions giving rise to the correct relic abundance. It turns out that such regions can also have σ_{SI} much lower than the Z-boson mediated expectation. One example is Ref. [72], which computed a posterior probability conditioned on the correct relic abundance and direct detection limits at the time. As of now, this range has largely been excluded but with some posterior volume remaining between the current exclusions and the 'neutrino fog' (see section 12.5). In Bayesian language, we have updated our prior on SUSY. A more recent SUSY scan is Ref. [55], which found the frequentist preferred region, conditioned on direct detection and relic abundance. Much of this region has very low cross section and lies below the 'neutrino fog'. Low cross-section models in SUSY provide motivation to continue the search for WIMPs.

12.4.1 Detectors

Some of the most sensitive WIMP direct detection experiments are noble gas scintillators. When a scintillating material is hit by an incoming particle, it absorbs kinetic energy which it then later re-emits (scintillates) in the form of measurable

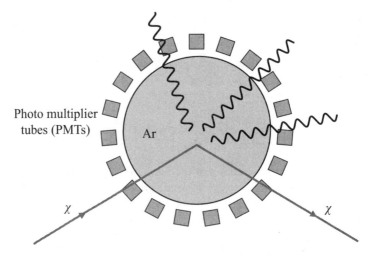

Figure 12.6 DEAP-3600–DM particles collide with 1 tonne of liquid argon and excite the nuclei causing them to emit scintillation photons. These photons are then detected by photomultiplier tubes.

light. Like all direct detection experiments, their sensitivity scales with the mass of detector material. Modern experiments use on the order of 1 tonne of target material.

One example is DEAP-3600 which, as the name suggests, uses a total of 3600 kg of liquid argon. Of this, 1000 kg make up the 'true' detectable region. Radiation emitted by the target is detected by an array of 255 photomultiplier tubes that surround it. The detector is cooled down to 87 K and encased in shielding to reduce background noise. As well as this, the detector was placed 2 km below the Earth's surface to further protect it from background radiation such as cosmic rays. A sketch of the experiment design is shown in Fig. 12.6.

A different detector design is used by XENON1T. This experiment uses a total of 3.2 tonnes of xenon with a fiducial volume of around 2 tonnes. As with similar experiments such as LUX and PandaX, it uses a two-phase detector using liquid xenon with a region of gaseous xenon above it as shown in Fig. 12.7. In this setup, scintillation events that occur in the liquid region produce electrons that drift upwards due to an applied field. Once these electrons reach the liquid gas boundary, a second scintillation event occurs. The photons produced by both events are detected by photomultiplier tubes above and below the target xenon.

Lastly, the Homestake experiment used germanium. A modern incarnation of a germanium-based detector is SuperCDMS. SuperCDMS was located at the Soudan mine in Minnesota, USA, and its successor will be housed at SNOLAB in Canada. SuperCDMS makes use of a large number of stacks of germanium crystals housed in a cryogenic environment at sub-1 K temperatures. The ultimate aim is around 200 kg of germanium target material.

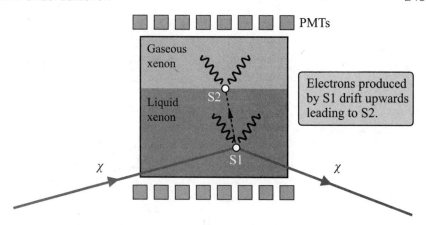

Figure 12.7 XENON1T–DM particles collide with 2 tonnes of liquid xenon causing a scintillation event. This produces an electron which drifts upwards due to an applied field causing an additional scintillation event. Both events produce photons which are then detected using photomultiplier tubes. See colour insert.

12.4.2 Backgrounds

Xenon and DEAP both use liquid noble gases as targets, but the designs are significantly different. This is because they have different approaches to distinguishing between their signal and their background.

Backgrounds are events that can be measured in the detector that share some characteristics with a DM recoil event. Examples of sources of backgrounds are:

- Cosmic rays giving rise to γ-rays.
- Radiogenic or cosmogenic neutrons.
- Solar, radiogenic or cosmogenic neutrinos.
- Radioactive decays of nuclei within the detector.

Distinguishing between signal and background is done by measuring some other characteristic of the event which indicates whether the signal comes from, for example, radiation in the equipment or DM. Each type of experiment has its own strengths and weaknesses.

In a germanium detector such as those used in SuperCDMS, the ionization yield of the events is measured. The yield for a dark matter scattering event is expected to be significantly lower than a typical background event as shown in the left panel of Fig. 12.8. A low-ionization event at high recoil energy would be known to be DM and not background.

The single-phase setup used by DEAP allows them to measure the 'pulse shape' and charge-to-light ratio of events. As we see in the middle panel of Fig. 12.8, using both variables leads to very clearly separated signal and background regions, but the signal region itself is quite small. This means that such a detector can be very confident of signal events.

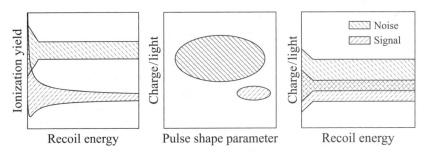

Figure 12.8 WIMP detector backgrounds—a sketch comparison of the signal properties with the background for a germanium detector (left), a single-phase liquid argon detector (middle) and a two-phase xenon detector (right). Figure inspired by Ref. [73].

The two-phase detectors like XENON1T measure the charge-to-light ratio of each event as a function of recoil energy only, without pulse shape discrimination. Thus, the separation between signal and background is not so good, but the signal region is much larger. Therefore, while they cannot differentiate between the signal and noise quite so well, this is compensated by having access to a potentially much larger number of signal events.

If we have a known background level with a mean b and an expected signal with mean μ then we can use *Poisson statistics* to calculate the probability of seeing n events via the equation

$$P(n|\mu) = \frac{(\mu + b)^n e^{-(\mu+b)}}{n!}. \tag{12.17}$$

Imagine we have very carefully designed an experiment such that we have no background ($b = 0$); however, we also observe no events ($n = 0$). By setting $P = 0.1$, we can use Eq. (12.17) to constrain our prediction to the 90% confidence interval. The mean signal μ will be given by the rate times the duration of the observation, and so we can set an exclusion on the DM cross section and mass. In real examples of data analysis the statistical analysis is more involved (see e.g. the statistics section of the PDG [37]), but Eq. (12.17) should give a sense of the basics.

12.5 THE NEUTRINO FOG

Constraints on the WIMP-nucleon cross section can always be improved by building a bigger detector or by simply running our existing detectors for a longer amount of time. It is then natural to ask at what point do we stop? Will we be forever building bigger and better detectors to rule out more and more of the WIMP parameter space? One potential stopping point comes from the *neutrino fog*.

Currently, we don't know the WIMP-proton cross section (if WIMPs exist at all). However, we do know the cross section for neutrinos interacting with protons. Additionally, we know that neutrinos are extremely abundant on Earth as

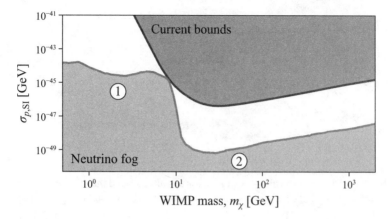

Figure 12.9 Neutrino fog—comparison between WIMP bounds from XENON-1T and the potential limit of WIMP direct detection which arises from the neutrino fog. Region 1: for low WIMP masses ($m_\chi \lesssim 10$ GeV) the fog is primarily made up of solar neutrinos. Region 2: for higher masses ($m_\chi \gtrsim 10$ GeV) it is atmospheric and 'diffuse supernova neutrinos' which make up the fog.

they are emitted from the Sun, the atmosphere and distant supernovae. Importantly, these neutrinos will also pass through our WIMP direct detection experiments and, while their interaction cross section is very small, WIMP detectors will soon be large enough to detect them (think for comparison of the size of the Super-Kamiokande water Cherenkov neutrino detector, which is dozens of kilotonnes). This is problematic as these neutrinos will have a differential recoil rate that mimics dark matter.

Fig. 12.9 sketches the neutrino fog (see Ref. [74] for a precise definition) and representative contemporary exclusions (in this case, XENON-1T [75], from 2018). The neutrino fog is not so far from the current exclusions and will soon be an issue for detecting WIMPs of masses around 10 GeV. Eventually, the entire mass range will be affected.

All hope will not be lost once experiments reach the neutrino fog as there are some ways to differentiate between a dark matter signal and a neutrino signal. One key approach comes from the annual and daily modulation in the dark matter signal. The velocity of the dark matter wind we have considered so far ($v \approx 200$ km s^{-1}) is the velocity at which the Sun orbits through the Milky Way. The orbit of the Earth around the Sun gives the relative dark matter velocity a small ($\sim 10\%$) yearly modulation. This means that in June, when the Earth's velocity is in the same direction as the Sun's, we expect that slightly more dark matter will pass through our detector every second than in December when the opposite is true. We therefore would expect any dark matter signal to oscillate over one year, peaking in June. Since the neutrino background and most other background sources are expected to be constant, this modulation would be a 'smoking gun' for the presence of dark matter. As well as this, we would also expect a much smaller daily modulation due to the rotation of the Earth about its axis.

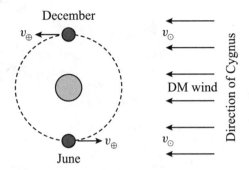

Figure 12.10 Dark matter wind—due to the Earth's orbit around the Sun, the mean velocity of the dark matter wind oscillates over the year, peaking in June and falling to a minimum in December.

Unfortunately, the amplitude of this modulation is small and some backgrounds may also exhibit some kind of annual modulation. A similar approach is to build detectors that can measure the direction of the incoming particles. Again, due to the Sun's orbit through the Milky Way, the distribution of WIMP-induced nuclear recoils should be strongly peaked in the opposite direction to the Sun's motion in the galaxy. Other background sources would either have no directional dependence or one which is distinctly different, since sources unrelated to DM would not be sensitive to the standard of rest with respect to the DM halo. For example, the solar neutrino spectrum would naturally be peaked in the direction of the Sun, which is a different direction to the direction of the Sun's motion (the Sun is never in the constellation Cygnus towards which we are moving).

Lastly, different target nuclei experience different cross-section 'fogs' due to the neutrino background. This can be used to differentiate between a signal from neutrinos and WIMPs by comparing the results of detectors using different nuclei.

QUIZ

i. What is the approximate energy scale of nuclear recoils for a 100 GeV WIMP?
   a. 10s of keV.
   b. 10s of eV.
   c. 10s of GeV.
   d. The weak scale.

ii. The galactic escape velocity plays an important role in WIMP detection because:
   a. High-mass WIMPs are pulled to the centre of the galaxy and evade detection.
   b. Low-mass WIMPs escape the galaxy and evade detection.
   c. A maximum velocity implies recoil energy thresholds can't be passed at low WIMP mass.
   d. The escape velocity implies a maximum energy for any WIMP, excluding high masses.

iii. The DM nucleon cross section (select multiple):
   a. Is always spin independent.
   b. Depends on the mass of the 'portal' boson.
   c. Is equal to the total annihilation cross section.
   d. Can be much smaller than the total annihilation cross section.
   e. Is excluded for couplings with the Z.
   f. Is ruled out for all possible WIMPs.

iv. Consider Fig. 12.9. Assign the features to the causes. Feature 1: rise of the red curve at low mass. Feature 2: rise of the red curve at large mass. Feature 3: the brown-shaded region.
   a. Caused by neutrinos interfering with the WIMP signal.
   b. Caused by a minimum threshold recoil energy.
   c. Caused by the reduced number density of WIMPs.

v. Consider Fig. 12.8. Which statement refers to which detector type?
   a. Best signal/background discrimination.
   b. Largest signal region.
   c. Worst signal/background discrimination.

Answers

i. a; ii. c; iii. b,d,e; iv. a-3,b-1,c-2; v. argon, xenon, xenon

Chapter Thirteen

WIMP Astrophysics and Indirect Detection

13.1 WIMP MASS LOWER BOUND

The physics of galaxy formation restricts theories of DM. We will meet such restrictions for all of our theories: WIMP-like DM, axion-like DM and PBHs. There are two types of bound, the first is *local* (i.e. independent of the cosmological history), arising from the density distribution of DM within galaxies observed at present, related to what we learned in chapters 3 and 4 about DM density profiles. The second type of bound is *cosmological*, arising from the theory of structure formation we developed in chapter 5.

13.1.1 Fermions and the Tremaine–Gunn Bound

The SUSY WIMP candidate, the neutralino, is a fermion. A lower bound on fermionic DM is arrived at by considering the *phase space density* and the *Pauli exclusion principle*. The number density, n, is the integral of the distribution function over momenta:

$$n = \int f(\vec{p}) d^3 p. \tag{13.1}$$

Note the units: f is dimensionless, so n has units $[p]^3 = [E]^3 = [V]^{-1}$. The exclusion principle states that no state can be multiply occupied, that is, $f(\vec{p}) \leq 1$. The total mass of a halo, M, is given by the particle mass, m_f (where subscript f is for 'fermion'), times the halo volume, V, times the number density. Since $f \leq 1$, we have that

$$M = m_f \times V \times n \lesssim m_f \times V \times \int d^3 p \sim m_f \times \left(\frac{4}{3}\pi R^3\right) \times (m_f v)^3, \tag{13.2}$$

where v is the characteristic velocity of particles in the halo, which we have approximated the integral $d^3 p$ to be peaked at. For v we use the *virial velocity* from chapter 2:

$$v = \sqrt{\frac{3}{5}\frac{GM}{R}}. \tag{13.3}$$

Plugging this into Eq. (13.2) and rearranging we arrive at an inequality for m_f known as the *Tremaine–Gunn bound*:

$$m_f \gtrsim \left[\frac{4}{3}\pi \left(\frac{3}{5}GR\right)^3 M\right]^{-1/8}. \tag{13.4}$$

If we observe a galaxy with inferred DM halo mass M and radius R, then fermionic DM must be heavier than the Tremaine–Gunn bound: if it is lighter, then the Pauli exclusion principle forbids a DM halo with the given properties from forming. Note that the $\mathcal{O}(1)$ constants in Eq. (13.4) differ in different references depending on the estimate of v used and the level of approximation taken on f.

The Tremaine–Gunn bound is strongest (largest lower limit) when applied to the smallest observed galaxies dominated by DM (lowest M and smallest R); these are the *dwarf spheroidal galaxies*. Roughly, these have masses on the order of $10^8 M_\odot$ and half-light radii of around 500 pc. Plugging these values in leads to the bound

$$\boxed{m_f \gtrsim 120 \text{ eV}} . \tag{13.5}$$

A more precise bound can be arrived at by solving the Jeans equation for the velocity profile, and the exclusion principle is applied by demanding that the Fermi velocity remains lower than the local escape velocity. The precise bound (see e.g. Ref. [76]) is very close to our estimate (although this is really a cheat from the choices we made for M and R).

13.1.2 Warm Dark Matter

A more generic bound (in the sense that it applies for bosons as well as fermions) arises for any DM that was relativistic when it was produced, that is, with $T \gtrsim m$. This occurs for thermal production but also for some non-thermal channels. The bound applies, for example, to *sneutrinos*, *gravitinos*, *KK photons* and *standard model and sterile neutrinos*.

If DM freezes out while $T > m$, then it *free streams* until $T \sim m$. Such relativistic DM can move across the whole cosmic horizon (since $v \sim c$), and no cosmic structure can form. The horizon size when $T = m$ is

$$R_H \sim [aH(T = m)]^{-1} . \tag{13.6}$$

Anticipating that structure formation constraints probe relatively late cosmic epochs, let's adopt $g_\star \sim g_\star(T_0) = 3.38$. Structure formation binds galaxies together, with typical separations of order 1 Mpc (this is roughly the distance between our MW galaxy and our galactic neighbour Andromeda, M31). This is the largest quasi-linear scale of structure formation sensitive to primordial free streaming. Requiring $R_H \lesssim 1$ Mpc and substituting the Friedmann equation for H leads to the very approximate bound

$$\boxed{m_{\text{thermal}} \gtrsim 100 \text{ eV}} . \tag{13.7}$$

Using structure formation data from the *Lyman-α forest flux power spectrum*, a stronger bound on thermal DM can be found. The Lyman-α forest probes smaller separations but requires finding the non-linear power spectrum, $P(k)$, and gas dynamics by numerical simulation. The scale probed is roughly $k = 4\ h\text{Mpc}^{-1}$. Deriving even an approximate limit requires solving the linearised Boltzmann equations for warm DM: some discussion of the physics can be found in Ref. [77].

The linear power spectrum in WDM is suppressed relative to CDM as $P(k)_{\mathrm{WDM}} = T^2_{\mathrm{WDM}}P(k)_{\mathrm{CDM}}$ with 'transfer function' $T_{\mathrm{WDM}}(k) \approx [1 + (\alpha k)^{2\mu}]^{-5\mu}$, $\mu = 1.12$, $\alpha = 0.052(m/\mathrm{keV})^{-1.15}$. Setting k to the largest scale probed by Lyman-α and demanding $T(k) > 0.5$ at this scale (since the observations are consistent with CDM) allows you to get an approximate sense of bounds. A recent bound derived from a full likelihood analysis including non-linear simulations is [78]:

$$\boxed{m_{\mathrm{thermal}} > 5.3 \text{ keV}} .$$ (13.8)

You will notice that the parameter space of WIMP-like DM is now bounded from above and below. The range between 1 GeV and the unitarity bound, Eq. (9.28), around 50 TeV, is accessible to traditional direct detection, such as nuclear recoil. The range from 1 keV to 1 GeV is referred to as 'light DM'. Due to the low mass, the recoil energies are too small for this type of DM to be probed by the nuclear recoil direct detection techniques we discussed in chapter 12. Thus, this is the subject of intense exploration currently, exploring new direct detection techniques that often use novel materials and aim to be sensitive to very low recoil energies (see e.g. Ref. [79]).

13.2 ANNIHILATIONS AND INDIRECT DETECTION

The production of canonical WIMPs requires initial chemical equilibrium maintained by annihilations. After freeze out, annihilations become rare, but they can still occur, particularly if the DM density is high. We can thus try to observe the products of this annihilation occurring in the local Universe. Eventually, we observe photons as the final product of many different annihilation channels (see Fig. 13.1), resulting in a spectrum of photon energies. A typical DM mass is 1 GeV, and many decays eventually involve pions, with mass larger than 100 MeV. Thus typical photons in the spectrum will have energies far in excess of 100 keV, that is, in the *gamma ray energy range*.

13.2.1 Gamma Ray Flux Production

Consider DM annihilation in our galactic neighbourhood, for example in a dwarf satellite of the MW or in the galactic centre. We would observe gamma rays produced in these dense dark matter environments, which must travel through the interstellar medium to reach us. The most intense sources are the densest regions of DM. We know the NFW DM profile:

$$\rho(r) = \frac{\rho_s}{(r/r_s)(1 + r/r_s)^2} ,$$ (13.9)

which peaks towards $r = 0$. The densest regions are the *MW galactic centre, and the centres of dense satellite galaxies*.

We need to solve the Boltzmann equation for the DM and the photons. In the local environment we can neglect the expansion of the Universe. Furthermore, the

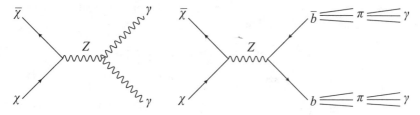

Figure 13.1 Two DM particles χ annihilate and eventually produce photons. Note that DM is often its own anti-particle, $\bar{\chi}$. *Left*: annihilation via a Z-boson, which decays directly to two photons. *Right*: annihilation via a Z-boson decaying to b quarks. The b quarks hadronise to pions in a shower. The pions themselves eventually decay to photons.

local plasma photons have $E \ll m_\chi$ and so cannot themselves create DM. Thus, the DM Boltzmann equation is

$$\frac{dn_\chi}{dt} = - \overbrace{\frac{n_\chi^2}{2}}^{\substack{\text{two DM particles}\\\text{required}}} \underbrace{\langle \sigma v \rangle}_{\text{average over local distribution}} \overbrace{\times 2}^{\substack{\text{lose two DM}\\\text{particles in the}\\\text{annihilation}}} , \qquad (13.10)$$

where σ is the total annihilation cross section. Now we need the Boltzmann equation for the photons. If the annihilation into photons occurs via many channels i, with cross section σ_i, each of which creates N_i photons, then the total Boltzmann equation is

$$\frac{dn_\gamma}{dt} = \sum_i \frac{n_\chi^2}{2} \langle \sigma_i v \rangle N_i . \qquad (13.11)$$

For example, the process on the right of Fig. 13.1 via b hadronization will create a gamma ray shower which can be solved numerically with Monte Carlo event generators like those used to compute LHC processes.

We observe a point in the sky and all of the photons produced by annihilation along the line of sight to that point. The line of sight is a radial direction, ℓ, from us to the source, inclined at an angle ψ from the galactic plane. The DM density is specified by the radial coordinate r relative to the galactic centre. The flux $d\Phi$ (which has units of rate per solid angle) in an energy band dE_γ in direction ψ is

$$\frac{d\Phi}{dE_\gamma} = \underbrace{\overbrace{\frac{A}{4\pi}}^{\text{detector area}}}_{\text{unit solid angle}} \underbrace{\left[\int d\ell \rho^2 (r(\ell, \psi)) \right]}_{\text{``J-factor''}} \overbrace{\frac{1}{2m_\chi^2}}^{n=\rho/m} \sum_i \langle \sigma_i v \rangle \frac{dN_i}{dE_\gamma} . \qquad (13.12)$$

Note the appearance of ρ^2 in the integral: this comes from the n_χ^2 dependence in Eq. (13.11), since the annihilation process involves two DM particles.

Eq. (13.12) defines the famous 'J-factor', which absorbs all of the astrophysical modelling. Objects with a high J-factor are good targets to observe gamma rays from DM. The integral in J is performed along the line of sight, ℓ. J has units GeV/cm^5. Some typical values are

$$J \sim 10^{19-20} \text{ GeV}^2 \text{ cm}^{-5} \quad \text{dwarf spheroidals}, \tag{13.13}$$

$$J \sim 10^{20} \text{ GeV}^2 \text{ cm}^{-5} \quad \text{Andromeda}, \tag{13.14}$$

$$J \sim 10^{22-25} \text{ GeV}^2 \text{ cm}^{-5} \quad \text{MW galactic centre}. \tag{13.15}$$

Note that J *is measured* (albeit with some error) by fitting ρ to, for example, the rotation curve or velocity dispersion in an object. The galactic centre clearly has the largest J factor, but it also has a lot of *backgrounds* from the messy environment. Dwarf galaxies have smaller J factors but are much cleaner to observe.

The great advantage of DM indirect detection is that $\langle \sigma v \rangle$ in Eq. (13.11) (when summed over all channels) *is the exact same total annihilation cross section that appears in the WIMP relic density!* This is distinct from direct detection, where dominant annihilation to b's could suppress the nuclear recoil cross section, σ_p, by orders of magnitude compared to the total annihilation cross section, which occurs for example in the so-called 'Higgs portal'.

If $\langle \sigma v \rangle$ has strong velocity dependence, it may need to be taken inside the J-factor integral, introducing model dependence. Recall that we expand

$$\langle \sigma v \rangle \sim \sigma_0 + \sigma_1 x^{-1} + \cdots, \tag{13.16}$$

where $x = m/T$, σ_0 is the 's-wave annihilation', and σ_1 is the 'p-wave annihilation'.

13.2.2 Diffusion-Loss Equation[1]

Gamma rays do not propagate freely from source to observer. We need to model how they evolve passing through the galaxy. We consider the number density of gamma rays in a phase space cell labelled E and x as depicted in Fig. 13.2. The flux in and out of the cell occurs in the x-direction or E-direction and is labelled $\phi_{x,E}$. With these definitions, the number density evolves as

$$\frac{d}{dt} N dE dx = [\phi_x(x) - \phi_x(x+dx)] \, dE + [\phi_E(E) - \phi_E(E+dE)] \, dx + Q(E,x) dE dx, \tag{13.17}$$

where the first two terms are the free evolution, and the last Q term is injection from a source, such as DM annihilation. We define the following quantities:

$$\phi_x = -D \frac{\partial N}{\partial x}, \tag{13.18}$$

$$\frac{\partial \phi_E}{\partial E} = -\frac{\partial}{\partial E}(Nb), \tag{13.19}$$

[1] This section inspired by notes from Malcolm Fairbairn.

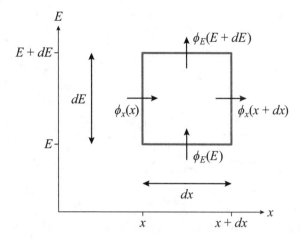

Figure 13.2 Flux in and out of a phase space cell.

where D is the diffusion coefficient and $b = -\partial E/\partial t$ is the energy loss. Generalizing from one dimension (x) to three we arrive at the diffusion-loss equation:

$$\frac{dN}{dt} = \underbrace{D\nabla^2 N}_{\text{diffusion}} + \overbrace{\frac{\partial}{\partial E}(Nb)}^{\text{loss}} + \underbrace{Q}_{\text{injection}} .$$

(13.20)

Examples of energy loss of photons in the galaxy include *synchrotron radiation in magnetic fields* and *Compton scattering of free electrons*. Thus, solving the diffusion-loss equation requires a model for the galactic magnetic field and free electron density in order to compute b. This is done numerically using codes such as GALPROP [80].

13.2.3 Limits from Gamma Ray Telescopes

The *Fermi Large Area Telescope* ('Fermi-LAT', a gamma ray satellite) observed 25 dSphs in the Milky Way for a total of six years. None were detected in diffuse gamma ray emission characteristic of DM annihilation, which allows us to set upper bounds on $\langle \sigma v \rangle$. The bounds depend on the channel of annihilation. The total emission summing over channels is

$$\frac{dN}{dE_\gamma} = \sum_i B_i \frac{dN_i}{dE_\gamma} ,$$

(13.21)

where B_i is the branching ratio (computable by hand using Feynman diagrams), and the individual spectra dN_i/dE_γ would normally be computed numerically using, for example, PYTHIA. The spectra dN_i/dE_γ are independent of the branching ratios, since they only involve decays of Standard Model particles. Thus, exclusions on

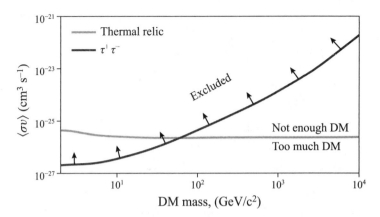

Figure 13.3 Exclusion on WIMP parameter space in the $\tau^+\tau^-$ channel from Fermi-LAT data [81].

$\langle\sigma v\rangle$ can be presented assuming, for example, one channel is dominant over the others. An example of such an exclusion is shown in Fig. 13.3. Note that the Fermi constraints disappear at low values of $m_\chi \lesssim E_{min} \approx 500$ MeV, where the energy of the produced photons would be too low to be detected by this telescope.

Constraints are derived using Poisson statistics assuming zero total flux over six years. The total flux is found by integrating Eq. (13.12):

$$\Phi = \frac{A}{4\pi}\frac{\langle\sigma v\rangle}{2m_\chi^2}\int_{E_{min}}^{E_{max}} dE\frac{dN}{dE}\times J. \tag{13.22}$$

Typical spectra $E^2 dN/dE$ rise to a peak at $E \lesssim m_\chi$, with a sharp cut-off for $E > 2m_\chi$ from energy conservation. The spectra units are GeV cm^{-2} s^{-1} (think: N is photons per unit volume, and we restore one factor of c to get a rate per unit area) with peak values near 10^{-6-7} for $m_\chi \approx 100$ GeV. An approximate fit for the shape is

$$\frac{dN}{dE} \propto E^{-\alpha}\exp\left[-\left(\frac{E}{E_c}\right)^\beta\right], \tag{13.23}$$

where E_c is a cut-off energy that depends on the channel, and $\alpha \sim 1$ and $\beta \sim 1$ are fitted exponents. Sketches of this function are shown in Fig. 13.4. The peak of the spectrum moves to larger energies as we increase the WIMP mass. The spectrum also depends strongly on the dominant annihilation mode (whether it be b-quarks, τ-leptons, etc.); see the references for more details.

Inspecting Fig. 13.3, we notice something very important: we can derive a *lower limit on the thermal WIMP mass*:

$$\boxed{m_\chi \gtrsim 10-100 \text{ GeV}} \quad \text{(WIMP mass lower limit from Fermi)}, \tag{13.24}$$

where the exact limit depends on the branching ratios. The lower limit exists because if the WIMP is lower mass it requires a lower cross section to match the Fermi

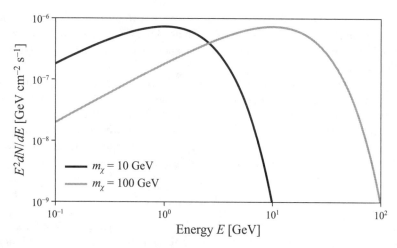

Figure 13.4 Sketch of the energy spectrum, Eq. (13.23).

constraints; however, with a lower cross section, the WIMP would produce *too much DM* and so is excluded. Thus low-mass WIMPs produced by annihilation and freeze out in the hot early Universe cannot be even a small fraction of the DM: there are no values of the cross section consistent with the Fermi data and the maximum value of the relic density allowed by the CMB.

The above argument fails, however, at very low cross sections where the WIMP is no longer in thermal equilibrium in the early Universe. At this point the dependence of the relic density on the cross section (visible in Fig. 9.3) turns over: there is a region at low cross sections that does not produce too much DM and is also allowed by the Fermi limits. This opens up a new region of parameter space at very low $\langle \sigma v \rangle$ known as feebly interacting massive particles (FIMPs).

13.2.4 Other Indirect Detection Probes

We close this chapter with a brief overview of some other ways that WIMPs can be probed by their annihilations or decays into high-energy particles and the effects of these particles on astrophysical observables. If WIMPs annihilate or decay into Standard Model particles, those particles inject energy into the population that is already present or create an entirely new source of high-energy particles.

The CMB offers one powerful probe of the thermal state of the Universe. The CMB is observed to be an almost perfect black body. Any injection of photons by DM annihilations or decays can distort this black body spectrum, if the photons do not have time to thermalize. CMB spectral distortions are tightly constrained by the spectrum measured by COBE/FIRAS in the 1990s [82]. DM annihilations and decays can also have an indirect effect on the CMB by heating the gas between us and the CMB. Hot gas can become ionized, and the CMB is sensitive to the overall number density of free electrons via the optical depth parameter, τ. Ionized electrons also lead to polarisation of the CMB, and the polarization anisotropies are in fact sensitive to the whole shape of the ionization history. The measured polarization anisotropies are consistent with an almost step-like reionization of the Universe at a redshift $z \approx 10$, which is consistent with ionization caused by starlight in the first galaxies.

Decaying or annihilating WIMPs lead to a different shape to the ionization history, in particular causing ionization at very high redshifts, $z > 100$, and this can be used to constrain DM annihilations. The CMB is sensitive to a combination of parameters $f_{\text{eff}} \langle \sigma v \rangle / m$, where f_{eff} is an efficiency factor, which ranges from around 0.2 to 0.6 depending on the dominant annihilation channel (it is almost zero for annihilation to neutrinos). The *Planck* CMB anisotropies require:

$$f_{\text{eff}} \frac{\langle \sigma v \rangle}{m} < 3.2 \frac{10^{-26} \text{ cm}^3 \text{ s}^{-1}}{100 \text{ GeV}} . \tag{13.25}$$

The thermal relic cross section is excluded for masses around 30 GeV if the energy injection is efficient. Like the limit from Fermi, this places a lower bound on the thermal relic mass.

The other indirect detection constraint we mention is from cosmic rays. Cosmic rays are high-energy particles from outer space. For example, the Alpha Magnetic Spectrometer on the International Space Station measures positron flux at energies up to 350 GeV. The observations show a smoothly varying function, which can be explained by astrophysical sources such as supernova explosions. Annihilating DM would produce a spiked spectrum peaked somewhere near the DM mass. The DM-induced spike should not exceed the observed spectrum, and this allows an upper limit to be placed on $\langle \sigma v \rangle$. The sort of limits found again resemble those from Fermi (Fig. 13.3), this time giving the strongest limit at $m \approx 200$ GeV for annihilations dominantly to $e^+ e^-$.

QUIZ

i. Which of the following statements contribute to the lower bound on WIMP-like particle mass arising from structure formation alone? (Select multiple.)
   a. For fermions, the Pauli exclusion principle.
   b. For bosons, the temperature of Bose–Einstein condensation.
   c. The temperature at which DM becomes non-relativistic.
   d. The unitarity constraint on the cross section.
   e. The non-observation of nuclear recoil events.
   f. The absence of supersymmetry at the LHC.

ii. What is the resulting lower bound on WIMP-like particle mass from structure formation?
   a. About 1 keV.
   b. About 1 GeV.
   c. About 100 GeV.

iii. The 'J-factor' for annihilation is (select multiple):
   a. Largest in regions of low DM density.
   b. Known from observations of dwarf galaxies.
   c. Given by the line of sight integral of the DM density.
   d. Given by the line of sight integral of the squared DM density.
   e. Largest in regions of high DM density.
   f. Largest in dwarf galaxies.

iv. What causes the high energy cut-off in the annihilation gamma ray spectrum, dN/dE?
   a. The total cross section.
   b. Unitarity.
   c. The maximum energy resolution of the experiment.
   d. Conservation of energy.

v. Annihilation of DM into gamma rays:
   a. Has been observed by Fermi-LAT.
   b. Places an upper limit on the thermal relic particle mass around 1 TeV.
   c. Places a lower limit on the thermal relic particle mass around 10 GeV.
   d. Can be searched for at the LHC.

Answers

i. a,c; ii. a; iii. b,d,e; iv. d; v. c

Chapter Fourteen

Axion Direct Detection

Recall that axion dark matter is produced in the early Universe by the classical motion of the axion field and as such can be modelled as

$$\phi(x, t) = \phi(x) \cos(\omega t), \tag{14.1}$$

where, in the non-relativistic limit,

$$\omega = m_a + \frac{1}{2} m_a v^2. \tag{14.2}$$

Additionally, recall that the density is given by

$$\rho_a = \frac{1}{2}\dot{\phi}^2 + \frac{1}{2}m^2\phi^2. \tag{14.3}$$

It can hence be shown that the spatial component of our field is given by

$$\phi(x) = \sqrt{\frac{2\rho_a}{m_a^2}}. \tag{14.4}$$

In the *standard halo model*, the velocity v is drawn from the local Maxwell–Boltzmann distribution with dispersion σ_v as shown previously in Fig. 12.3. The local density and dispersion are known to be

- $\rho_a \approx 0.4$ GeV cm^{-3}.
- $\sigma_v = v_0/\sqrt{2} \approx 160$ km s$^{-1} \approx 10^{-3}c$.

The dispersion is therefore

$$\frac{\Delta\omega}{\omega} = \sigma_v^2 \approx 10^{-6}. \tag{14.5}$$

Therefore, we can look for axionic matter by searching for an oscillating field of a known amplitude and a central frequency of

$$v = \frac{m_a}{2\pi} \approx \frac{1\text{Hz}}{2\pi} \frac{m_a}{10^{-15}\text{eV}}. \tag{14.6}$$

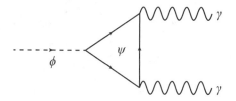

Figure 14.1 A loop of electromagnetically charged fermions gives rise to the axion-photon coupling.

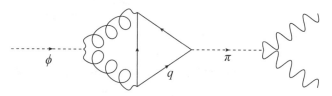

Figure 14.2 The QCD axion by definition has a coupling to two gluons, which mediates an axion-photon coupling via the pion.

14.1 THE AXION-PHOTON COUPLING

14.1.1 Lagrangian for Axion-Photon Coupling

How does the axion couple to particles of the Standard Model? The axion Lagrangian includes the following term for the coupling between axions and photons:

$$\mathcal{L} \supset -\frac{1}{4}\phi g F_{\mu\nu}\tilde{F}^{\mu\nu}, \tag{14.7}$$

where g is the axion-photon coupling (often also denoted $g_{a\gamma}$). This coupling arises automatically for the QCD axion and is a generic coupling for ALPs. Consider the Feynman diagram in Fig. 14.1. This diagram gives an effective interaction Lagrangian term

$$\mathcal{L} \propto \mathcal{E}\phi F\tilde{F}, \tag{14.8}$$

where $\mathcal{E}$ is the *electromagnetic anomaly*. This arises if the axion couples to electromagnetically charged fermions of the Standard Model. For the QCD axion, the process can also be mediated through the defining axion-gluon coupling, as shown in Fig. 14.2.

This second contribution for the QCD axion arises from mixing with the pion, and below the QCD confinement scale the effective Lagrangian is

$$\mathcal{L} \propto \mathcal{C}\phi F\tilde{F}, \tag{14.9}$$

where $\mathcal{C}$ is the *colour anomaly* introduced briefly in chapter 10. The constant of proportionality is well known in the literature and was computed in detail by

Srednicki [83]. The total axion-photon coupling is given by

$$g = \frac{\alpha}{2\pi f_a}\left[\frac{\mathcal{E}}{\mathcal{C}} - \frac{2}{3}\frac{4 + m_u/m_d}{1 + m_u/m_d}\right]. \qquad (14.10)$$

QCD axion models differ from one another depending on the ratio $\mathcal{E}/\mathcal{C}$. Note that for ALPs, where $\mathcal{C} = 0$, this formula should be replaced by simply

$$g = \frac{\alpha\mathcal{E}}{2\pi v_a}. \qquad (14.11)$$

Combining Eqs. (14.10) and (10.86) we see that for the QCD axion $g \propto m_a$. This defines a clear experimental target to try to reach:

$$g_{\text{QCD}} = 10^{-15}\left(\frac{m_a}{5 \times 10^{-6}\ \text{eV}}\right)\text{GeV}^{-1}\mathcal{C}_{\phi\gamma}, \qquad (14.12)$$

where $\mathcal{C}_{\phi\gamma} = -1.92, 0.75$ (given the quantity in square brackets in Eq. (14.10)) for the KSVZ and DFSZ models respectively.

Now we know the field amplitude from Eq. (14.4), the frequency from Eq. (14.6) and the order of magnitude of the coupling from Eq. (14.12). How can we actually find the axion?

14.1.2 Axion-Maxwell Equations

The effective Lagrangian density for photons and axions is given by

$$\mathcal{L} = -\frac{1}{4}F_{\mu\nu}F^{\mu\nu} + \frac{1}{2}\partial_\mu\phi\partial^\mu\phi - \frac{1}{2}m_a^2\phi^2 - \frac{g}{4}\phi F_{\mu\nu}\tilde{F}^{\mu\nu}, \qquad (14.13)$$

where we have used the metric signature $(+, -, -, -)$. The coupling term modifies the Euler–Lagrange equations for the electromagnetic field. As a result, the axion appears in Maxwell's equations:

$$\nabla \cdot \vec{B} = 0, \qquad (14.14)$$

$$\nabla \times \vec{E} + \dot{\vec{B}} = 0, \qquad (14.15)$$

$$\nabla \cdot \vec{E} = \rho_e - g\vec{B} \cdot \nabla\phi, \qquad (14.16)$$

$$\dot{\vec{E}} - \nabla \times \vec{B} = -\vec{J}_e - g(\vec{B} \cdot \dot{\phi} - \vec{E} \times \nabla\phi), \qquad (14.17)$$

$$\Box\phi - m_a^2\phi = -g\vec{E} \cdot \vec{B}. \qquad (14.18)$$

In the DM halo at leading order, we can take our axion field discussed at the start of the chapter as a fixed background solution and plug it in as a source on the right-hand side of Maxwell's equations. The axion field then acts as a source of 'anomalous' charge and current density.

14.2 AXION CAVITY 'HALOSCOPE'

Now consider turning on a source of $\vec{B}$ in the lab; let's find an equation of motion for the response $\vec{E}$ field. When we apply a magnetic field in the presence of the axion field, an electric field is induced parallel to the magnetic field. This is in contrast to the ordinary form of Maxwell's equations in which $\vec{E}$ and $\vec{B}$ are always perpendicular to each other.

The axion field has a de Broglie wavelength given by

$$\lambda_{dB} = \frac{2\pi}{m_a v}, \tag{14.19}$$

where again v is the velocity of the axion particle and, as usual, $\hbar = 1$. Taking this velocity to be $v \approx 270$ km s^{-1} from the standard halo model, we find that

$$\lambda_{dB} \approx 1\text{km} \left(\frac{10^{-6}\text{eV}}{m_a} \right). \tag{14.20}$$

When the length scales we are interested in are much *smaller* than the de Broglie wavelength then we can assume $\nabla\phi \approx 0$: the field is smooth on the de Broglie scale. In this approximation, the induced electric field has the equation of motion:

$$\ddot{\vec{E}} - \nabla^2\vec{E} = \frac{\omega^2}{m}\sqrt{2\rho_a g}\vec{B}\cos\omega t. \tag{14.21}$$

We recognise this as the equation for a *driven harmonic oscillator*. This gives us the possibility of driving the oscillator to resonance. In 1D, we can achieve this by confining our electric field between two mirrors as shown in Fig. 14.3. This is known as a *cavity*.

Due to the reflecting boundary conditions of the cavity, the electromagnetic field inside it has quantised modes given by

$$E_n \propto \sin\left(\frac{n\pi x}{L}\right), \tag{14.22}$$

where L is the length of the cavity and n is an integer. For these modes, we find that

$$\nabla^2 E = -\frac{n^2\pi^2}{L^2}E = \omega_n^2 E. \tag{14.23}$$

This means that our system will be in resonance when $n\pi/L = \omega$.

If we have a system with no damping, the amplitude of the induced field diverges on resonance, that is, it grows without bound and tends towards infinity over time. However, in reality, this can never be the case since real cavities suffer from energy losses. For example, imperfectly conducting mirrors lead to energy 'leaking out'. These losses are quantified by the *resonance quality factor*:

$$Q = \frac{\text{energy stored}}{\text{energy loss per radiation}} = \frac{\omega}{\Gamma}, \tag{14.24}$$

where, in the uniform loss approximation, the loss term Γ is given by a single number.

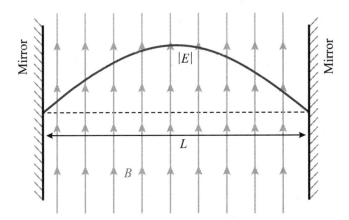

Figure 14.3 1D cavity. The magnetic field is applied along the z-direction, which is the direction of the induced electric field. Mirrors along the x-direction confine the electric field and force it to take solutions given by the normal modes $E_n \propto \sin(n\pi x/L)$ for some integer n.

The solution for a driven harmonic oscillator with losses is (you can find this solution using Green's functions):

$$E = \text{Re}\left[e^{-i\omega t} \frac{-A}{\omega^2 + \frac{i\omega\omega_n}{Q} - \omega_n^2} \right], \tag{14.25}$$

which has a 'Lorentzian' or 'Breit–Wigner' shape resonance with bandwidth of $\Delta\omega = \omega/Q$, and A is the driving amplitude from the right-hand side of Eq. (14.21) given by

$$A = \frac{\omega^2}{m}\sqrt{2\rho g}B. \tag{14.26}$$

On resonance, $\omega_n = \omega$, the E-field amplitude is $|E| = AQ/\omega^2$.

The energy that leaks out of the cavity is something we can detect, typically using an antenna. The power is given by

$$P(t) = \text{force} \times \text{velocity}, \tag{14.27}$$

and the power lost is given by the integral of the power over time over one cycle:

$$\begin{aligned}
\langle P_{\text{loss}} \rangle &= \frac{\omega}{2\pi} \int_0^{2\pi/\omega} P(t)\mathrm{d}t \\
&= \frac{\omega}{Q} \times \text{(energy stored)} \\
&= \frac{\omega}{Q} \times \frac{1}{2}|E|^2 V,
\end{aligned} \tag{14.28}$$

where we have used the definition of the quality factor from Eq. (14.24) and the fact that the energy density of an electric field is given by $|E|^2/2$. We can therefore substitute in the field amplitude from Eq. (14.25) to calculate the power we expect to detect.

We will now calculate the power we would detect on resonance ($\omega = \omega_n$) for the largest mode volume ($n = 1$). Additionally, we will assume that the effective quality factor is smaller than the spread of frequencies of the axion source, that is,

$$Q < \frac{\omega}{\Delta \omega} \approx \sigma_v^{-2} \sim 10^6. \tag{14.29}$$

This enables us to treat the axion source as a delta function in frequency. We can therefore consider a single frequency for the axion field, avoiding the necessity of convolving the Breit–Wigner resonance of the electric field with the power spectrum of the axion field.

Substituting this in, we find that the output power is finally

$$P = \frac{g^2 B^2 V \rho_{\mathrm{DM}}}{m} Q. \tag{14.30}$$

This, however, is only true for one-dimensional cavities. Real cavities instead tend to be cylinders as shown in Fig. 14.4. Therefore, we have to modify the power given by Eq. (14.30) to include a form factor $C \sim \mathcal{O}(1)$: this accounts for the real-life problems of solving Maxwell's equations in some given geometry and is computed numerically using engineering software. Additionally, in the case when the quality factor is of the order the inverse square velocity dispersion ($Q \sim \sigma_v^{-2} \sim 10^6$), we then also have to take into account the spectrum of the axion velocity distribution. This can be roughly accounted for by making the replacement $Q \rightarrow \min[Q, 10^6]$. Additionally, our antenna to detect the induced E-field is not perfectly efficient and has some coupling to the E-field in the cavity $\kappa < 1$. Finally, one should use the *loaded quality factor* which accounts for the additional extraction of energy by the antenna. Taking into account these effects, our modified equation for the induced power is

$$P = \kappa C \frac{g^2 B^2 V \rho_{\mathrm{DM}}}{m} \min[Q, \sigma_v^{-2}]. \tag{14.31}$$

The additional terms compared to our simple estimate are only rather small $\mathcal{O}(1)$ corrections.

14.2.1 Axion Dark Matter Experiment (ADMX)

To date, ADMX is the most successful and well-known cavity haloscope experiment. The cavity is a cylinder, as shown in Fig. 14.4, with a form factor of $C = 0.69$, a volume of around 100 litres and a quality factor $Q \approx 3 \times 10^4$. This cavity is subjected to a ~ 7.5 T magnetic field. The antenna setup is able to measure electromagnetic fields with powers as low as $P \approx 10^{-22}$ W. Substituting into Eq. (14.31) (taking $\kappa = 1$), we find that for axion masses around 3×10^{-6} eV ADMX would be able to detect the axion with a coupling strength as small as $g \approx 10^{-15}$ GeV^{-1}, which is right around where the QCD axion of this mass is predicted to be according to Eq. (14.12). This is of course why ADMX is designed with such specifications.

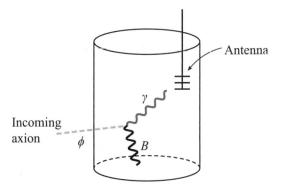

Figure 14.4 Haloscope cavity—sketch of a typical axion haloscope. The magnetic field B resonantly converts axions ϕ to photons which are then detected by the antenna.

For axion mass 3×10^{-6} eV the cavity frequency is in the GHz range, and the diameter length scale of the cavity is 30 cm. This is a nice human-sized cavity that fits inside commercial magnets and can be cooled down to 1 K to reduce background noise. The problem is that we don't currently know what the axion mass is. Therefore we cannot build a detector with a dedicated cavity width, L, to sit right on the correct resonant frequency. Instead, the cavity has to be able to 'scan' its resonant frequency. As shown in Fig. 14.5, one design ADMX has used achieves this by using two tuning rods which can be moved to modify the effective cavity length, that is, to modify the resonant frequency.

We want to scan for the axion without missing any frequency. Therefore, the effective length change should be such that we move a single bandwidth $\Delta\omega$. Using the relationship between bandwidth and quality factor for the Lorentzian lineshape, and the central frequency relation $\omega \sim 1/L$, we find that the relative change in length is related to the quality factor by

$$\frac{\delta L}{L} \sim \frac{1}{Q}. \tag{14.32}$$

Therefore, for ADMX, $\delta L \sim 10^{-4}$ cm. By moving these rods, you change the resonant frequency ν_{res} of the cavity. This is done in steps of size

$$\Delta\nu \sim \frac{\nu_{\text{res}}}{Q}, \tag{14.33}$$

from ν_{min} to ν_{max}. For a fixed cavity, there is not much difference between the maximum and minimum values. For ADMX in a given run such as Ref. [84] we may have, for example, $\nu_{\text{min}} \sim 0.7$ GHz and $\nu_{\text{max}} \sim 0.8$ GHz relating to axion masses $m_a \in [2.8, 3.3] \mu$ eV.

When conducting an experiment like this, we need to decide how long to collect power at each frequency to constrain the axion-photon coupling to a particular level. The signal-to-noise ratio (SNR) is given by the *Dicke radiometer equation*

$$\text{SNR} = \frac{P_{\text{s}}}{P_{\text{N}}} \sqrt{\Delta t \Delta\nu}, \tag{14.34}$$

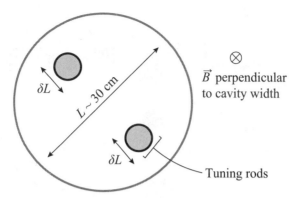

Figure 14.5 ADMX tuning–sketch of how a haloscope such as ADMX tunes the length of the cavity to change the resonant frequency.

where P_s and P_N are the signal and noise power respectively, Δv is the bandwidth, and Δt is the measuring time. Noise in a haloscope is described by a total system noise temperature, T_{sys}. Modern haloscopes aim for noise close to the quantum limit imposed by zero point fluctuations, $T_{sys} \sim v$. The noise power is given by the noise temperature times the measurement bandwidth:

$$P_N = T_{sys} \Delta v. \tag{14.35}$$

Substituting into Eq. (14.34), the signal-to-noise ratio can be written as

$$\text{SNR} = \frac{P_s}{T_{sys}\sqrt{\Delta v}}\sqrt{\Delta t}. \tag{14.36}$$

We see that our signal is improved by reducing our resonance step size Δv or increasing the measuring time Δt.

ADMX uses a dilution refrigerator which is able to reduce the system temperature to 300 mK. This is safely above the 'standard quantum limit' given by

$$T_{sys} \approx m_a \approx 10^{-2} \text{K} \left(\frac{m_a}{10^{-6} \text{ eV}} \right). \tag{14.37}$$

This means that we can use $T_{sys} = 300$ mK in our calculation for the system noise.

Typically for detection, we require a signal-to-noise ratio of 3. As mentioned previously, ADMX can measure power as low as $P \approx 10^{-21}$ W as well as a bandwidth of $\Delta v = 10^{-5}$ GHz. Substituting these into Eq. (14.36) we find that it would only take 100 s to detect/exclude the axion if you knew where to look.

However, again, we don't know the mass of the axion and each scan only excludes a range of frequencies of width Δv. Therefore, we need to scan $\mathcal{O}(Q)$ times to cover a $\mathcal{O}(1)$ range of frequency. As a result, experimental campaigns normally take on the order of years to complete.

So far, ADMX has excluded the QCD axion in the mass range of $2.7 \lesssim m_a \lesssim 3.3\mu$eV for DFSZ axions and $2 \lesssim m_a \lesssim 3.7\mu$eV for KSVZ axions. The reason why

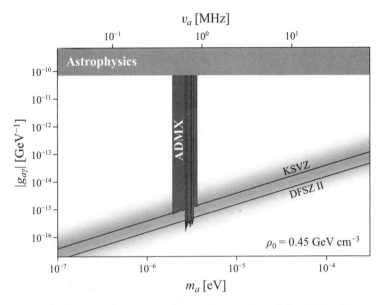

Figure 14.6 ADMX constraints—constraints on the coupling of the QCD axion coupling to photons from the ADMX haloscope experiment. Generated using AxionLimits [85]. See colour insert.

the range is much wider for KSVZ axions is that this model predicts larger axion-photon couplings $g_{a\gamma}$ and therefore less time is required to constrain to this level. To date, only ADMX and some other haloscopes have been able to make any successful bound on the QCD axion.

14.3 OTHER AXION EXPERIMENTS AND COUPLINGS

14.3.1 Other Couplings

While the axion-photon coupling tends to be the main focus of axion dark matter experiments, the axion also has other couplings we could try to detect. As we saw in chapter 10, these include the nucleon coupling in which the gradient of the axion field couples to the axial nucleon current

$$(\partial_\mu \phi)\bar{N}\gamma^\mu \gamma_5 N, \tag{14.38}$$

the electron coupling

$$(\partial_\mu \phi)\bar{e}\gamma^\mu \gamma_5 e, \tag{14.39}$$

and of course the direct coupling to gluons that we needed for the axion to solve the strong-CP problem

$$\frac{\phi}{f_a} G_{\mu\nu}\tilde{G}^{\mu\nu}. \tag{14.40}$$

For the nucleon and electron couplings in the non-relativistic limit, the axion gradient couples to the *fermion spin*. The spin coupling contributes to the Hamiltonian, resulting in a shift given by

$$\Delta H = g_i \nabla \phi \cdot \sigma_i, \qquad (14.41)$$

where σ is the spin vector and $i = e$ or N for electrons or nucleons respectively. This means that the axion gradient $\nabla \phi$ looks like an *effective magnetic field* capable of driving a spin resonance. We can resonantly couple to this if the spin precession frequency of the fermion itself is equal to the oscillation frequency of the axion. The spin precession frequency of the fermion is its Larmor frequency

$$\omega_L = \gamma B, \qquad (14.42)$$

where γ is the gyromagnetic ratio and B is the applied magnetic field. Recall that the gradient of the axion field is given by

$$\nabla \phi = m_a \vec{v} \sqrt{\frac{2\rho_a}{m^2}} \cos \omega t, \qquad (14.43)$$

where $\vec{v}$ is the velocity of the axion dark matter. This means that the coupling is proportional to the velocity of the dark matter which, as mentioned previously, we know to be around 230 km s^{-1} due to the orbit of the Sun through the Milky Way's dark matter halo. This is sometimes thought of as a 'dark matter wind' and, for this reason, this coupling is known as the *axion wind coupling*.

We now briefly discuss two experiments looking for axion-fermion couplings. The first, QUAX, searches for the axion-electron coupling, while the second, CASPEr-Wind, searches for the axion-nucleon coupling.

14.3.2 QUAX

The QUAX experiment is based in Italy at the Institute for Nuclear Physics in Padova. It uses the coupling to the Larmor frequency of the electron inside a few hundred cubic millimetres of a ferromagnetic material called yttrium iron garnet (YIG) as shown in Fig. 14.7. The effective magnetic field from the axion wind causes the YIG spheres to resonate at the electron Larmour frequency, creating a magnetization. In order to detect this magnetization, the YIG spheres are placed inside a microwave cavity. The resulting electromagnetic field produced by the YIG couples to the modes of the cavity hybridizing into so-called Kittel modes. The resonantly excited field in the cavity can then be measured using a microwave antenna in a similar way to an ADMX-style haloscope.

The QUAX experiment is sensitive to axion masses $\approx 50 \ \mu$eV although there are plans to increase the range. Like with ADMX, they do this by scanning at different resonant frequencies. However, from Eq. (14.42), we see that instead of changing the cavity length, here we alter the resonant frequency by changing the applied magnetic field B.

Currently, the constraints from this experiment are weaker than those from astrophysical observations. However, by using hundreds of YIG spheres, they may be able to test for DFSZ axions in the future.

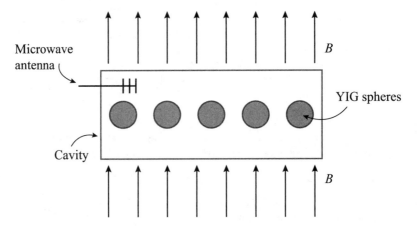

Figure 14.7 QUAX experiment—uses ferromagnetic spheres in an applied magnetic field in an attempt to measure the induced field inside the spheres due to the axion wind.

14.3.3 CASPEr

14.3.3.1 CASPEr-Wind

The Cosmic Axion Spin Precession Experiment (CASPEr) includes a number of different experiments to try to constrain different axion-SM couplings. One of those is CASPEr-Wind based in Mainz in Germany. Like QUAX, CASPEr-Wind tries to detect the effective magnetic field due to the axion field gradient but for the nuclear spin. This is simply *nuclear magnetic resonance* (NMR) which is well known, among other things, for its applications in medicine.

The experiment has access to a very wide range of axion masses from 10^{-6} eV down to 10^{-12} eV. This width comes from the fact that it is much easier to tune the resonance with a magnetic field. In principle, when off-resonance, the experiment can also go down to very low AC frequencies, approaching DC, allowing it to place constraints on axions with masses as low as 10^{-22} eV. This is done by using samples of hyper-polarized liquid xenon with a volume of around 1 cm^3.

As shown in Fig. 14.8, CASPEr-Wind is unlikely to reach the level of sensitivity required to constrain the QCD axion. However, it will increase existing bounds on axion-like particles by a factor of more than 10^4.

14.3.3.2 CASPEr-Electric

As mentioned, the last coupling, given by Eq. (14.40), is the one that solves the strong-CP problem. Since ϕ oscillates around zero, the neutron electric dipole moment is on average equal to zero. However, most of the time, it is not equal to zero but instead oscillates as given by

$$d_n = 2.4 \times 10^{-16} \frac{\phi}{f_a} e \text{ cm},$$

$$\equiv g_d \phi,$$

(14.44)

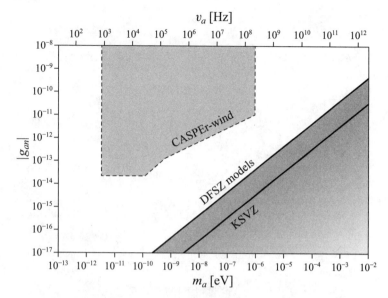

Figure 14.8 CASPEr-Wind constraint—projected constraint from phase 1 of the CASPEr-Wind experiment on the axion-neutron coupling constant. Generated using AxionLimits [85].

where we have defined the coupling g_d between the axion and the *nuclear electric dipole moments*. This also leads to nuclear spin resonance. Another wing of the CASPEr team based in Boston, USA, is trying to detect this resonance in an experiment called CASPEr-Electric. They also use NMR but with ^{207}Pb. The group aim to be sensitive to QCD axion coupling values in the mass range from 10^{-12} eV up to around 10^{-8} eV.

Both CASPEr-Wind and CASPEr-Electric have made some preliminary measurements and aim to begin their full experimental campaigns in the coming years.

QUIZ

i. If the axion has a mass of order 1 μeV, what is the oscillation frequency?
   a. kHz.
   b. MHz.
   c. GHz.
   d. THz.

ii. In the presence of a background magnetic field B in the z-direction, the axion sources:
   a. An electric field in the x-y plane.
   b. An electric field in the z-direction.
   c. No electric field.
   d. Electron-positron pair production.

iii. In a cavity, the electric field is resonant with the axion of mass m when:
   a. The length of the cavity is approximately $1/m$.
   b. The length of the cavity is approximately m.
   c. The magnetic field in the cavity is approximately m.
   d. The cavity volume is equal to the axion coherence volume.

iv. The power of a cavity haloscope (select multiple):
   a. Increases with quality factor.
   b. Decreases with quality factor.
   c. Increases linearly with magnetic field strength.
   d. Increases with cavity resonant frequency.
   e. Increases quadratically with magnetic field strength.
   f. Is equal to $(\omega/Q)\times$(energy stored).

v. The axion gradient couplings (select multiple):
   a. Resonate with fermion Larmour frequencies.
   b. Give rise to the time dependence of the neutron EDM.
   c. Come from the 'axion wind' velocity relative to Earth.
   d. Can be interpreted as an effective magnetic field.
   e. Are constrained by ADMX.
   f. Are constrained by QUAX.

vi. Existing axion DM searches:
   a. Exclude the QCD axion completely.
   b. Exclude only axions more weakly coupled than the QCD axion.
   c. Exclude a narrow range of the QCD axion mass.
   d. Disprove the axion solution of the strong-CP problem.

Answers

i. c; ii. b; iii. a; iv. a,c,d,f; v. a,e,f; vi. c

Chapter Fifteen

Astrophysical Axion Bounds

In this chapter we consider ways that axions are constrained by astrophysics. First we consider bounds on ultralight axions based on the ways in which such axions do not behave as CDM, but rather as so-called fuzzy DM. These limits are of fundamental significance since they provide an absolute lower bound on the DM particle mass and precision limits on the allowed composition of DM if it is multi-component.

Then we consider two independent bounds from black holes and from stellar evolution. These two bounds actually have nothing to do with axions as a DM candidate and apply independently of the relic density, but they provide useful information on axion masses and couplings that are relevant to the DM problem. The true equivalents of WIMP indirect detection for axions are considered briefly at the very end of this chapter, from axion DM decays and axion-photon conversion in neutron star magnetospheres. Presently the limits from neutron stars suffer from large uncertainties, while the limits from decays apply at large mass, outside of the range accessible to the direct detection techniques we discussed in the last chapter.

15.1 ULTRALIGHT AXIONS AND FUZZY DARK MATTER

15.1.1 Bounds from Structure Formation

The first bound we will derive is an important lower bound on the ALP. Recall that axions start to behave as DM when

$$H(a_{\rm osc}) \approx m_a. \tag{15.1}$$

Before this epoch the axion fluid has equation of state $w = -1$: it is like the cosmological constant and cannot cluster. Thus structure formation is suppressed for all modes within the cosmological horizon at this time, that is, all modes smaller than the comoving distance:

$$R_H^{-1} \sim a_{\rm osc} H(a_{\rm osc}). \tag{15.2}$$

For ALPs, let's take $m_a = $ constant. How small can m_a be given what we know about cosmic structures on scales R?

We have already derived a similar bound for thermal DM, so we can skip a step and identify:

$$R_H \sim 1 \text{ Mpc} \Rightarrow T_{\rm osc} \sim m_{\rm thermal} \sim 100 \text{ eV}. \tag{15.3}$$

So the equivalent ALP mass bound is

$$H(T = 100 \text{ eV}) = \left(\frac{\pi^2}{90} g_\star \frac{T^4}{M_{pl}^2} \right)^{1/2} = m_a, \qquad (15.4)$$

which implies

$$\boxed{m_a \gtrsim 2.5 \times 10^{-24} \text{ eV}}. \qquad (15.5)$$

As for warm DM, there is a stronger bound that can be derived from the Lyman-α forest. The transfer function for ultralight axions must be computed with full cosmological perturbation theory; it can be approximated as $T_{\text{ULA}}(k) = \cos x_J^3 / (1 + x_J^8)$ with $x_J = 1.61 (m/10^{-22} \text{ eV})^{1/18} k/k_{J,\text{eq}}$, $k_{J,\text{eq}} = 9(m/10^{-22} \text{ eV})^{1/2}$ Mpc^{-1}, where $k_{J,\text{eq}}$ is the Jeans scale at matter-radiation equality, which we touched on in section 10.3.

The linear theory transfer function is useful for estimating the bound from an observation consistent with CDM at some scale and for working out where deviations from CDM might occur. A full analysis, again using simulations, gives the bound [86]

$$\boxed{m_a > 2 \times 10^{-20} \text{ eV}}. \qquad (15.6)$$

The first bound Eq. (15.5) is very conservative and only relies on linear theory. It can also be derived rigorously from the CMB anisotropies and galaxy clustering. The second bound Eq. (15.6) relies on non-linear physics and modelling of gas.

We can derive an intermediate and relatively intuitive bound by considering *dwarf spheroidal galaxies* (dSph):

$$R_{\text{dSph}} \approx 1 \text{ kpc}, \quad M_{\text{dSph}} \approx 10^8 M_\odot \Rightarrow \sigma_v \approx 20 \text{ km s}^{-1}. \qquad (15.7)$$

Demanding that the reduced ALP de Broglie wavelength ($\bar{\lambda} = \lambda/2\pi = 1/mv$ in natural units) is smaller than R_{dSph} gives the bound

$$\boxed{m_a \gtrsim 10^{-22} \text{ eV}}. \qquad (15.8)$$

If the de Broglie wavelength is too large then the DM halo cannot form since the ALPs cannot 'fit inside' the halo. This is the bosonic equivalent of the Tremaine–Gunn bound.

Some dwarf galaxies are known to have 'density cores' with $\rho \sim$ const. on kpc scales. ALPs sitting exactly on or around the bound Eq. (15.8) are called *fuzzy dark matter*. The large de Broglie wavelength causes the density of the ALPs to be smoothed out on small scales (due to axion star formation and wave physics that we discussed in section 10.3). This may improve consistency with the observations and may also help alleviate other *small-scale problems of CDM*. The mass scale 10^{-22} eV is, however, apparently inconsistent with the bounds from the Lyman α forest (Eq. (15.6)). The Lyman-α forest bounds are cosmological and might be avoided by changing other aspects of cosmology and/or the early Universe physics of the ALP. So let's consider another purely local effect of ultralight ALPs that can be used to place a lower limit on the particle mass.

15.1.2 'Quasiparticle Heating'

As derived in section 10.3 bosonic DM is coherent on the de Broglie wavelength and on the time scale:

$$\tau_{\text{coh.}} \approx \frac{\bar{\lambda}}{v} = \frac{1}{m_a v^2}. \tag{15.9}$$

We can think of an axion DM halo as composed of *short-lived quasiparticles* with mass:

$$m_{\text{qp}} \sim \bar{\lambda}^3 \bar{\rho}_{\text{DM}}, \tag{15.10}$$

where $\bar{\rho}_{\text{DM}}$ is the local DM density.

The quasiparticles interact with each other and with other objects in the halo (stars, gas) via gravitational scattering. Gravitational *two-body relaxation* is the diffusion of a body's velocity caused by continual scattering by close encounters (see Ref. [5], 1.2.1). The relaxation time is

$$t_{\text{relax}} = 0.1 \frac{R}{v} \frac{M}{m \log \Lambda}, \tag{15.11}$$

where M is the host mass (the object containing the bodies responsible for the majority of the scattering), R is the host radius, m is the scatterer mass, v is the characteristic velocity of the scattered particle, and $\log \Lambda$ is the 'Coulomb logarithm' which regulates the integral over impact parameters, $\Lambda = R_{\text{max}}/R_{\text{min}}$. The Coulomb logarithm is typically a quantity of order unity. For quasiparticles scattering off each other we replace

$$m = m_{\text{qp}}, \quad R_{\text{min}} = \bar{\lambda} = \frac{1}{m_a v}. \tag{15.12}$$

It is now left as an exercise to plug in the numbers and arrive at the relaxation time:

$$t_{\text{relax}} \sim \frac{10^{10}}{\log \Lambda} \text{ yr} \left(\frac{m_a}{10^{-22} \text{ eV}}\right)^3 \left(\frac{v}{100 \text{ km s}^{-1}}\right)^2 \left(\frac{R}{5 \text{ kpc}}\right)^4. \tag{15.13}$$

Over the time scale t_{relax} a distribution of test particles will evolve away from its initial conditions (for example, some typical equilibrium distribution).

The relaxation time itself does not depend on the mass of the test particle being scattered. What matters is whether the test particle is more or less massive than the scatterer. If the scatterer is heavier than the test particle, two-body relaxation leads to heating. If the scatterer is lighter than the test particle, two-body relaxation leads to cooling. For stars in a fuzzy dark matter halo we thus have

$$m_\star \ll m_{\text{qp}} \Rightarrow \text{heating}, \tag{15.14}$$

$$m_\star \gg m_{\text{qp}} \Rightarrow \text{cooling}. \tag{15.15}$$

A stellar distribution which is cooled will contract, while one which is heated will expand.

We can estimate a bound on axions by considering the heating of the *old star cluster in Eridanus-II*. The DM velocity dispersion and central density of the Eridanus II DM halo are

$$\sigma_v = 7 \text{ km s}^{-1}, \quad \bar{\rho} = 0.15 \, M_\odot \text{ pc}^{-3}, \tag{15.16}$$

leading to a quasiparticle mass:

$$m_{\text{qp}} \sim 3 \, M_\odot \left(\frac{10^{-19} \text{ eV}}{m_a} \right). \tag{15.17}$$

The central star cluster has the following properties:

$$R_h = 13 \text{ pc}, \quad \tau_{\text{orb}} = 10^6 \text{yr}, \quad \text{Age} \sim 10^{10} \text{ yr}. \tag{15.18}$$

The quasiparticle mass implies that stars in the star cluster are heated by the quasiparticles for $m_a \lesssim 10^{-19}$ eV. If the heating time scale is shorter than the age of the star cluster, then it should not exist: it is unstable on the time scale of its age. The star cluster is stable to heating if the relaxation time is larger than the estimated age. It is also stable if $\tau_{\text{coh}} > \tau_{\text{orb}}$: a coherence time longer than the orbital period leads only to adiabatic perturbations. Imposing these two limits, we find that axions are excluded in the mass range:

$$\underbrace{10^{-21} \text{ eV}}_{\tau_{\text{coh}} > \tau_{\text{orb}}} \lesssim m_a \lesssim \underbrace{10^{-19} \text{ eV}}_{t_{\text{relax}} > \text{Age}}. \tag{15.19}$$

Axion masses in this range destabilise the central star cluster of Eridanus II. At lower masses, the star cluster can reside entirely in the central axion star/soliton of the Eridanus II DM halo (see section 10.3). If the outer quasiparticle region of the halo is tidally stripped, then this soliton might be stable and free of oscillations on large time scales and as such would not destabilise the star cluster, which allows for some possibility that lower axion masses are compatible with the Eridanus II star cluster dynamics [87, 88], although such low masses will struggle to form the Eridanus II DM halo itself and other MW satellites [89, 90].

15.1.3 Bounds on ULA Fraction

The bounds we have considered so far can be generalised to the case that DM contains only a fraction of ultralight ALPs. Precision cosmology generally is sensitive to the range

$$10^{-33} \text{ eV} \lesssim m_a \lesssim 10^{-20} \text{ eV}, \tag{15.20}$$

where the mass limits come from the length scales involved (the lower limit being the Hubble scale, and the upper limit corresponding to length scales probed by the Lyman-α forest). Current limits are at around the 1–10% level, while future bounds could improve this by some orders of magnitude and possibly test the axion relic density predicted by GUT scale symmetry breaking. There is possibly some room for fuzzy dark matter to be all of the DM in the mass range 10^{-22} eV $\lesssim m_a \lesssim$ 10^{-21} eV depending on some uncertainties in the structure formation of satellites like Eridanus II and some modification of the early Universe dynamics of the axion to allow consistency with the Lyman-α forest. Finally we remind ourselves that for $m_a < H_0 \approx 10^{-33}$ eV the axion field is not oscillating today and so is not classified as DM (at such low masses it is in fact a dark energy candidate).

15.2 BLACK HOLE SUPERRADIANCE

A spinning BH is known as a Kerr BH and has a known metric specified by the mass, M_{BH}, and dimensionless spin, $a_J \in [0, 1]$, which you can find in any textbook on GR. The axion field on Kerr obeys the Klein–Gordon equation:

$$\Box\phi - m_a^2\phi = 0, \tag{15.21}$$

where $\Box$ should be computed for the Kerr metric following the definition in Eq. (10.3) and thus depends on M_{BH} and a_J. We expand the axion field in the eigenfunctions of spheroidal symmetry, the spheroidal harmonics $S_{\ell\mu}$ (deformed spherical harmonics reflecting the azimuthal symmetry of a rotating BH spacetime):

$$\phi = \sum_{\ell,\mu} e^{-i\omega_{\ell\mu}t + i\mu\varphi} S_{\ell\mu}(\theta)\psi_{\ell\mu}(r) + \text{h.c.}, \tag{15.22}$$

where μ, ℓ are the eigenvalues (quantum numbers) and φ, θ are the spheroidal coordinates. This leads to an eigenvalue problem: to find $\omega_{\ell\mu}$ for the radial eigenfunction ψ in an effective potential (the non-relativistic limit is not unlike the one we used to arrive at the Schrödinger–Poisson equations for scalar field structure formation). The approximate shape of the effective potential is sketched in Fig. 15.1 for the so-called tortoise coordinate $r_\star$, which has $r_\star = -\infty$ at the event horizon.

The occupation number of the mode ℓ, μ grows *exponentially* if Im$\omega \neq 0$. Such growth occurs when the bound state wavefunction can tunnel through the potential barrier into the 'ergoregion'. In this scenario, part of the wavefunction crosses from the ergoregion into the event horizon, while part of it escapes back to the bound region, triggering the 'Penrose process', displayed in Fig. 15.2. The wavefunction in the ergoregion co-rotates, while the part that falls in has negative energy. Conservation of energy now dictates

$$E_1 = E_2 + E_3. \tag{15.23}$$

Since $E_2 < 0$ we must have $E_3 > E_1$. The larger energy after the process leads to a larger occupation number of the states in the bound region, that is, an increase in amplitude of the mode $\psi_{\ell\mu}$. The gain in energy is *balanced by a reduction in the angular momentum of the BH*, that is, a reduction in a_J. If the particle is confined by a mirror, then the process repeats over and over. For a bosonic field on Kerr, a natural 'mirror' is provided by the bound state solution energy barrier, and thus the process naturally self-perpetuates in a process known as *black hole superradiance* (for a review, see Ref. [91]).

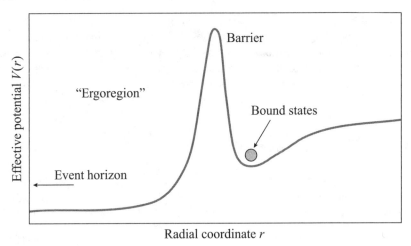

Figure 15.1 Effective scalar field potential on Kerr spacetime. The radial coordinate is the Wheeler tortoise coordinate, for which the event horizon resides at $r_\star = -\infty$.

Tunnelling is efficient when the radial scale of the bound state wavefunction is of order the radial scale in the potential, that is, when the axion Compton wavelength is of order the horizon size:

$$\lambda_C = \frac{1}{m_a} \sim R_H \sim GM_{BH} \Rightarrow GM_{BH}m_a \sim 1. \qquad (15.24)$$

The growth of ϕ reduces a_J. Following the detailed evolution it can be shown that this reduction is eventually shut off when the ergoregion becomes too small at some minimal value of a_J. We can think of this shut-off as occurring when the overlap of the ergoregion with the boson wavefunction becomes too small.

This logic implies that a certain region of the three-dimensional parameter space (m_a, M_{BH}, a_J) is excluded: in this region axions spin down BHs and do not allow for large values of a_J. The region is only truly excluded if the time scale for the evolution of the BH due to other factors is shorter than the superradiant time scale: if the BH evolves due to, for example, accretion, then these processes compete with superradiance, and the exclusion cannot be made so cleanly. Since BHs can have their parameters (M_{BH}, a_J) measured, we can arrive at an exclusion on m_a.

Let's estimate what these bounds might look like by making some simple assumptions about BHs. First, let's assume there is almost blanket coverage of BH observations in the stellar and supermassive mass ranges:

$$7M_\odot < M_{BH} < 30M_\odot \quad \text{(stellar)}, \qquad (15.25)$$

$$10^6 M_\odot < M_{BH} < 10^9 M_\odot \quad \text{(supermassive)}. \qquad (15.26)$$

Next assume all of these BHs are known to have large spins, $a_J > 0.5$, with high confidence. If this were the case, then, setting $GM_{BH}m_a = 1$ at the extreme ends,

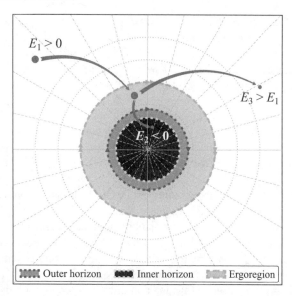

Figure 15.2 Schematic of the Penrose process. A particle with energy E_1 enters the Kerr ergoregion where it is forced to co-rotate with the BH. It then splits in two, with one particle falling into the event horizon with $E_2 < 0$ and another escaping to infinity with $E_3 > E_1$, with the balance of energy paid by a reduction in the BH spin and a shrinking of the ergoregion. If the released particle is confined by a 'mirror', then this process repeats over and over, spinning the BH down. Reproduced from "Superradiance in string theory" by Viraf M. Mehta et al. (*Journal of Cosmology and Astroparticle Physics*, Vol. 2021) © 2021 IOP Publishing Ltd and Sissa Medialab. See colour insert.

within our very rough approximations we would exclude axions with

$$4 \times 10^{-12} < m_a < 2 \times 10^{-11} \text{ eV}, \qquad (15.27)$$

$$1 \times 10^{-19} < m_a < 1 \times 10^{-16} \text{ eV}. \qquad (15.28)$$

The lower limit of Eq. (15.28) meets our ULA bounds from Eridanus II and closes a gap in axion parameter space. There is a gap between Eq. (15.28) and Eq. (15.27) caused by the non-observation of intermediate-mass BHs, leading to a window for axions around $m_a = 10^{-15}$ eV, that is, frequencies $\nu = 1$ Hz. This region can be probed in the lab by NMR experiments, as we discussed in section 14.3. Finally, note that Eq. (15.27) covers the QCD axion for

$$10^{17} \text{ GeV} \lesssim f_a \lesssim 10^{18} \text{ GeV}. \qquad (15.29)$$

Decay constants above the Planck scale run into various problems with quantum gravity, which we do not have time to explain. Thus find that *superradiance gives a lower bound on the QCD axion mass, $m_{\text{QCD}} \gtrsim 10^{-11}$ eV.*

Our superradiance bound estimates deserve a number of caveats. BH observations are not complete in the stated mass ranges. As our knowledge of BHs becomes more detailed, it is possible that we could find statistically significant gaps in the observations. Such a gap could be interpreted as evidence for superradiance, if BHs in a certain window never reach high spin and line up on 'superradiant trajectories'. Finally, axion self-interactions, parameterised by f_a, can also lead to processes in the superradiant cloud that compete with the growth of $\psi_{\ell\mu}$. Exclusions are then rightly cast in the full space $(m_a, f_a, M_{\mathrm{BH}}, a_J)$. Practically, this means that the ranges Eq. (15.28) and Eq. (15.27) each apply only down to minimum values for f_a of 10^{15} GeV and 10^{12} GeV respectively.

15.3 BOUNDS FROM STELLAR PRODUCTION OF AXIONS

We are now going to find an upper bound on the QCD axion mass and generic upper bounds on axion couplings to the Standard Model.

Axions can be produced inside stars via interactions with the plasma. For example, consider the process in Fig. 15.3, known as the Primakoff process. This process is allowed as long as $T_\gamma > m_a$. For main sequence stars, the internal temperature is of order 1 keV, which means axions can be produced inside these stars if $m_a \lesssim 1$ keV. Since, after their production, axions interact only very feebly with the stellar constituents, this implies that they have a *long mean free path and constitute an energy loss channel for the star.*

The rate of axion production via the Primakoff process (see Fig. 15.3) is

$$\Gamma_{\gamma \to a} = \frac{g^2 T \kappa_s^2}{32\pi} \left[\left(1 + \frac{\kappa_s^2}{4E^2} \right) \ln \left(1 + \frac{4\kappa_s^2}{E^2} \right) - 1 \right], \tag{15.30}$$

where g is the axion-photon coupling constant, and κ_s is the screening length caused by charge redistribution in the plasma, given by

$$\kappa_s^2 = \frac{4\pi\alpha}{T} \left(n_e + \sum_j Z_j^2 n_j \right), \tag{15.31}$$

with fine structure constant α, free electron density n_e and free ion densities n_j of ions of charge Z_j. To know n_e, n_j we need to specify a stellar model and integrate over the volume.

After doing this for the standard solar model, we find that the flux of axions at Earth can be fit by

$$\frac{d\Phi}{dE} = 6 \times 10^{10} \text{ cm}^{-2} \text{ s}^{-1} \text{ keV}^{-1} \left(\frac{g}{10^{-10} \text{ GeV}^{-1}} \right)^2 \left(\frac{E}{\text{keV}} \right)^{2.481} e^{-E/1.201 \text{ keV}}. \tag{15.32}$$

The same models, when integrated over energy, give the total stellar luminosity in axions:

$$L_a = 1.85 \times 10^{-3} L_\odot \left(\frac{g}{10^{-10} \text{ GeV}^{-1}} \right)^2, \tag{15.33}$$

with $L_\odot$ the solar luminosity. A very simple bound can be arrived at by the simple fact that if the Sun gave off an $\mathcal{O}(1)$ fraction of its energy in an invisible particle not

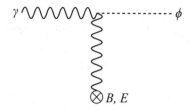

Figure 15.3 The Primakoff process. Photons convert into axions in the presence of ambient electric and magnetic fields (virtual photons) sourced by the stellar plasma.

in the Standard Model, we might think this would have been pretty obvious since at least the time in history of the understanding of the theory of nuclear fusion as the fuel of the Sun. Therefore, demanding $L_a < 1L_\odot$, we arrive at

$$g \lesssim 10^{-9} \text{ GeV}^{-1}. \tag{15.34}$$

If axions are emitted from the Sun, we can try to observe them on Earth, using the inverse process of Fig. 15.3, as depicted in Fig. 15.4 where the magnetic field is supplied inside the laboratory. Such an 'axion solar telescope' is known as a *helioscope*. The most sensitive helioscope to have been constructed to date is the CERN Axion Solar Telescope (CAST), which consists of a large LHC dipole magnet pointed at the Sun. The non-observation of keV (X-ray) photons produced via axion-photon conversion in CAST leads to the bound

$$\boxed{g < 6 \times 10^{-11} \text{ GeV}^{-1}} \quad \text{axion-photon coupling bound from CAST.} \tag{15.35}$$

Reconversion is inefficient for high-mass axions, and the bound ceases to apply for $m_a \gtrsim 1$ eV. Future plans for helioscopes include the International Axion Observatory (IAXO), which could improve the CAST bound on g by more than an order of magnitude (or, of course, IAXO may observe solar axions!).

15.3.1 Horizontal Branch Stars

If axion emission is significant then it affects stellar burning and thus affects the evolution of stars along the *Hertzsprung–Russell diagram*. Stars in globular clusters are all around the same age. Axion emission, if present, changes the ratio

$$R_{\text{HB}} = \frac{\#\text{Horizontal Branch}}{\#\text{Red Giant Branch}}. \tag{15.36}$$

Axion emission reduces the lifetime of stars on the horizontal branch, and thus a larger g reduces R_{HB}. In the Standard Model, R_{HB} is approximately given by

$$R_{\text{HB}} = 6.26 Y_{\text{He}} - 0.12. \tag{15.37}$$

The measured value of the primordial helium abundance is $Y_{\text{He}} \approx 0.25$ which is consistent with the measured value of $R_{\text{HB}} \approx 1.38$. Various models predict R_{HB} in

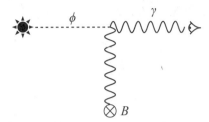

Figure 15.4 Axion helioscope. Axions produced in the Sun convert into visible photons in the presence of an applied magnetic field.

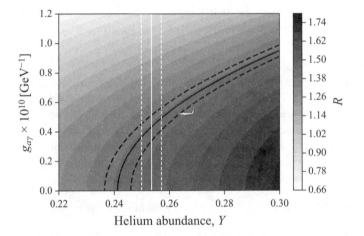

Figure 15.5 The effect of axions and helium abundance on the ratio of horizontal branch to red giant branch stars. Contours indicate the measured values, which lead to a slight preference for non-zero g.

the presence of axions. The impact of this coupling has to be calculated numerically, and results in the literature are not in total agreement. However, consider the following fit given from numerical stellar evolution models of Ref. [93]:

$$R_{\mathrm{HB}} = 6.26 Y_{\mathrm{He}} - 0.41 g_{10}^2 - 0.12, \tag{15.38}$$

where Y is the helium abundance and $g_{10} = g/10^{-10}\ \mathrm{GeV}^{-1}$. If we take the helium abundance and R-parameter as measured to be $Y_{\mathrm{obs}} = 0.2535 \pm 0.0036$ and $R_{\mathrm{obs}} = 1.39 \pm 0.04$ respectively, then we can find a preferred value for g, as shown in Fig. 15.5.

The preference for non-zero g from R_{HB} is only at a few standard deviations (and we note that there is some disagreement on the precise form of $R(g)$ in the literature). It is thus more conservative to set a 95% upper limit, which is approximately

$$\boxed{g < 6.6 \times 10^{-11} \text{ GeV}^{-1}}$$ axion-photon coupling bound from HB stars.

$$(15.39)$$

If the axion also couples directly to electrons then this further affects stellar evolution and one can arrive at the bound in the plane (g, g_e). The bound is almost exactly the same as the CAST bound in terms of g but extends all the way to the typical internal temperature of main sequence stars, $m_a \approx 1$ keV (the CAST limit cuts off due to requiring the reconversion of axions to photons inside the telescope).

15.3.2 SN1987A

Supernova 1987A was a core collapse supernova observed in great detail due to its proximity to the Earth. There was an accompanying neutrino burst observed in a number of neutrino detectors. The *neutrino burst was delayed with respect to the SN1987A lightcurve peak*. The standard explanation for this is that collapse leads to a proto-neutron star forming, which gravitationally traps the neutrinos and delays their emission.

   Axions are also produced in SN1987A via the process

$$N + N \longrightarrow N + N + a, \tag{15.40}$$

where N is a nucleon, which is allowed as long as the supernova internal temperature, $T \approx 30$ MeV, exceeds the axion mass. The process is mediated by the axion-neutron and axion-proton couplings:

$$\frac{C_n}{f_a} \partial_\mu \phi \bar{n} \gamma^\mu \gamma_5 n \quad \text{(axion-neutron coupling)}, \tag{15.41}$$

$$\frac{C_p}{f_a} \partial_\mu \phi \bar{p} \gamma^\mu \gamma_5 p \quad \text{(axion-proton coupling)}, \tag{15.42}$$

which leads to a total axion-nucleon effective coupling:

$$C_N^2 = Y_n C_n^2 + Y_p C_p^2, \tag{15.43}$$

where Y_n and Y_p are the relative proton and neutron abundances. These are $Y_p \approx 0.3$ and $Y_n = 1 - Y_p$. Axion emission by SN1987A would cause the proto-neutron star to *cool more rapidly, producing fewer neutrinos in the process and thus shortening the length of the observed neutrino burst*.

   The axion emissivity is given by

$$\varepsilon_a = \frac{1}{\rho} \left(\frac{C_N}{2f_a} \right)^2 \frac{n_N}{4\pi^2} \int_0^\infty \omega^4 S_\sigma, \tag{15.44}$$

where S_σ is the spin structure function of the proto-neutron star. This is difficult to calculate, especially in the presence of axion emission. The result can be approximated by (Ref. [37])

$$\varepsilon_a \approx \left(\frac{C_N}{2f_a}\right)^2 \frac{T^4}{m_N \pi^2}. \tag{15.45}$$

Simulations of SN1987A find that the neutrino burst emission is too short if

$$\varepsilon_a > 10^{19} \text{ erg g}^{-1} \text{ s}^{-1} \sim 7.2 \times 10^{-18} \text{ eV}. \tag{15.46}$$

Taking $T = 30$ MeV in Eq. (15.45) leads to the bound

$$\boxed{\frac{C_N}{f_a} < 6 \times 10^{-10} \text{ GeV}^{-1}} \quad \text{(SN1987A axion-nucleon coupling bound)}.$$
$$\tag{15.47}$$

Eq. (15.47) is very significant because the QCD axion *must* have non-zero C_N to solve the strong-*CP* problem. Thus the bound is a lower bound on f_a and *an upper bound on the QCD axion mass*. A reference model takes $C_n = 0$ and $Y_p = 0.3$ proton, giving the limit $f_a \gtrsim 4 \times 10^8$ GeV and thus

$$\boxed{m_{\text{QCD}} \lesssim 2 \times 10^{-2} \text{ eV}} \quad \text{(SN1987A QCD axion mass bound)}. \tag{15.48}$$

Various modern studies have refined this bound, but the overall scale is consistent with Eq. (15.48). Due to large theoretical uncertainties, and the sparse neutrino data on SN1987A, the bound is only indicative.

Our astrophysical considerations of superradiance, Eq. (15.27), and SN1987A, Eq. (15.48), lead to a *finite range of parameter space for the QCD axion*:

$$\boxed{2 \times 10^{-11} \text{ eV} \lesssim m_{\text{QCD}} \lesssim 2 \times 10^{-2} \text{ eV}} \quad \text{(QCD axion mass range)}. \tag{15.49}$$

15.4 DECAY AND CONVERSION OF AXION DM

The astrophysical limits we have just considered on axions, from superradiance and from stellar evolution, are very powerful in bounding the mass range of the QCD axion and limiting the allowed couplings of generic ALPs. They are powerful because they do not rely on the assumption that these particles are DM, but this is a book on DM, and so we should ask if there are any more astrophysical constraints on axion couplings if the axion is indeed DM. The answer is that, because of just how weakly coupled the axion is, there are surprisingly few such 'indirect detection' limits.

15.4.1 X-Rays and Axion Decays

The axions we have met so far have generally been very light, with $m_a \ll 1$ eV, but axions don't have to be so light. Heavier axions can be produced by vacuum realignment and also by thermal processes like freeze out just the same way WIMPs can. This is a relevant channel that can lead to DM composed of axions with masses of order keV.

The axion-photon coupling allows an axion to decay to two photons with lifetime

$$\tau = \frac{64\pi}{m_a^3 g^2}.$$ (15.50)

For keV mass and above, the axion lifetime can be short enough that we might try to observe axion decays. If all of the DM were composed of axions, then this decay would produce a monochromatic line visible in any galaxy at frequency $\omega = m_a/2$. Such a line has been searched for in X-ray observations and is not observed. Between approximately 1 keV and 10^4 keV the limit on the lifetime of any DM particle inferred from the absence of such a line is

$$\tau \gtrsim 10^{29} \text{ s}, \Rightarrow g \lesssim 10^{-18} \text{ GeV}^{-1} \left(\frac{10 \text{ keV}}{m_a} \right)^{3/2}.$$ (15.51)

This is a very strong limit on the axion-photon coupling at high mass, under the assumption that the keV axion is all of the DM. This limit is applicable to any keV decaying DM, most notably sterile neutrinos.

15.4.2 Neutron Star Magnetospheres

Axion-photon conversion in haloscopes happens in the presence of a magnetic field and a resonance driven by the haloscope geometry. Magnetized plasmas provide a natural haloscope, with the resonance condition being provided by the photon plasma frequency:

$$\omega_p = \sqrt{\frac{4\pi \alpha n_e}{m_e}},$$ (15.52)

where n_e is the free electron density, α is the fine structure constant, and m_e is the mass of the electron. Some of the largest magnetic fields in the Universe are present around neutron stars with $B \sim 10^{11}$ T. Due to the ionized atmosphere around them, neutron stars also happen to have plasma frequencies of order 10^{-6} eV in their magnetospheres and so can resonantly convert axions in this mass range into photons. Neutron stars thus provide hugely powerful astrophysical haloscopes, which will convert axion DM around them into radio frequency photons which we can try to observe.

Assuming the DM axions around the neutron star are non-relativistic, then in a region of length L over which $\omega_p = m_a$ the conversion probability from axions to photons is

$$P \sim B^2 g^2 L^2.$$ (15.53)

To compute the exact photon flux, one solves the coupled axion-photon equations of motion using a model for the neutron star magnetosphere, for which it is common to use the so-called Goldreich–Julian model.

The most sensitive limits on the axion-photon coupling were obtained recently from observing the galactic centre with the 'Breakthrough Listen' survey by the

Greenbank Telescope (which was designed to search for extraterrestrials). No radio lines consistent with axion-photon conversion were observed (and no signals from little green men either), which allows a limit to be set on g:

$$g \lesssim 10^{-11} \text{ GeV}^{-1}, \quad (17\,\mu\text{eV} \lesssim m_a \lesssim 32\,\mu\text{eV}). \tag{15.54}$$

The relatively narrow range in mass is fixed by the receiver bandwidth on the radio telescope. The limit relies not only on careful modelling of axion-photon conversion, but also of the population of neutron stars in the galactic centre.

The signal considered here is caused by the ambient DM density in the galactic centre and would lead to a continuously visible radio line. It is possible that a stronger signal from axion-photon conversion in neutron star magnetospheres might be seen if the axion DM is clumped in so-called miniclusters. These high-density DM halos form if SSB happens after inflation. If a minicluster collides with a neutron star, it will create a very bright *transient* radio line. Due to computational challenges, much is unknown about the exact distribution of miniclusters inside galaxies, making the signal of their collision with neutron stars hard to predict, and no conclusive search for transient radio lines due to this effect has been carried out so far. Miniclusters may also have other observational consequences, such as gravitational microlensing and on axion direct detection, which you can find references for in the appendix.

Chapter Sixteen

Constraints on PBHs

16.1 HAWKING RADIATION

Despite their name, black holes are not perfectly 'black'. Instead, they can give off radiation from their event horizons known as *Hawking radiation*. This radiation arises due to vacuum fluctuations occurring on curved space time. Matter anti-matter particle pairs can be formed in a vacuum by essentially 'borrowing' energy from the Universe. As shown in Fig. 16.1, there are three different possible outcomes when such a pair is created near a black hole:

1. The matter anti-matter pair immediately recombine and annihilate each other and the debt to the Universe is repaid.
2. Both particles fall into the black hole. In this scenario, the black hole simultaneously gains the mass of both particles, while losing the same amount of mass by repaying the energy debt. Therefore, there is no net change in the black hole's mass.
3. Sometimes, only one of the two particles falls into the black hole and the other particle is able to escape. Now the black hole has only gained the mass of one of the particles but must still pay the energy debt for both particles. As a result, the black hole loses mass. The particle which escapes is known as Hawking radiation.

This emission of particles is *thermal* and gives the black hole a temperature

$$T_{\mathrm{BH}} = \frac{1}{8\pi GM} \approx 100 \left(\frac{10^{14}\mathrm{g}}{M} \right) \ \mathrm{MeV}, \tag{16.1}$$

where $GM = R_{\mathrm{Sch}}$ is the *Schwarzchild radius*. Importantly, we see that the temperature of a black hole is inversely proportional to its mass. This means that while solar mass black holes are relatively cold, much lighter primordial black holes can have much higher temperatures. This process also means that, in the absence of any material for the black hole to consume, its mass will decrease until it disappears.[1] We can therefore ask: how long can a black hole with some initial mass

---

[1] Quantum gravitational effects might mean black holes lose mass until they reach some minimum mass given by, for example, the string scale in string theory. In the final stages of particle emission, the

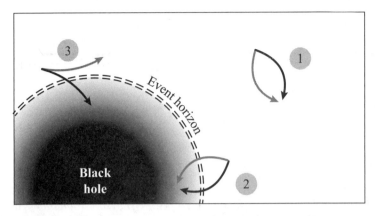

Figure 16.1 Vacuum fluctuations lead to the production of particle-anti-particle pairs. Near a black hole: (1) the pair can immediately recombine and annihilate, (2) both particles can fall into the black hole or (3) one particle can fall into the black hole, providing the black hole with negative energy and hence reducing its mass.

survive? It can be shown that the change in mass per unit time is given by

$$\frac{dM}{dt} = -5.34 \times 10^{-3} \left(\frac{10^{14} \text{ g}}{M}\right)^2 f(M) \text{ g s}^{-1}, \qquad (16.2)$$

where $f(M)$ is determined by considering all of the particles with a mass smaller than the temperature of the black hole, that is, $m < T_{\text{BH}}$.[2] This sum can be written as

$$f(M) = \sum_{i[m_i < T(M)]} f(s_i), \qquad (16.3)$$

where s_i is the spin of particle i and $f(s_i)$ is a value that depends on the particle's spin as summarised in Table 16.1.

Integrating over time while using $f(M)$ for the Standard Model, the total lifetime of a black hole of mass M is given by

$$\tau_{\text{BH}} \approx \frac{2.7 \times 10^{14}}{\langle f(M) \rangle} \left(\frac{M}{10^{14} \text{ g}}\right)^3 s, \qquad (16.4)$$

where $\langle f(M) \rangle$ is the weighted average of $f(M)$ which is approximately equal to the initial value $f(M_i)$. We hence see that, since larger black holes have a smaller temperature, they live longer.

---

momentum gained from each emission becomes significant compared to this mass, leaving behind a small fast-moving remnant that may behave like warm or hot dark matter.

[2] This can be thought of as the black hole quantity analogous to $g_\star$ in the Friedmann equation.

Table 16.1 Values relating to the emission of particles of different
spins as Hawking radiation.

s_i	Example(s)	$f(s_i)$
0	Higgs boson, axion	0.267
1	Photon, Z- and W-bosons	0.060
3/2	Gravitino	0.020
2	Graviton	0.007
1/2 (neutral)	Neutron, neutrino	0.147
1/2 (charge e)	Proton, electron	0.142

If PBHs are dark matter, then their lifetime must be greater than the age of the Universe (i.e. $\tau_{\mathrm{BH}} \gtrsim 13.8 \times 10^9$ years). Substituting this into Eq. (16.4) we find the requirement that

$$M_{\mathrm{PBH}} \gtrsim 5 \times 10^{14} \left(\frac{f(21 \text{ MeV})}{1.9} \right)^{1/3} \text{ g,} \tag{16.5}$$

or equivalently in solar masses ($M_\odot = 2 \times 10^{30}$ kg)

$$M_{\mathrm{PBH}} \gtrsim 2.5 \times 10^{-19} M_\odot. \tag{16.6}$$

Any black hole lighter than this shouldn't be able to survive long enough to be dark matter.

This has consequences for our ability to directly detect PBHs on Earth. Assuming that black holes, each with a mass equal to $2.5 \times 10^{-19} M_\odot$, make up all of the dark matter, we can calculate that we would expect around 10^{-7} collision events each year. Additionally, the Schwarzschild radius is given by

$$R = GM \sim 10^{-16} \left(\frac{M}{10^{-12} M_\odot} \right) \text{ m.} \tag{16.7}$$

This means we would have to look for objects with sizes on the scale of femtometers which only pass through the Earth around once every 10 million years. This makes attempts to detect PBHs *directly* (i.e., on Earth) pretty hopeless.

However, we might not need a PBH to collide directly with our detector. It is thought that if a PBH collided with the Earth it would cause an earthquake with a magnitude of around 4. Heavier PBHs would cause even bigger earthquakes, and therefore it is lucky for life on Earth that the expected collision rate goes down as PBH mass goes up! Furthermore, PBHs with masses that are still unconstrained by observations we will discuss later are found to have collision rates lower than one per Earth lifetime.

Since we (fortunately) are unlikely to be able to study PBHs on Earth, we need to find ways of constraining them with astrophysical observations.

16.2 LOWER BOUND FROM GAMMA RAYS

If PBHs have survived to the present day, they will still be emitting Hawking radiation which we can try to detect. The photon emission rate is

$$\frac{dN}{dE} \propto \begin{cases} E^3 M^3, & \text{for } E < T_{\text{BH}}, \\ E^2 M^2 e^{-E/T}, & \text{for } E > T_{\text{BH}}. \end{cases} \tag{16.8}$$

We see that the emission rate increases quickly for particles with energies less than the black hole temperature but that it is exponentially suppressed for energies greater than this. The emission, therefore, peaks for particle energies around the black hole temperature. The maximum intensity is given by

$$I(T_{\text{BH}}) \propto T_{\text{BH}}^2 f(M) \Omega_{\text{PBH}}. \tag{16.9}$$

Naturally, this spectrum cannot exceed the observed spectrum, particularly if this is understood due to known astrophysical processes. There are two main energy ranges that are of importance to us. The first is X-rays with energies in the range of 3–500 keV. Photons in this range are either measured using balloon-based experiments or with telescopes such as Chandra. The second range is gamma rays with energies of around 1 MeV as measured using telescopes such as EGRET.

As shown in Fig. 16.2, DM composed of a PBH with a mass $M \sim 10^{-16} M_\odot$ produces a spectrum of photons with a maximum intensity which just about reaches the observed value. However, as the mass of the PBH decreases, the spectrum is shifted to higher energies. These PBH masses can therefore not make up all of the dark matter since they would produce photons at intensities that exceed the observed spectrum. Instead, these masses can only make up a smaller fraction of all dark matter. Therefore, if we want PBHs to make up all of the dark matter, we have the lower bound

$$M_{\text{PBH}} \gtrsim 10^{-16} M_\odot. \tag{16.10}$$

For masses smaller than this we get strong bounds on the allowed value of Ω_{PBH}.

16.3 UPPER BOUND FROM GALAXY FORMATION

The simplest bound on PBHs comes from the masses of dwarf spheroidal galaxies (dSphs). These objects have masses $M \sim 10^8 M_\odot$ and radii $R \sim 0.5$ kpc. We also know that these galaxies are dominated by dark matter. For this to be the case we naturally require that the PBH mass is much smaller than the dwarf galaxy's total mass so that the PBHs can be treated as a large number of point masses making up a virialised halo:

$$M_{\text{PBH}} \ll 10^8 M_\odot. \tag{16.11}$$

You can think of this like the PBH version of the Tremaine-Gunn bound for fermions or the bound we derived on axions by considering their de Broglie wavelength relative to the size of a dSph.

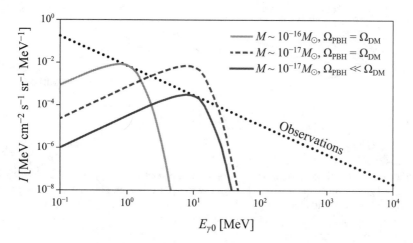

Figure 16.2 Sketch of emission spectra of PBHs due to Hawking radiation (inspired by Ref. [94]). For PBHs of mass $\sim 10^{-16}M_\odot$ the peak of the spectrum is just about consistent with the observed background spectrum while making up all of the dark matter. However, at lower masses, the PBH must make up a smaller fraction of the dark matter to be consistent with observations.

We can obtain a stronger upper bound on the PBH mass by considering the survival of ultra faint dwarfs such as Eridanus II. The presence of PBHs would cause a two-body relaxation which heats the star cluster just like the axion 'quasiparticles' we considered in section 15.1.2. We can even use our result from axion quasiparticles to save us from repeating the same calculation.

Recall that the axion quasiparticle mass is given by

$$m_{qp} = \frac{\rho_{DM}}{(m_a v)^3}. \tag{16.12}$$

From this, we estimated that for Eridanus II the quasiparticle mass was

$$m_{qp} = 3M_\odot \left(\frac{10^{-19} \text{eV}}{m_a} \right)^3. \tag{16.13}$$

By considering the heating of the cluster by these quasiparticles, we found that we required $m_a \gtrsim 10^{-19}$ eV if the axion was to make up all of the dark matter. This meant that we needed lighter quasiparticles to not disrupt the star cluster. Since in our comparison of the two-body relaxation time to the age of the star cluster, we didn't care what the quasiparticles were, the same limit applies to PBHs, that is:

$$M_{PBH} \lesssim 3M_\odot. \tag{16.14}$$

For masses greater than this, we get strong bounds on the allowed value of Ω_{PBH}.

We now have a lower bound at $10^{-16}M_\odot$ from X-ray and gamma-ray measurements and an upper bound at $3M_\odot$ from the survival of Eridanus II. In the next

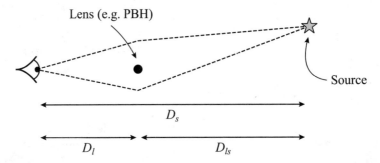

Figure 16.3 Microlensing–light from a source at a distance D_s away from the observer is bent by a gravitational lens at a distance D_l away from the observer. This enables two separate images of the source to reach the observer. However, in microlensing, these separate images cannot be resolved and are instead seen as an amplification of the unlensed source intensity.

section, we will see that most of this intermediate region is excluded by gravitational microlensing.

16.4 MICROLENSING

As discussed in chapter 4, due to general relativity, when light travels in the presence of a gravitational field its path is bent much like when it passes through a lens as shown in Fig. 16.4. We, therefore, know the process as gravitational lensing. This lensing can often enable us to see multiple images of the same object. However, when the lensing is very small, we cannot resolve these separate images. Instead, we observe an amplification of the original source magnitude. This is known as *gravitational microlensing*.

Recall that we write the lens equation in terms of two-dimensional vectors in the lens plane as

$$\vec{y} = \vec{x} - \vec{\alpha}, \tag{16.15}$$

where $\vec{y}$ and $\vec{x}$ are the 2D positions of the source and image respectively and where $\vec{\alpha}$ is the deflection angle. This deflection angle is given by the two-dimensional gradient of the lensing potential Ψ:

$$\vec{\alpha} = \vec{\nabla}_{2d}\Psi, \tag{16.16}$$

where the lensing potential is the two-dimensional projection of the gravitational potential onto the lensing plane. Taking the derivative of components of the source position, y_i, by the components of the image position, x_i, gives

$$\frac{\partial y_i}{\partial x_j} = A_{ij} = \delta_{ij} - \frac{\partial^2 \Psi}{\partial x_i \partial x_j}. \tag{16.17}$$

Recall also that the magnification is given by the inverse determinant of A:

$$\mu = \frac{1}{\det A}. \tag{16.18}$$

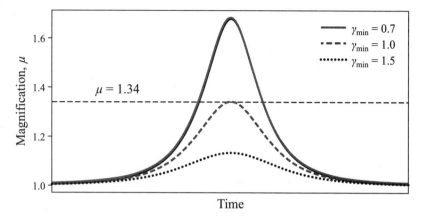

Figure 16.4 Magnification of a light source by a point lens for different minimum impact parameters $y_{\min}$.

However, as mentioned already, in microlensing we have two *unresolved* images.

For a point mass, we naturally have cylindrical symmetry on the lens plane. We can therefore write our source and image positions each as a single value ($\vec{x} \to x$ and $\vec{y} \to y$). The coordinate y of the source is the same as the impact parameter between it and the lens along the line of sight. It is then possible to derive the lens equation:

$$y = x - \frac{1}{x}. \tag{16.19}$$

Rearranging for x, we then obtain the location of the two images given by

$$x_{\pm} = \frac{1}{2}\left[y \pm \sqrt{y^2 - 4}\right]. \tag{16.20}$$

We then calculate $\det A$ for each of these two images as given by

$$\det A = \frac{\partial y}{\partial x}. \tag{16.21}$$

The magnification of each of these images is given by

$$\mu_{\pm} = (\det A)^{-1} = \left(1 - \left(\frac{1}{x_{\pm}}\right)^4\right)^{-1}, \tag{16.22}$$

with a total magnification then given by the sum of these two different values

$$\mu_{\text{tot}} = |\mu_+| + |\mu_-|. \tag{16.23}$$

Therefore, using Eqs. (16.22) and (16.20), the total magnification is given by

$$\mu_{\text{tot}} = \frac{y^2 + 2}{y\sqrt{y^2 + 4}}. \tag{16.24}$$

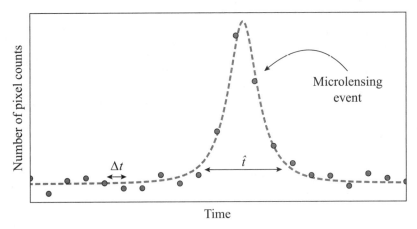

Figure 16.5 Example magnification event which lasts for time $\hat{t}$ and is observed with a cadence Δt.

We choose to measure all distances in units of the 'Einstein' radius R_{E}, which is given by

$$R_{\mathrm{E}} = \sqrt{\frac{4GM}{c^2} \frac{D_{\mathrm{L}} D_{\mathrm{LS}}}{D_{\mathrm{S}}}}. \tag{16.25}$$

In these units, when the impact parameter $y = 1$, a total magnification of $\mu = 1.34$ is produced as shown in Fig. 16.4.

A microlensing survey observes a large number of stars, returning to each one with a cadence of Δt. By doing so the survey can build up light curves for each star. Any magnifications above $\mu = 1.34$ are flagged as lensing events if their shape is the same as the predicted curve as shown in Fig. 16.5.

To use microlensing to calculate constraints on the abundance of PBHs we need to predict the rate of lensing events that we expect given our DM halo model. We also need to bear in mind that planets and stars are also capable of creating microlensing signals. In fact, microlensing is a tool used to discover new exoplanets. These additional events compose a background to our PBH search, particularly when looking through messy regions of space such as the galactic centre.

The rate of microlensing events is the rate of PBHs crossing the *microlensing tube* radius R_{E} between the observer and source. The total number of events over a survey is given by

$$N_{\mathrm{exp}} = E \int_0^{\mathrm{inf}} \frac{d\Gamma}{d\hat{t}} \epsilon(\hat{t}) d\hat{t}, \tag{16.26}$$

where E is the exposure in star years, that is, the number of stars observed multiplied by the total observation time, $\epsilon(\hat{t})$, is the efficiency of observing an event of time $\hat{t}$, an example of which is shown in Fig. 16.6, and $d\Gamma/d\hat{t}$ is the rate of events with time $\hat{t}$.

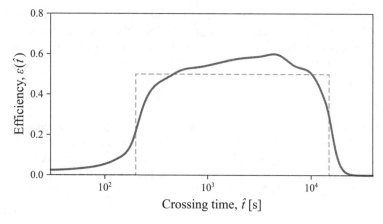

Figure 16.6 Example of a lensing survey efficiency of observing an event with a crossing time $\hat{t}$. Reproduced from data in Ref. [95].

The rate is found using the DM density profile and Maxwellian velocity of PBHs (the derivation can be found in Ref. [96]),

$$\frac{d\Gamma}{d\hat{t}} \approx \frac{32D_s}{\hat{t}^4 v_c^2} \frac{1}{M_{PBH}} \int_0^{x_h} dx \rho(x) R_E(x)^4 e^{-Q(x)}, \tag{16.27}$$

where

$$Q(x) \approx \frac{4R_E(x)^2}{\hat{t}^2 v_c^2}, \tag{16.28}$$

where v_c is the local circular speed of the Earth in the Milky Way ($v_c \approx 220$ km s^{-1}) and $\rho(x)$ is the DM density along the line of sight.

16.4.1 EROS and MACHO

The EROS and MACHO surveys observed stars within the Large Magellanic Cloud (LMC). The LMC is 50 kpc away from the Earth at a galactic coordinate $(l, b) = (280°, -33°)$ placing it within the MW halo. This means that we only need to consider the DM density of the MW when calculating the DM density along the line of sight

$$\rho(x) = \rho_{MW}(x). \tag{16.29}$$

The analysis in Ref. [97] modelled the density profile of the MW as a cored isothermal sphere given by

$$\rho_{iso}(r) = \rho_0 \frac{R_0^2 + R_\odot^2}{R_0^2 + r^2}, \tag{16.30}$$

where $R_\odot = 8.5 \times 10^3$ pc is the Sun's orbital radius around the MW, $R_c = 5 \times 10^3$ pc is the scale radius and $\rho_0 = 0.0079$ Msol pc^{-3} is the scale density.

The EROS survey had an exposure of $E = 3.68 \times 10^7$ star years, a cadence of around 1 day and a total viewing time of 1000 days. This long cadence means that the survey misses events with a short crossing time $\hat{t}$. As a result, they are unable to constrain low-mass PBHs due to their small Einstein radii.

16.4.2 Subaru-HSC

The Subaru-HSC survey instead observed stars within M31, also known as the Andromeda Galaxy. M31 is 770 kpc away from Earth, at a galactic coordinate of $(l, b) = (121.2°, -21.6°)$. Unlike the LMC, M31 has its own DM halo which makes a significant contribution to the total line of sight density. We, therefore, have to consider the sum of the MW and M31 DM densities

$$\rho(x) = \rho_{MW}(x) + \rho_{M31}(x). \tag{16.31}$$

Unlike EROS, the Subaru-HSC survey chose to model both Andromeda and the MW as NFW profiles given by

$$\rho_{NFW}(r) = \frac{\rho_0}{\frac{r_s}{r}\left(1 + \frac{r_s}{r}\right)^2}, \tag{16.32}$$

where the scale densities ρ_0 are $4.88 \times 10^6 M_\odot$ kpc^{-3} and $4.96 \times 10^6 M_\odot$ kpc^{-3} and scale radii r_s are 21.5 kpc and 25 kpc for the MW and M31 respectively. The masses of the MW and M31 are very similar at around $10^{12} M_\odot$ and $1.6 \times 10^{12} M_\odot$ respectively.

The campaign has a cadence of 2 minutes and a total observation time of 7 hours. As a result it was sensitive to much shorter events than EROS but was also insensitive to the much longer events that were potentially visible to EROS.

16.4.3 Estimating Constraints

For both of these surveys, in order to calculate the line of sight density profile, we need to calculate the radial coordinate r as a function of the line of sight distance $x = D/D_s$. For the MW, we can calculate this using the galactic coordinates of the source as

$$r_{MW}(D) = \sqrt{R_\odot^2 - 2R_\odot \cos l \cos b + D^2}. \tag{16.33}$$

The radial coordinate for M31, or any other galaxy we might be studying, is then simply given by

$$r_{M31} = D_s - D. \tag{16.34}$$

Since neither EROS, MACHO nor HSC saw any microlensing events, we can once again use Poisson statistics to calculate the resulting constraints. Recall that Poisson statistics tell us that if we don't observe any events, any model that predicts more than three events ($N_{exp} \geq 3$) is excluded at a 95% confidence level.

To estimate the PBH exclusion, we need to estimate the $\hat{t}(M)$ typical time scale for a microlensing event. We could make the slightly crude estimate that

$$v_c \hat{t} = 2R_E, \tag{16.35}$$

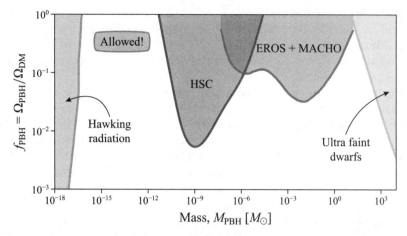

Figure 16.7 Compilation of key constraints on the fraction of DM composed of PBHs including microlensing (HSC and EROS), γ and X-ray measurements and the survival of ultra faint dwarf galaxies. Data collated by B. Kavanagh as part of the PBHbounds project, https://github.com/bradkav/PBHbounds (Ref. [181]). See colour insert.

remembering that R_E is maximized when $D_L = D_S/2$ (x = 1/2). However, we can do slightly better by noting that the rate $r = d\Gamma/d\hat{t}$ is maximized when $\partial r/\partial \hat{t} = 0$. Therefore, differentiating Eq. (16.27) with respect to $\hat{t}$, we see that

$$\int dx \rho(x) R_E^4 e^{-Q(x)} \left[-\frac{4}{\hat{t}^5} + \frac{8R_E}{\hat{t}^7 v_c^2} \right] = 0. \qquad (16.36)$$

Then, since the terms outside of the square brackets are non-zero, we therefore require that

$$-\frac{4}{\hat{t}^5} + \frac{8R_E}{\hat{t}^7 v_c^2} = 0. \qquad (16.37)$$

Rearranging we then find that our rate is maximized when

$$\hat{t}^2 = 2\frac{R_E^2}{v_c^2}. \qquad (16.38)$$

Setting $x = 1/2$ for a lens half way to the source, and using reference distance to the LMC and PBH mass of 1 $M_\odot$, we can finally write this in the slightly more useful form

$$\hat{t} \approx 10^7 \, \text{s} \left(\frac{M}{M_\odot} \right)^{1/2} \left(\frac{D_s}{50 \, \text{kpc}} \right)^{1/2}. \qquad (16.39)$$

By setting $\hat{t}$ to the maximum and minimum observable crossing times of the EROS and Subaru-HSC surveys, we can estimate the constrained masses for each. For EROS we find that masses are constrained in the region $10^{-4} \lesssim M_{PBH}/M_\odot \lesssim 10^3$, and for Subaru-HSC the range is $10^{-11} \lesssim M_{PBH}/M_\odot \lesssim 10^{-6}$.

Performing the full integral given in Eq. (16.27) for each of these surveys, while taking into consideration their respective sensitivities as well as wave effects,[3] one can calculate their constraints of the PBH mass fraction as shown in Fig. 16.7. Numerical calculations for these constraints can be found in the accompanying JUPYTER notebooks.

Fig. 16.7, and summaries like it, are rather remarkable. PBHs are bounded extremely strongly at both very high and very low mass, where they cannot make up more than 0.1% of the DM. There is only one window, between about $10^{-11} M_\odot$ and $10^{-15} M_\odot$, where a PBH of a single mass is allowed to compose all of the DM. Naturally, this region is the target of researchers searching for evidence of PBHs or seeking to exclude them. A caveat, however, is that our limits have assumed all of the DM is composed of PBHs of a single mass, while realistic formation mechanisms typically lead to a spread of masses: a mass function like that for DM halos. Lensing limits in particular must be calculated taking this spread of masses into account. This may allow PBHs to compose all of the DM at higher masses if the mass function is wide but naturally makes them more elusive.

---

[3] For very small masses, the Schwarzschild radius of the PBH is smaller than the typical wavelength of the observed light. This leads to suppression in the magnification which prevents us from being able to see these low-mass objects.

QUIZ

i. Which gives off higher energy photons of Hawking radiation?
   a. Light BHs.
   b. Heavy BHs.
   c. BHs cannot radiate.

ii. Hawking radiation (select multiple):
   a. Makes the BHs visible like stars, rather than dark.
   b. Excludes masses for which PBHs evaporate on time scales shorter than the age of the Universe.
   c. Excludes some masses due to X-ray and gamma ray emission in the sky, which is not observed.
   d. Excludes some masses due to radio emission in the sky, which is not observed.
   e. Excludes PBH DM because of the 'black hole information paradox'.
   f. Excludes PBHs with $M < 10^{-16} M_\odot$.
   g. Excludes PBHs with $M > 10^8 M_\odot$.

iii. PBHs of $M \approx 1 M_\odot$ heat the old star cluster in Eridanus-II by:
   a. Hawking radiation.
   b. Tidal heating.
   c. Two-body relaxation.

iv. Microlensing:
   a. Leads to an observable shift in location between source and image.
   b. Leads to an amplification of light from a source.
   c. Is only caused by black holes.

v. For an impact parameter of fixed physical distance from the line of sight:
   a. Heavier lenses cause more magnification.
   b. Lighter lenses cause more magnification.
   c. All lenses cause the same magnification.

vi. The expected microlensing event duration in the EROS survey was 70 days. If the PBH mass decreased by four orders of magnitude, what would the expected duration be?
   a. 0.7 days.
   b. Unchanged.
   c. 7 days.
   d. 700 days.

vii. What type of microlensing survey has the best constraints on low-mass PBHs?
   a. Short cadence, long observation time.
   b. Long cadence, long observation time.
   c. Short cadence, short observation time.
   d. Long cadence, short observation time.

Answers

i. a; ii. b,c,f; iii. c; iv. c; v. b; vi. a; vii. a

Problems on Testing Dark Matter

Problem III.1. For many years the DAMA experiment has seen an 'annual modulation' signal, which is claimed to be consistent with the expected June–December annual modulation of DM nuclear recoils. The group used 250 kg of radio pure NaI(Ti) scintillator and can detect recoil energies in the range of 2 to 6 keV.

a) Assuming direct elastic recoil, show that

$$(E_R - m_N v_i^2)m_\chi^2 + E_R m_\chi m_N + E_R m_N^2 = 0. \qquad (16.40)$$

b) Hence calculate the maximum and minimum possible WIMP mass detectable by DAMA assuming that recoils from Na (sodium) nuclei dominate.

c) Following the example of section 12.3 but for a *general* WIMP with $\sigma_p = 2 \times 10^{-40}$ cm^2, estimate the expected:

   i. Spin independent cross section

   ii. Recoil rate per nucleus

   iii. Total detector recoil rate

   for the maximum and minimum WIMP mass calculated in **b**.

d) Around the same time, the XENON100 experiment, which used a target of 62 kg of liquid Xenon, reported no evidence of a WIMP signal.

   Calculate the expected recoil rate for Xenon100 for the minimum and maximum masses calculated in **b**. Is this compatible with the DAMA findings?

e) XENON100 recorded data for a total of 100.9 days. Using Poisson statistics, calculate the constraint on $\langle \sigma v \rangle$ at a 90% confidence interval for the minimum and maximum masses calculated in **b**.

Problem III.2. Consider scaling ADMX from GHz to THz.

a) What is the cavity length, L, for resonance, in cm? Taking $V \sim L^3$, what is V in litres?

b) What is the axion mass, in eV, corresponding to 1 THz? For the QCD axion, what f_a does this correspond to in GeV? What is the coupling to photons, g, in GeV^{-1}?

c) Assuming the cavity quality factor, Q, and magnetic field, B, are the same as for ADMX, compute the power output of the cavity in watts.

d) Compute the standard quantum limit noise temperature in kelvin at 1 THz.

e) What is the scan time to reach the QCD axion?

f) For your value of the quality factor, with what precision does L have to be tuned to scan a bandwidth v/Q?

g) Using what you have found in a–f, comment on the problems facing such a proposed THz experiment. What are the main limiting factors? How might they be overcome?

Problem III.3. Consider scaling ADMX from GHz to MHz.

a) What is the cavity length, L, for resonance, in cm? Taking $V \sim L^3$, what is V in litres?

b) What is the axion mass, in eV, corresponding to 1 MHz? For the QCD axion, what f_a does this correspond to in GeV? What is the coupling to photons, g, in GeV^{-1}?

c) Assuming the same magnetic field as for ADMX, what is the total magnetic energy stored in the cavity, in joules? Why might this be a problem for the proposed experiment?

d) Assuming the same quality factor and form factor as ADMX, what is the power output in Watts?

e) What is the standard quantum limit noise temperature at 1 MHz? Why might this be a problem?

f) Think about the identified problems in part c and part e. What is the root cause? How might you overcome this?

Problem III.4. Derive the limit on PBH mass from Eridanus-II explicitly using the two-body relaxation time and the parameters for Eridanus-II given above.

Problem III.5. Verify the PBH bounds from microlensing quoted above, using an estimate for the typical $\hat{t}$ for a given M.

Chapter Seventeen

Epilogue: The DM Candidate Zoo

Throughout this book we have rather indiscriminately used the term CDM—'cold dark matter'—without investigating in detail what it means. Cosmologists normally have three adjectives in mind subsumed under the single term 'cold': namely, 'cold', 'collisionless' and 'pressureless'. Such an idealised fluid is easy to model in GR (having equation of state $w = 0$ and defining the synchronous gauge where it has a vanishing velocity field) and makes theoretical life as easy as possible. Breaking any of these assumptions corresponds to a departure from canonical CDM. The idea that DM may not be canonical CDM, but rather something lying just at the edge of our astrophysical and cosmological knowledge of its properties, has always been alive and well in the community and drives forward progress by demanding we push our theories and observations further. Each of our three main DM candidates, WIMPs, axions and PBHs, could be reclassified as 'particle-like', 'wave-like' and 'macroscopic'. Under this reclassification, each candidate has a distinctive way that it can depart from the 'cold' limit, and we have actually met each one, although only very briefly. In this chapter, we expand somewhat on these ideas and meet some other members of the DM candidate zoo.

For macroscopic DM, like PBHs, the departure from the cold and pressureless limits occurs when there are simply too few PBHs to allow a thermodynamic limit. This leads us to the very rough bound that such macroscopic DM candidates should be much lower in mass than any system where we know that we can use an extended distribution to model the DM, for example, much lower in mass than any system where we observe a flat rotation curve. We met a more precise version of this limit when we considered the old star cluster in Eridanus-II being heated by PBHs: the absence of such heating places an upper limit on the mass of a macroscopic DM component of $M \lesssim 3\,M_\odot$.

For particle-like DM, the key variable is temperature. If such DM is produced (freezes out) while it is relativistic, then it starts life with $E \sim T \gg m$. As the Universe cools, it may become non-relativistic but with a non-negligible value for T/m. Such is the case for warm DM, where the small thermal velocities the DM particle had during the early epoch of structure formation cause a suppression in the matter power spectrum relative to canonical CDM, which is incompatible with the Lyman-α forest flux power spectrum unless $m \gtrsim 5$ keV.

For wavelike DM, we encountered a departure from canonical CDM also in the low mass limit, where in this case the departure is in the form of an effective pressure. For wave DM, the ratio H/m plays a similar role to the ratio T/m for particle-like DM. At early times defined by $H \gg m$ the effective fluid description of the bosonic field has an equation of state $w = -1$, which corresponds to a large negative pressure inhibiting structure formation. On smaller scales, and at late times ($H \ll m$), gradient energy in the Klein–Gordon equation acts to oppose the inward force of gravity and gives wavelike DM a de Broglie wavelength: it is fuzzy, rather than cold. Such fuzzy DM, as we have seen, is excluded unless $m \gtrsim 2 \times 10^{-20}$ eV. The disastrous effects of too much pressure on structure formation underlie the cosmological reason that we rejected a Universe dominated by baryons and requiring DM: the presence of non-linear power on 1 Mpc scales and the lack of large acoustic features in $P(k)$ on this scale (recall Fig. 6.4).

The contrast between DM and baryons can be used to help us think about other constraints on the possible properties of DM. One of our key pieces of evidence for DM was the striking visual representation of the Bullet Cluster. We saw a separation between the DM and baryons, driven by the fact that baryons are collisional and drag on one another, leading to pressure and the formation of the visible wake in the distribution of baryonic gas inferred from the X-ray observations. Clearly, if DM also had interactions amongst its constituents, then a similar phenomenon would be observed in the mass distribution inferred from gravitational lensing. The possibility of such self-interacting DM (SIDM, reviewed in Ref. [99]) can be constrained: the approximate scale of a cross section where departures from the collisionless limit can be observed in astrophysics and cosmology is $\sigma/m \sim \mathrm{cm}^2\,\mathrm{g}^{-1} \sim 10^{-24}\mathrm{cm}^2\,\mathrm{GeV}^{-1}$.

Warm DM, fuzzy DM and SIDM have all in different circumstances and by different authors been invoked as solutions to perceived problems with canonical CDM. Such problems arise always on the smallest observable scales[1] where computations are pushed to their limits by non-linearities and the astrophysics of star formation begins to have significant effects on gravitational potentials. A possible simplistic change in our assumptions about DM is then attractive to bring observations and theory back into line with one another. On the other hand, our holistic view of DM, requiring consistency across many scales, tends to push such ideas into 'catch-22' situations, unable to offer a one-size-fits-all solution. So far, perceived discrepancies of the CDM paradigm have fallen away either as data improves (in the case of 'missing satellites') or as the physical models of star formation in simulations get closer to the truth (when 'cold dark matter heats up' [100]). Nonetheless, these frontiers of our knowledge remain incredibly important in the advancement of DM science.

Continuing our tour of the frontiers of DM research, let's return to the beginning and our first candidate: the humble neutrino. Is there a way to restore it to centre stage? There is one such way, which goes under the name of the sterile neutrino. Sterile neutrinos are defined as fermions with no charge under the gauge group of the Standard Model, but which mass (and thus flavour) mix with ordinary neutrinos. They are defined by their mass and their mixing angle (normally written as $\sin^2 2\theta$),

---

[1] Cosmological problems on large scales, for example the low quadrupole of the CMB temperature anisotropies, or the currently discrepant measurements of H_0 from different probes, typically require solutions in the initial conditions or the late time dark energy-dominated phases of the Universe and are not conducive to solution by modification of the DM sector.

and the relic abundance produced by flavour mixing with the ordinary neutrinos can be fit approximately by $\Omega_\nu h^2 = 0.12(m/\text{keV})^2(\sin^2 2\theta/t.3 \times 10^{-8})^{1.23}$ [101] (the 'Dodelson–Widrow' mechanism: other mechanisms also exist but do not substantially change the conclusions). Sterile neutrinos can radiatively decay to X-rays (for masses of order 1 keV), and the non-observation of such emission from galaxies constrains the mass from above. Being in the keV range, the Lyman-α forest also constrains the mass from below. This leaves a relatively small and bounded region of parameter space allowed for sterile neutrino models, and they can be probed by direct detection in similar ways to WIMPs. For a comprehensive review, see Ref. [102].

Our canonical particle candidates for DM, the WIMP and the axion, covered only two possibilities for the fundamental property of quantum mechanical spin: the axion with spin 0 and the neutralino with spin 1/2. This far into this epilogue, you may no longer be surprised to learn that there are DM candidates that go beyond these assumptions. Spin-1 candidates are known either as vector DM or 'dark photons'. Being bosonic, the phenomenology both cosmological and in direct detection is quite similar to what we learned for axions, but with three polarisation states. A lower bound on the mass of the dark photon arises similarly to fuzzy DM, $m \gtrsim 10^{-22}$ eV, if it dominates the relic density. Increasing in spin once more, SUSY again gives us a candidate: the spin 3/2 gravitino, the supersymmetric partner of the graviton, which is present in supergravity and string theory and whose mass is related to the scale of SUSY breaking. One limit on the mass of the gravitino arises by demanding that it does not contribute too much warm DM and is therefore more massive than a few keV. Finally, it is even possible for so-called bimetric massive gravity to provide a spin-2 DM particle, the mass of which as a boson is also bounded from below by fuzzy DM-like limits.

The possible existence of fundamental particles with spin larger than 2 is beyond the scope of the present discussion (the idea goes by the name 'Vasiliev theory' [103]). However, composite particles can certainly have higher spin, like the resonances of QCD, which leads us onto our last stop in the zoo: MACROs. MACROs are a type of macroscopic DM of much lower mass than a PBH and much larger mass than a fundamental particle (which cannot have mass larger than M_{pl} without being equivalent to a BH due to having a Schwarzschild radius larger than its Compton wavelength). A humorous example is the spherical cow, with mass $M_{sc} \approx 10^3$ kg $\approx 10^{-27}M_\odot \approx 10^{12}M_{pl}$. DM composed of such spherical cows would be a perfectly cold and collisionless fluid, consistent with the phenomenology of DM on galactic scales. The problem arises that we cannot create a population of primordial spherical cows by freeze out, vacuum realignment or gravitational collapse of a radiation fluid. Nonetheless, models for kg and larger scale MACROs do exist, for example as composite particles related to QCD (the historical example being Witten's quark nuggets [104], and for more recent models see Refs. [105, 106]). Such MACROs also allow DM to be strongly interacting, where the large cross section is compensated for by the relatively low number density.

It is now time to end this book, and we do so by returning to the assertion made in the introduction that the coming decades will see the microscopic identification of DM. The problem of DM has pushed forward theoretical physics in ways we may never have imagined, to think outside of every box. The problem of DM has also pushed the boundaries of experimental physics and observational cosmology, forcing levels of precision never before conceived. Right now there is a huge zoo of DM candidates, but we know the locations of all the lampposts and have tools to search under them. Soon we will know which DM candidates (and there may be many) in the hypothetical zoo exist in the real pantheon of nature.

Acknowledgements

This book began life as a lecture course in Dark Matter taught by DJEM at the University of Göttingen during the summer term of 2020—the beginning of the coronavirus pandemic. At that stage it was sheets of handwritten notes intended to be presented on a blackboard alongside interactive JUPYTER notebooks, created by DE, for tutorials. Then, with VMM on board, the long journey from notes to book began and has taken nearly four years—including a large stretch of lockdown time—to complete. Jens Niemeyer supported the initial idea to teach this course, and has been helpful in many ways throughout. DJEM is grateful for the support and enthusiasm of the faculty for subsequent delivery of the course at King's College London. We are grateful to all the students who have taken the Dark Matter course at Göttingen and King's College London, whose enthusiasm for the subject has spurred the development and refinement of this book and the JUPYTER notebooks, and who have helped find numerous typos in draft versions. We are grateful to Bobby Acharya, Malcolm Fairbairn, Stan Lai, Nick Mavromatos, and Lukas Witkowski for sharing lecture notes with us used in early development of the topics. We are grateful to Ciaran O'Hare for a reading of the manuscript, and discussions on recoil rates and the halo model, and to Federico Carta for clarifying subtleties about the torus and string theory. This work made use of open source software (in addition to that cited in the text and Appendix): NUMPY, SCIPY, MATPLOTLIB, JUPYTER, and we are very grateful to python developers around the world for their constant hard work. Finally, we would like to thank Pedro Ferreira, who suggested that this project become a textbook rather than simply a set of notes, and put us in contact with Ingrid Gnerlich at Princeton.

Appendix A

Reading List

A.1 BOOKS AND REVIEW ARTICLES

As noted in the introduction, this book does not attempt to provide a complete technical background in cosmology or particle physics, and readers would do well to have access to books on each. There are also advanced topics, on general relativity and quantum field theory, which while not necessary for a first reading are certainly necessary if you want to take many of the ideas here further into research. Here is a short list of some books I recommend on these and other topics.

- *The Early Universe*, by Kolb and Turner [49] (KT), is excellent for all things thermal Universe and also contains good technical particle physics background.
- *Quantum Field Theory in a Nutshell*, by Zee [58], is an excellent introduction to the concepts of QFT and is brilliant for building intuition about concepts that may feel buried in the details in more technical books. I particularly like the treatment of the chiral anomaly. This kind of intuition is about the perfect level to accompany this book.
- *Cosmological Physics*, by Peacock [107], covers just what it says in the title. It is a great resource for large-scale structure, and the introduction to particle physics and QFT is also useful for non-experts to get started.
- *Modern Cosmology*, by Dodelson [22], is where we get our treatment of cosmological perturbations. Go here for the cosmic microwave background, matter power spectrum and cosmological statistics. It is also my favourite treatment of the collisional Boltzmann equation.
- *Physical Foundations of Cosmology*, by Mukhanov [36], has a more theoretical treatment of cosmology, including inflation, and a unique analytical approach to the CMB anisotropies.
- *Galactic Dynamics*, by Binney and Tremaine [5] (BT), covers the statistical model for dark matter and stars that we rely on heavily throughout Part I.
- *Galaxy Formation and Evolution*, by Mo, van den Bosch and White [108], is an excellent and modern treatment of the theory and numerics of galaxy formation.
- *Astroparticle Physics: Theory and Phenomenology*, by Sigl [3], is a vast and technical overview of most everything an astroparticle physicist needs to know at a high level.
- *A First Course in General Relativity*, by Schutz [109], provides a very accessible introduction without getting bogged down in too much abstract mathematics.

- *Spacetime and Geometry*, by Carroll [18], covers GR in a more formal way, defining manifolds and differential forms, but still keeps a physicist's approach.
- *Quarks and Leptons: An Introductory Course in Modern Particle Physics*, by Halzen and Martin [110], is an accessible and practical book for QFT computations using Feynman diagrams in the Standard Model. Learn the rules, and turn the crank.
- *Introduction to Elementary Particles*, by Griffiths [111], is an overview of particle physics background and the Standard Model.
- *The Standard Model in a Nutshell*, by Goldberg [112]. This is exactly what it says in the title. You will quickly get familiar with the nuts and bolts of the Standard Model.
- *An Introduction to Quantum Field Theory*, by Peskin and Schroeder [20], is formal in its development and the go-to reference for many serious QFT practitioners. It includes useful cross-section computations and reference sections on Feynman rules and conventions.
- *Quantum Field Theory*, by Srednicki [59], has a formal development based on the path integral. I found the chapters related to axions extremely useful in their technical detail.
- *The Search for Ultralight Bosonic Dark Matter*, eds Kimball and van Bibber [63]. This book covers everything about wave dark matter including axions and related particles, with an emphasis on experimental techniques.

As a modern and fast developing subject, review articles and lecture notes are invaluable for the study of dark matter. Here are some that we used often in the preparation of this book:

- Baumann's book and lecture notes on cosmology [113, 114].
- Bertone, Hooper and Silk [50] (BHS).
- Axion cosmology, Marsh [64].
- Particle Data Group (PDG) [37].
- Lisanti's lecture on DM [21].

A.2 EVIDENCE FOR DARK MATTER

A.2.1 Galactic Scales

Notes and resources:

- BHS Sec. 2.1, 2.2, 2.5.
- Sigl Ch. 3.2.5, 3.7, 14.1.
- BT Ch. 1, 2.
- 'Mass Distribution and Rotation Curve in the Galaxy', Sofue [6], de Salas and Widmark [12], McMillan [115].

- 'Introduction to Gravitational Lensing', Meneghetti [17]. These lecture notes are a good supplement to the older review by Bartlemann and Schneider [16].
- 'Rotation Curves of Spiral Galaxies', Sofue and Rubin [116]. This is from a more astrophysical perspective.

Original papers:

- 'Clusters', 1933. Zwicky [2].
- 'Rotation Curves', 1970. Rubin and Ford [4].
- 'Dark Matter in the Milky Way'. Iocco et al. [117], de Salas et al. [118], Eilers et al. [13], and application to the DM 'standard halo model' by Evans et al. [11].
- 'Rotation Curves and Modified Gravity'. Begemann et al. [119].
- 'The Bullet Cluster'. Clowe et al. [15].

A.2.2 Cosmology

Notes and resources:

- BHS Sec. 2.3.
- Dodelson Ch. 2 (for GR), 3 (thermodynamics), 4, 5, 7, 8 (perturbation equations of motion and solutions).
- KT Ch. 3 (thermodynamics).
- Sigl Ch. 3, 4.
- BT Ch. 9.1.
- Review of CMB anisotropies, Hu & Dodelson [39].
- Accessible and concise review of inflation and perturbations, Langlois [120].
- Review of the halo model, Cooray and Sheth [121].

Original papers:

- The classic on cosmological perturbation theory, Ma and Bertschinger [23].
- Peebles began the gravitational instability paradigm and worked out the basis of the cosmological Standard Model. Ref. [38] predicts the CMB amplitude from galaxy clustering, which turned out to be correct when measured ten years later, by presuming DM exists.
- Efstathiou et al. (1992) [44] on the CMB establishing the instability paradigm.
- Wilkinson Microwave Anisotropy Probe (WMAP) three-year CMB (2007) [122]. WMAP firmly established the cosmological Standard Model based on the acoustic peaks. Earlier papers like this can be easier to digest, since they assume less prior knowledge. Sec. 3.3 briefly discusses the role of CDM versus MOND (see below). Galaxy surveys 2dF and SDSS are also featured.
- The *Planck* (2018) summary paper [123] includes a historical view of the beyond ΛCDM parameters and a compilation of matter power spectrum data. The cosmological parameters are given in Ref. [43] (including standard parameters as well as, for example, WIMP indirect detection, inflation and much more).

- CAMB, Python code for computing cosmological observables like C_ℓ and $P(k)$ https://camb.readthedocs.io/en/latest/.

The Nail in the Coffin of MOND

This is some additional reading for the keen student. Cosmology is the real reason that we must have dark matter and that no 'modified gravity' will do. Original papers:

- Original paper by Beckenstein on TeVeS [124].
- 'The Real Problem with MOND', Dodelson and Ligouri [125, 41].
- The CMB with MOND, Skordis et al. [126] (see in particular Fig. 3 in Skordis' review [45]), does not fit the third and higher peaks measured in the power spectrum.
- Review of all modified gravity theories (with more of a focus on dark energy and cosmology) [127].

I hope that this reading, in the context of the gravitational instability paradigm of chapter 6, will convince you that there is no theory without DM that can explain the CMB and large-scale structure (which, remember, is actually an incredibly simple linear system). Dodelson points out that the galaxy power spectrum goes wrong in a simple no-DM theory. Skordis et al. found that some particular version of TeVeS can get around this due to vector perturbations. The crucial point is that, even in this case, in the absence of DM the CMB acoustic peaks after the second have the wrong heights. This was the case approximately in the WMAP-3 data in 2006 and is even more pronounced in the exquisitely measured peaks of *Planck*, ACT and SPT. Varying all of the free parameters in the 'TeVeS' relativistic realisation of MOND still only gives a very poor fit [128]. In any case, the 'S' in TeVeS is key. To quote Sean Carroll:[1]

[Proponents of MOND] can't just wave [their] hands and say that a mysterious 'forcing term' will help explain the CMB. If there is no non-baryonic dark matter, there is no way that even-numbered peaks can be different from odd-numbered peaks; the configuration of baryons is precisely analogous. You can mimic the situation in TeVeS (although the numbers don't seem to work out) because you've introduced an independently propagating scalar degree of freedom whose energy density doesn't follow the baryons. *You can give that scalar whatever name you like, but it is 'non-baryonic dark matter.' A particularly contrived version, but that's what it is.*

You can't explain the third peak without a source for gravity that propagates independently of the baryons.

---

[1] https://www.preposterousuniverse.com/blog/2012/05/09/dark-matter-vs-modified-gravity-a-trialogue/

If it walks like a duck and quacks like a duck, it's a duck. If it is a weakly interacting, propagating degree of freedom that gets the early formation of structure in the Universe correct, it's DM.

A.3 THEORIES OF DARK MATTER

A.3.1 The Standard Model and Quantum Field Theory

Notes and resources:

- PDG 9 (QCD), 10 (electroweak model), 11 (Higgs boson).
- Griffiths Ch. 4.7.11 (note Griffiths uses the 'Dirac', rather than chiral, representation of the γ-matrices).
- Sigl Ch. 1.6, 2.4.1, 2.4.3, 2.4.4, 2.5, 2.6, 2.7.
- Langacker's notes on the Standard Model [129]. These are more advanced than we need but a good reference.
- Srednicki Ch. 22, 24, 34, 54, 69, 84–89. This is more advanced than we need but gives all the gory details.

A.3.2 WIMPs, SUSY and Other Thermal Relics

Notes and resources:

- KT Ch. 5.2 (freeze out).
- BHS, Sec. 1.5 (relic density), 3.2, 3.3, 3.4 (particle models). Appendix A, B for neutralino masses and cross sections.
- Lisanti's lecture on DM [21], Sec. 2.2 (freeze out).
- PDG 12 (CKM matrix), 14 (neutrino mass), 110 (SUSY), 26 (dark matter).
- Logic of our cross-section estimates, Peskin and Schroeder Ch. 1.
- Olive's DM lectures [130], Sec. 2.2.2 (neutrinos as DM), 3 (SUSY).
- Review of supersymmetric DM, Jungman, Kamionkowski and Griest [131].

Original literature:

- Complete neutralino annihilation cross-section analytic expressions in minimal supergravity [51, 52].
- Nihei et al. (2001) computation of neutralino freeze out and relic density [53].
- DarkSUSY http://www.darksusy.org/ [54], a code people use to make WIMP calculations 'in the real world'.
- GAMBIT 'global fits' to the MSSM [55] using DarkSUSY. Note Fig. 4 on the relic density and neutralino mass.

A.3.3 Axions

Notes and resources:

- 'Axions for Amateurs' [132].
- Marsh, Sec. 2, 3, 4.
- KT Ch. 10.
- Quinn [133], a non-technical article on the strong-CP problem.
- Di Luzio et al. [134] review of QCD axion models. Sec. 2 provides a detailed model description, and Sec. 3 discusses cosmology and relic density.
- The book, *Aspects of Symmetry* by Coleman [57], Ch. 7, contains a nice description of the Θ-vacua and 'instantons', and a proof that the vacuum energy goes as $E_{vac} \propto -\cos\theta$.
- Cosmological perturbation theory for axions is covered in Chapter 3 of Ref. [63], on arXiv as Ref. [135].
- Reviews of ultralight DM physics including the Schrödinger–Poisson equations by Niemeyer [65] and Hui [67].
- Srednicki Ch. 93, 94 (advanced).
- The 'String Axiverse' [136], Arvanitaki et al. (2009). There are more details in Svrcek and Witten [137] on axions in string theory (advanced).

Original literature:

- Borsanyi et al. [47], numerical computation of the QCD topological susceptibility.
- Peccei–Quinn: first solution to the strong-CP problem [138, 139] (note that they assume Φ is the Higgs, which was quickly disproven and gives no axion DM).
- Kim [140], Shifman, Vainshstein, Zakharov [141] (KSVZ) axion model, which is what we base our treatment on. The other historical model is the Dine, Fischler, Srednicki [142], Zhitnitsky [143] (DFSZ) model, which uses the Higgs and Standard Model quarks in a different way.
- Simulations of the Schrödinger–Poisson equations in cosmology by Schive et al. [66] and in the kinetic regime by Levkov et al. [144].

A.3.4 Primordial Black Holes

Notes and resources:

- Sasaki et al. [71], Sec. 2, 'Formation Mechanisms'.
- Reviews of inflation by Langlois [120] and Baumann [145] (in particular Sec. 5 overview of inflation, Sec. 11 quantizing the harmonic oscillator, Sec. 12.1–12.2 quantization of curvature during inflation).
- Review of PBHs by Green and Kavanagh [68].

Original literature:

- Hawking (1971) [146].
- Carr and Hawking (1974) [147].
- Carr (1975) [148].

- Hertzberg and Yamada (2017) [70], how to reverse engineer the inflaton potential necessary for PBH DM.

A.4 TESTS OF DARK MATTER

A.4.1 WIMPs

Notes and resources:

- Lisanti [21], Sec. 3 (direct detection), Sec. 4 (indirect detection).
- Undagoitia and Rauch, review of WIMP direct detection with analysis details [73].
- Slatyer, review of indirect detection [149].
- Hooper's lectures on DM cover indirect detection very nicely, as well as many other topics [150].
- PDG 'Dark Matter' Sec. 26.6 on laboratory searches, references for almost all operating and planned direct searches.
- 'RESONAANCES' blog post from 2018 with an excellent overview of direct detection limits historically and in the context of models. Read the comments too. http://resonaances.blogspot.com/2018/05/wimps-after-xenon1t.html.
- Del Nobile, guide to DM direct detection computations [151].
- BHS, Sec. 4, 5, 6.

Original literature:

- Goodman and Witten (1985) on the first ideas for direct detection [152].
- The Homestake 1987 search for WIMPs [153].
- CDMS (2005) results [154]: a detailed look at how a direct search analysis is carried out.
- Trotta et al. [72], predictions of the CMSSM enforcing that WIMPs are all DM, direct detection bounds *not* imposed.
- GAMBIT global fits for the MSSM [55] allowing for sub-dominant fractions imposing direct limits.
- The neutrino floor: Billard et al. (2014), the neutrino floor [155].
- Fermi telescope indirect detection bounds [81].
- Leane et al. (2018), indirect detection lower bounds on WIMP mass [156].
- Cirelli et al. (2010), indirect detection 'cookbook' [157].
- Bergstrom et al. (2013), AMS limits from cosmic ray positron spectrum [158].

A.4.2 Axions

Notes and resources:

- Marsh, Sec. 5, 6, 8, 9.
- PDG 112 (axions and similar particles).

- Raffelt (2008) [159], review of bounds from stars.
- Grin et al. (2019) [160], summary of gravitational probes of ultralight axions.

Original literature:

- Sikivie (1983) [161], who invented many axion searches all in one paper.
- Recent ADMX constraints [84].
- The QUAX axion search [162].
- The CASPER axion search [163].
- Graham and Rajendran (2013) [164], new ideas for axion detection.
- O'Hare and Green (2017) [165], nice description and derivation of axion signal power spectrum, and discussion of astrophysical uses.
- Hui et al. (2017) [166], the case for fuzzy DM from observations.
- Marsh and Niemeyer (2019) [89], bounds on fuzzy DM from Eridanus-II.
- Hoof et al. (2019) [167], GAMBIT axion global fits.
- X-ray limits on decaying axions, Refs. [168, 169].
- Axion-photon conversion in neutron stars, Refs. [170, 171].
- Axion miniclusters, Refs. [172, 173, 174, 175].

A.4.3 Primordial Black Holes

Notes and resources:

- Carr et al. [176, 177].
- Sasaki et al. [71].
- Green and Kavanagh [68].

Original literature. Many of the below papers also contain summaries of other current limits:

- Carr et al. (2010), extragalactic gamma rays and X-rays [94].
- Brandt (2016) [178], PBH bounds from ultra faint dwarf galaxies.
- Green (2016) [179], a nice explanation of PBH microlensing bounds, including an extended mass function and combination with ultra faint dwarfs.
- EROS [97], MACHO [180] (and their combination [181]) and HSC [95], PBH microlensing constraints.
- 'Did LIGO Detect Dark Matter?', Bird et al. [182]. You'll notice many of the above papers come during or after 2015. Interest in PBHs was renewed by the first LIGO detection of gravitational waves from the binary inspiral of 30 $M_\odot$ BHs in 2015, which for a time were thought to possibly be PBHs. The combination of ultra faint dwarf galaxy heating and microlensing excludes this possibility for a monochromatic PBH mass spectrum.
- Kavanagh's PBH limits plotting package [98].

- Bounds in the open PBH window: Katz et al. [183] on femtolensing and why it gives no current bounds. Graham et al. (2015) [184], PBH bounds from white dwarf supernova triggers.
- Luo et al. (2012), consequences of PBH collision with Earth [185].
- Review of current status in the open window: Montero-Camacho et al. (2019) [186].

Bibliography

[1]   G. Bertone and D. Hooper, Rev. Mod. Phys. **90**, 045002 (2018), arXiv: 1605.04909.

[2]   F. Zwicky, Helv. Phys. Acta **6**, 110 (1933) [Gen. Rel. Grav. 41, 207 (2009)].

[3]   G. Sigl, *Astroparticle Physics: Theory and Phenomenology* (Springer, 2017), ISBN: 9789462392427.

[4]   V. C. Rubin and W. K. Ford, Jr., Astrophys. J. **159**, 379 (1970), arXiv: 10.1086/150317.

[5]   J. Binney and S. Tremaine, *Galactic Dynamics* (Princeton University Press, 1987), ISBN: 9780691130279.

[6]   Y. Sofue, *Mass Distribution and Rotation Curve in the Galaxy* Vol. 5 (Springer, 2013), p. 985, 10.1007/978-94-007-5612-0_19.

[7]   S. Z. Kam et al., The Astronomical Journal **154**, 41 (2017).

[8]   GRAVITY Collaboration et al., A&A **625**, L10 (2019), arXiv:1904.05721.

[9]   SDSS, M. Juric et al., Astrophys. J. **673**, 864 (2008), arXiv:astro-ph/0510520.

[10]   Event Horizon Telescope Collaboration et al., ApJLett **930**, L12 (2022).

[11]   N. W. Evans, C.A.J. O'Hare and C. McCabe, Phys. Rev. D. **99**, 023012 (2019), arXiv:1810.11468.

[12]   P. F. de Salas and A. Widmark, Rept. Prog. Phys. **84**, 104901 (2021), arXiv:2012.11477.

[13]   A.-C. Eilers, D. W. Hogg, H.-W. Rix and M. K. Ness, Ap.J. **871**, 120 (2019), arXiv:1810.09466.

[14]   E. Pouliasis, P. Di Matteo and M. Haywood, A&A **598**, A66 (2017), arXiv:1611.07979.

[15]   D. Clowe et al., Astrophys. J. **648**, L109 (2006), arXiv:astro-ph/0608407.

[16]   M. Bartelmann and P. Schneider, Phys. Rept. **340**, 291 (2001), arXiv:astro-ph/9912508.

[17]   M. Meneghetti, *Introduction to Gravitational Lensing: With Python Examples* (Springer, 2022), ISBN: 3030735826.

[18]   S. M. Carroll, *Spacetime and Geometry: An Introduction to General Relativity* (Addison Wesley, 2004), ISBN: 9781108770385.

[19]   O. Wantz and E.P.S. Shellard, Phys. Rev. D. **82**, 123508 (2010), arXiv: 0910.1066.

[20]   M. E. Peskin and D. V. Schroeder, *An Introduction to Quantum Field Theory* (Westview Press, 1995), ISBN: 9780367320560.

[21]   M. Lisanti, 'Lectures on Dark Matter Physics', in *Theoretical Advanced Study Institute in Elementary Particle Physics: New Frontiers in Fields*

and Strings (WSP, 2017), ISBN: 9789813149434, pp. 399–446, arXiv: 1603.03797.

[22] S. Dodelson, *Modern Cosmology* (Academic Press, 2003), ISBN: 9780 122191411.

[23] C.-P. Ma and E. Bertschinger, Astrophys. J. **455**, 7 (1995), arXiv:astro-ph /9506072.

[24] J. Martin, C. Ringeval and V. Vennin, Phys. Dark Univ. **5–6**, 75 (2014), arXiv:1303.3787.

[25] L. Kofman, A. D. Linde and A. A. Starobinsky, Phys. Rev. D. **56**, 3258 (1997), arXiv:hep-ph/9704452.

[26] W. H. Press and P. Schechter, ApJ **187**, 425 (1974).

[27] J. R. Bond, S. Cole, G. Efstathiou and N. Kaiser, ApJ **379**, 440 (1991).

[28] J. M. Bardeen, J. R. Bond, N. Kaiser and A. S. Szalay, Astrophys. J. **304**, 15 (1986).

[29] R. K. Sheth, H. J. Mo and G. Tormen, MNRAS **323**, 1 (2001), arXiv:astro-ph/9907024.

[30] J. F. Navarro, C. S. Frenk and S.D.M. White, Astrophys. J. **462**, 563 (1996), arXiv:astro-ph/9508025.

[31] J. F. Navarro, C. S. Frenk and S.D.M. White, Astrophys. J. **490**, 493 (1997), arXiv:astro-ph/9611107.

[32] M. Trenti and P. Hut, Scholarpedia **3**, 3930 (2008), revision #91544.

[33] J. Barnes and P. Hut, Nature **324**, 446 (1986).

[34] G. L. Bryan et al., ApJS **211**, 19 (2014), arXiv:1307.2265.

[35] Enzo community, https://enzo-project.org/.

[36] V. Mukhanov, *Physical Foundations of Cosmology* (Cambridge University Press, 2005).

[37] Particle Data Group, P. A. Zyla et al., PTEP **2020**, 083C01 (2020).

[38] P. Peebles, Astrophys. J. **263**, L1 (1982).

[39] W. Hu and S. Dodelson, Ann. Rev. Astron. Astrophys. **40**, 171 (2002), arXiv:astro-ph/0110414.

[40] A. Lewis, A. Challinor and A. Lasenby, ApJ **538**, 473 (2000), arXiv:astro-ph/9911177.

[41] S. Dodelson, Int. J. Mod. Phys. **D20**, 2749 (2011), arXiv:1112.1320.

[42] Planck, N. Aghanim et al., Astron. Astrophys. **641**, A5 (2020), arXiv: 1907.12875.

[43] Planck, N. Aghanim et al., Astron. Astrophys. **641**, A6 (2020), arXiv: 1807.06209.

[44] G. Efstathiou, J. R. Bond and S.D.M. White, Mon. Not. Roy. Astron. Soc. **258**, 1 (1992), 10.1093/mnras/258.1.1P.

[45] C. Skordis, Classical and Quantum Gravity **26**, 143001 (2009), arXiv:0903. 3602.

[46] B. W. Lee and S. Weinberg, Phys. Rev. Lett. **39**, 165 (1977).

[47] S. Borsanyi et al., Nature **539**, 69 (2016), arXiv:1606.07494.

[48] K. Griest and M. Kamionkowski, Phys. Rev. Lett. **64**, 615 (1990).

[49] E. W. Kolb and M. S. Turner, *The Early Universe* (Addison-Wesley, 1990), ISBN: 9780201626742.

[50] G. Bertone, D. Hooper and J. Silk, Phys. Rept. **405**, 279 (2005), arXiv: hep-ph/0404175.

[51] M. Drees and M. M. Nojiri, Phys. Rev. D. **47**, 376 (1993), arXiv:hep-ph/92 07234.

[52] A. Birkedal-Hansen and E.-h. Jeong, JHEP **02**, 047 (2003), arXiv:hep-ph/ 010041.

[53] T. Nihei, L. Roszkowski and R. Ruiz de Austri, JHEP **05**, 063 (2001), arXiv:hep-ph/0102308.

[54] P. Gondolo et al., JCAP **07**, 008 (2004), arXiv:astro-ph/0406204.

[55] GAMBIT, P. Athron et al., Eur. Phys. J. C. **77**, 879 (2017), arXiv:1705. 07917.

[56] J. M. Pendlebury et al., Phys. Rev. D. **92**, 092003 (2015), arXiv:1509. 04411.

[57] S. Coleman, *Aspects of Symmetry* (Cambridge University Press, 1988).

[58] A. Zee, *Quantum Field Theory in a Nutshell* (Princeton University Press, 2003), ISBN: 9780691140346.

[59] M. Srednicki, *Quantum Field Theory* (Cambridge University Press, 2007).

[60] C. P. Burgess, Ann. Rev. Nucl. Part. Sci. **57**, 329 (2007), arXiv:hep-th/0701053.

[61] F. Denef, Les Houches **87**, 483 (2008), arXiv:0803.1194.

[62] M. Demirtas, L. McAllister and A. Rios-Tascon, Fortsch. Phys. **68**, 200 0086 (2020), arXiv:2008.01730.

[63] D. F. Jackson Kimball, L. D. Duffy and D.J.E. Marsh, 'Ultralight Bosonic Dark Matter Theory' in *The Search for Ultralight Bosonic Dark Matter Theory* (Springer, 2023), ISBN: 9783030958510, pp. 31–72.

[64] D.J.E. Marsh, Phys. Rept. **643**, 1 (2016), arXiv:1510.07633.

[65] J. C. Niemeyer, Prog. Part. Nucl. Phys. **113**, 103787 (2019), arXiv:1912. 07064.

[66] H.-Y. Schive, T. Chiueh and T. Broadhurst, Nature Phys. **10**, 496 (2014), arXiv:1406.6586.

[67] L. Hui, Ann. Rev. Astron. Astrophys. **59**, 247 (2021), arXiv:2101.11735.

[68] A. M. Green and B. J. Kavanagh, J. Phys. G. **48**, 043001 (2021), arXiv: 2007.10722.

[69] A. D. Gow, C. T. Byrnes, P. S. Cole, and S. Young, JCAP **02**, 002 (2021), arXiv:2008.03289.

[70] M. P. Hertzberg and M. Yamada, Phys. Rev. D. **97**, 083509 (2018), arXiv:1712.09750.

[71] M. Sasaki, T. Suyama, T. Tanaka and S. Yokoyama, Class. Quant. Grav. **35**, 063001 (2018), arXiv:1801.05235.

[72] R. Trotta, F. Feroz, M. P. Hobson, L. Roszkowski and R. Ruiz de Austri, JHEP **12**, 024 (2008), arXiv:0809.3792.

[73] T. Marrodán Undagoitia and L. Rauch, J. Phys. G. **43**, 013001 (2016), arXiv:1509.08767.

[74] C.A.J. O'Hare, Phys. Rev. Lett. **127**, 251802 (2021), arXiv:2109.03116.

[75] XENON, E. Aprile et al., Phys. Rev. Lett. **121**, 111302 (2018), arXiv: 1805.12562.

[76]   J. Alvey et al., Mon. Not. Roy. Astron. Soc. **501**, 1188 (2021), arXiv:
        2010.03572.

[77]   P. Bode, J. P. Ostriker and N. Turok, Astrophys. J. **556**, 93 (2001), arXiv:
        astro-ph/0010389.

[78]   V. Iršič et al., Phys. Rev. D. **96**, 023522 (2017), arXiv:1702.01764.

[79]   Y. Kahn and T. Lin, Rept. Prog. Phys. **85**, 066901 (2022), arXiv:2108.03239.

[80]   A. W. Strong and I. V. Moskalenko, ApJ **509**, 212 (1998), arXiv:astro-
        ph/9807150.

[81]   Fermi-LAT, M. Ackermann et al., Phys. Rev. Lett. **115**, 231301 (2015),
        arXiv:1503.02641.

[82]   J. C. Mather et al., ApJ **420**, 439 (1994).

[83]   M. Srednicki, Nucl. Phys. B. **260**, 689 (1985).

[84]   ADMX, T. Braine et al., Phys. Rev. Lett. **124**, 101303 (2020), arXiv:
        1910.08638.

[85]   C. O'Hare, cajohare/axionlimits: Axionlimits, https://cajohare.github.io
        /AxionLimits/, 2020.

[86]   K. K. Rogers and H. V. Peiris, Phys. Rev. Lett. **126**, 071302 (2021), arXiv:
        2007.12705.

[87]   H.-Y. Schive, T. Chiueh and T. Broadhurst, Phys. Rev. Lett. **124**, 201301
        (2020), arXiv:1912.09483.

[88]   B. T. Chiang, H.-Y. Schive and T. Chiueh, Phys. Rev. D. **103**, 103019 (2021),
        arXiv:2104.13359.

[89]   D. J. Marsh and J. C. Niemeyer, Phys. Rev. Lett. **123**, 051103 (2019), arXiv:
        1810.08543.

[90]   DES, E. O. Nadler et al., Phys. Rev. Lett. **126**, 091101 (2021), arXiv:
        2008.00022.

[91]   R. Brito, V. Cardoso and P. Pani, Lect. Notes Phys. **906**, 1 (2015), arXiv:
        1501.06570.

[92]   V. M. Mehta et al., JCAP **07**, 033 (2021), arXiv:2103.06812.

[93]   A. Ayala, I. Domínguez, M. Giannotti, A. Mirizzi and O. Straniero, Phys.
        Rev. Lett. **113**, 191302 (2014), arXiv:1406.6053.

[94]   B. Carr, K. Kohri, Y. Sendouda and J. Yokoyama, Phys. Rev. D. **81**, 104019
        (2010), arXiv:0912.5297.

[95]   H. Niikura et al., Nat. Astron. **3**, 524 (2019), arXiv:1701.02151.

[96]   K. Griest, ApJ **366**, 412 (1991).

[97]   EROS-2, P. Tisserand et al., Astron. Astrophys. **469**, 387 (2007),
        arXiv:astro-ph/0607207.

[98]   B. J. Kavanagh, bradkav/pbhbounds: Release version, 2019.

[99]   S. Adhikari et al. (2022), arXiv:2207.10638.

[100]   A. Pontzen and F. Governato, Nature **506**, 171 (2014), arXiv:1402.1764.

[101]   B. Dasgupta and J. Kopp, Phys. Rept. **928**, 1 (2021), arXiv:2106.05913.

[102]   M. Drewes et al., JCAP **01**, 025 (2017), arXiv:1602.04816.

[103]   V. E. Didenko and E. D. Skvortsov (2014), arXiv:1401.2975.

[104]   E. Witten, Phys. Rev. D **30**, 272 (1984).

[105]   D. M. Jacobs, G. D. Starkman and B. W. Lynn, Mon. Not. Roy. Astron. Soc.
        450, 3418 (2015), arXiv:1410.2236.

[106]   A. Zhitnitsky, Mod. Phys. Lett. A. **36**, 2130017 (2021), arXiv:2105.08719.

[107]   J. A. Peacock, *Cosmological Physics* (Cambridge Astrophysics, 1999).

[108]   H. Mo, F. van den Bosch and S. White, *Galaxy Formation and Evolution* (Cambridge University Press, 2010).

[109]   B. Schutz, *A First Course in General Relativity* (Cambridge University Press, 2009).

[110]   F. Halzen and A. D. Martin, *Quarks and Leptons: An Introductory Course in Modern Particle Physics* (Wiley, 1984).

[111]   D. Griffiths, *Introduction to Elementary Particles* (Wiley, 2008).

[112]   D. Goldberg, *The Standard Model in a Nutshell* (Princeton University Press, 2017).

[113]   D. Baumann, PoS **TASI2017**, 009 (2018), arXiv:1807.03098.

[114]   D. Baumann, *Cosmology* (Cambridge University Press, 2022).

[115]   P. J. McMillan, MNRAS **465**, 76 (2017), arXiv:1608.00971.

[116]   Y. Sofue and V. Rubin, Ann. Rev. Astron. Astrophys. **39**, 137 (2001), arXiv: astro-ph/0010594.

[117]   F. Iocco, M. Pato and G. Bertone, Nature Phys. **11**, 245 (2015), arXiv: 1502.03821.

[118]   P. F. de Salas, K. Malhan, K. Freese, K. Hattori and M. Valluri, JCAP **2019**, 037 (2019), arXiv:1906.06133.

[119]   K. G. Begeman, A. H. Broeils and R. H. Sanders, MNRAS **249**, 523 (1991), 10.1093/mnras/249.3.523.

[120]   D. Langlois, Lect. Notes Phys. **800**, 1 (2010), arXiv:1001.5259.

[121]   A. Cooray and R. K. Sheth, Phys. Rept. **372**, 1 (2002), arXiv:astro-ph/0206508.

[122]   WMAP, D. N. Spergel et al., Astrophys. J. Suppl. **170**, 377 (2007), arXiv:astro-ph/0603449.

[123]   Planck Collaboration et al. (2018), arXiv:1807.06205.

[124]   J. D. Bekenstein, Phys. Rev. D. **70**, 083509 (2004).

[125]   S. Dodelson and M. Liguori, Phys. Rev. Lett. **97**, 231301 (2006), arXiv:astro-ph/0608602.

[126]   C. Skordis, D. F. Mota, P. G. Ferreira and C. Boehm, Phys. Rev. Lett. **96**, 011301 (2006), arXiv:astro-ph/0505519.

[127]   T. Clifton, P. G. Ferreira, A. Padilla and C. Skordis, Phys. Rept. **513**, 1 (2012), arXiv:1106.2476.

[128]   X.-d. Xu, B. Wang and P. Zhang, Phys. Rev. **D92**, 083505 (2015), arXiv: 1412.4073.

[129]   P. Langacker, 'Introduction to the Standard Model and Electroweak Physics', in *Proceedings of Theoretical Advanced Study Institute in Elementary Particle Physics on the Dawn of the LHC Era (TASI 2008): Boulder, USA, June 2–27, 2008* (2010), pp. 3–48, arXiv:0901.0241.

[130]   K. A. Olive, 'TASI Lectures On Dark Matter', in *Particle Physics and Cosmology: The Quest for Physics beyond the Standard Model(s). Proceedings, Theoretical Advanced Study Institute, TASI 2002, Boulder, USA, June 3–28, 2002* (2003), pp. 797–851, arXiv:astro-ph/0301505.

[131]   G. Jungman, M. Kamionkowski and K. Griest, Phys. Rept. **267**, 195 (1996).

[132] D.J.E. Marsh, Cont. Phys. **24** (2023), arXiv:2308.16003.

[133] H. R. Quinn (2001), arXiv:hep-ph/0110050.

[134] L. Di Luzio, M. Giannotti, E. Nardi and L. Visinelli, Phys. Rept. **870** (2020), arXiv:2003.01100.

[135] D.J.E. Marsh and S. Hoof, in *The Search for Ultralight Bosonic Dark Matter*, p. 73 (2023), arXiv:2106.08797.

[136] A. Arvanitaki, S. Dimopoulos, S. Dubovsky, N. Kaloper and J. March-Russell, Phys. Rev. D. **81**, 123530 (2009), arXiv:0905.4720.

[137] P. Svrcek and E. Witten, JHEP **6**, 51 (2006), arXiv:hep-th/0605206.

[138] R. Peccei and H. R. Quinn, Phys. Rev. Lett. **38**, 1440 (1977).

[139] R. Peccei and H. R. Quinn, Phys. Rev. D. **16**, 1791 (1977).

[140] J. E. Kim, Phys. Rev. Lett. **43**, 103 (1979).

[141] M. A. Shifman, A. I. Vainshtein and V. I. Zakharov, Nuclear Physics B. **166**, 493 (1980).

[142] M. Dine, W. Fischler and M. Srednicki, Phys. Lett. B. **104**, 199 (1981).

[143] A. R. Zhitnitsky, Sov. J. Nucl. Phys. **31**, 260 (1980).

[144] D. G. Levkov, A. G. Panin and I. I. Tkachev, Phys. Rev. Lett. **121**, 151301 (2018), arXiv:1804.05857.

[145] D. Baumann, 'Inflation', in *Theoretical Advanced Study Institute in Elementary Particle Physics: Physics of the Large and the Small*, pp. 523–686 (2011), arXiv:0907.5424.

[146] S. Hawking, Mon. Not. Roy. Astron. Soc. **152**, 75 (1971).

[147] B. J. Carr and S. W. Hawking, Mon. Not. Roy. Astron. Soc. **168**, 399 (1974).

[148] B. J. Carr, ApJ **201**, 1 (1975).

[149] T. R. Slatyer, 'Indirect Detection of Dark Matter', in *Theoretical Advanced Study Institute in Elementary Particle Physics: Anticipating the Next Discoveries in Particle Physics*, pp. 297–353 (2018), arXiv:1710.05137.

[150] D. Hooper, 'Particle Dark Matter', in *Theoretical Advanced Study Institute in Elementary Particle Physics: The Dawn of the LHC Era*, pp. 709–764 (2010), arXiv:0901.4090.

[151] E. Del Nobile, Lect. Notes Phys. **996** (2021), arXiv:2104.12785.

[152] M. W. Goodman and E. Witten, Phys. Rev. D. **31**, 3059 (1985).

[153] S. Ahlen et al., Phys. Lett. B. **195**, 603 (1987).

[154] CDMS, D. Akerib et al., Phys. Rev. D. **72**, 052009 (2005), arXiv:astro-ph/0507190.

[155] J. Billard, L. Strigari and E. Figueroa-Feliciano, Phys. Rev. D. **89**, 023524 (2014), arXiv:1307.5458.

[156] R. K. Leane, T. R. Slatyer, J. F. Beacom and K. C. Ng, Phys. Rev. D. **98**, 023016 (2018), 1805.10305.

[157] M. Cirelli et al., JCAP **03**, 051 (2010), arXiv:1012.4515 [Erratum: JCAP **10**, E01 (2012)].

[158] L. Bergstrom, T. Bringmann, I. Cholis, D. Hooper and C. Weniger, Phys. Rev. Lett. **111**, 171101 (2013), arXiv:1306.3983.

[159] G. G. Raffelt, Lect. Notes Phys. **741**, 51 (2008), arXiv:hep-ph/0611350.

[160] D. Grin et al., Bull. Am. Astron. Soc. **51**, 567 (2019), arXiv:1904.09003.

[161] P. Sikivie, Phys. Rev. Lett. **51**, 1415 (1983).

[162] N. Crescini et al., Eur. Phys. J. C. **78**, 703 (2018), arXiv:1806.00310 [Erratum: Eur. Phys. J. C. **78**, 813 (2018)].

[163] D. F. Jackson Kimball et al. (2017), arXiv:1711.08999.

[164] P. W. Graham and S. Rajendran, Phys. Rev. D. **88**, 035023 (2013), arXiv:1306.6088.

[165] C.A.J. O'Hare and A. M. Green, Phys. Rev. D. **95**, 063017 (2017), arXiv:1701.03118.

[166] L. Hui, J. P. Ostriker, S. Tremaine and E. Witten, Phys. Rev. D. **95**, 043541 (2017), arXiv:1610.08297.

[167] S. Hoof, F. Kahlhoefer, P. Scott, C. Weniger and M. White, JHEP **03**, 191 (2019), arXiv:1810.07192 [Erratum: JHEP **11**, 099 (2019)].

[168] D. Cadamuro and J. Redondo, JCAP **02**, 032 (2012), arXiv:1110.2895.

[169] J. W. Foster et al., Phys. Rev. Lett. **127**, 051101 (2021), arXiv:2102.02207.

[170] A. Hook, Y. Kahn, B. R. Safdi and Z. Sun, Phys. Rev. Lett. **121**, 241102 (2018), arXiv:1804.03145.

[171] J. W. Foster et al., Phys. Rev. Lett. **129**, 251102 (2022), arXiv:2202.08274.

[172] C. J. Hogan and M. J. Rees, Phys. Lett. B. **205**, 228 (1988).

[173] E. W. Kolb and I. I. Tkachev, Phys. Rev. D. **50**, 769 (1994), arXiv:astro-ph/9403011.

[174] D. Ellis et al., Phys. Rev. D. **106**, 103514 (2022), arXiv:2204.13187.

[175] T.D.P. Edwards, B. J. Kavanagh, L. Visinelli and C. Weniger, Phys. Rev. Lett. **127**, 131103 (2021), arXiv:2011.05378.

[176] B. Carr, F. Kuhnel and M. Sandstad, Phys. Rev. D. **94**, 083504 (2016), arXiv:1607.06077.

[177] B. Carr, K. Kohri, Y. Sendouda and J. Yokoyama, Rept. Prog. Phys. **84**, 116902 (2021), arXiv:2002.12778.

[178] T. D. Brandt, Astrophys. J. Lett. **824**, L31 (2016), arXiv:1605.03665.

[179] A. M. Green, Phys. Rev. D. **94**, 063530 (2016), arXiv:1609.01143.

[180] MACHO, C. Alcock et al., Astrophys. J. **461**, 84 (1996), arXiv:astro-ph/9506113.

[181] T. Blaineau et al., Astron. Astrophys. **664**, A106 (2022), arXiv:2202.13819.

[182] S. Bird et al., Phys. Rev. Lett. **116**, 201301 (2016), arXiv:1603.00464.

[183] A. Katz, J. Kopp, S. Sibiryakov and W. Xue, JCAP **12**, 005 (2018), arXiv:1807.11495.

[184] P. W. Graham, S. Rajendran and J. Varela, Phys. Rev. D. **92**, 063007 (2015), arXiv:1505.04444.

[185] Y. Luo, S. Hanasoge, J. Tromp and F. Pretorius, Astrophys. J. **751**, 16 (2012), arXiv:1203.3806.

[186] P. Montero-Camacho, X. Fang, G. Vasquez, M. Silva and C. M. Hirata, JCAP **08**, 031 (2019), arXiv:1906.05950.

Index